## Trends in Mathematics

*Trends in Mathematics* is a book series devoted to focused collections of articles arising from conferences, workshops or series of lectures.

Topics in a volume may concentrate on a particular area of mathematics, or may encompass a broad range of related subject matter. The purpose of this series is both progressive and archival, a context in which to make current developments available rapidly to the community as well as to embed them in a recognizable and accessible way.

Volumes of TIMS must be of high scientific quality. Articles without proofs, or which do not contain significantly new results, are not appropriate. High quality survey papers, however, are welcome. Contributions must be submitted to peer review in a process that emulates the best journal procedures, and must be edited for correct use of language. As a rule, the language will be English, but selective exceptions may be made. Articles should conform to the highest standards of bibliographic reference and attribution.

The organizers or editors of each volume are expected to deliver manuscripts in a form that is essentially "ready for reproduction." It is preferable that papers be submitted in one of the various forms of TEX in order to achieve a uniform and readable appearance. Ideally, volumes should not exceed 350-400 pages in length.

Proposals to the Publisher are welcomed at either:
Birkhäuser Boston, 675 Massachusetts Avenue, Cambridge, MA 02139, U.S.A.
or
Birkhäuser Verlag AG, 40–44 Viadukstrasse, CH-4051 Basel, Switzerland.

# Control and Modeling of Complex Systems

## *Cybernetics in the 21st Century*

Festschrift in Honor of Hidenori Kimura
on the Occasion of his 60th Birthday

Koichi Hashimoto
Yasuaki Oishi
Yutaka Yamamoto
Editors

Springer Science+Business Media, LLC

Koichi Hashimoto
Department of Information Physics
and Computing
The University of Tokyo
Hongo 7-3-1, Bunkyo-ku
Tokyo 113-8656
Japan

Yasuaki Oishi
Department of Mathematical Informatics
The University of Tokyo
Hongo 7-3-1, Bunkyo-ku
Tokyo 113-8656
Japan

Yutaka Yamamoto
Department of Applied Analysis
and Complex Dynamical Systems
Kyoto University
Yoshida-Honmachi, Sakyo-ku
Kyoto 606-8501
Japan

**Library of Congress Cataloging-in-Publication Data**

Control and modeling of complex systems : cybernetics in the 21st century : in honor of Hidenori Kimura on the occasion of his 60th birthday / Koichi Hashimoto, Yasuaki Oishi, Yutaka Yamamoto, editors.
p. cm. – (Trends in mathematics)
Includes bibliographical references.
ISBN 978-1-4612-6577-1 ISBN 978-1-4612-0023-9 (eBook)
DOI 10.1007/978-1-4612-0023-9
1. Control theory. 2. Cybernetics. I. Kimura, Hidenori, 1941- II. Hashimoto, Koichi. III. Oishi, Yasuaki. IV. Yamamoto, Yutaka. V. Series.

QA402.3 .C6268 2003
003'.5–dc21 2002038201
CIP

AMS Subject Classifications: 93-06, 93B30, 93B51, 93B52, 93C95, 93E11

Printed on acid-free paper.

Originally published by Birkhäuser Boston in 2003
Softcover reprint of the hardcover 1st edition 2003



ISBN 978-1-4612-6577-1 SPIN 10901656

Typeset by the editors.

9 8 7 6 5 4 3 2 1

Dedicated to Hidenori Kimura on the occasion of his 60th birthday

Hidenori Kimura

# Participants of the Kimura Festschrift Symposium

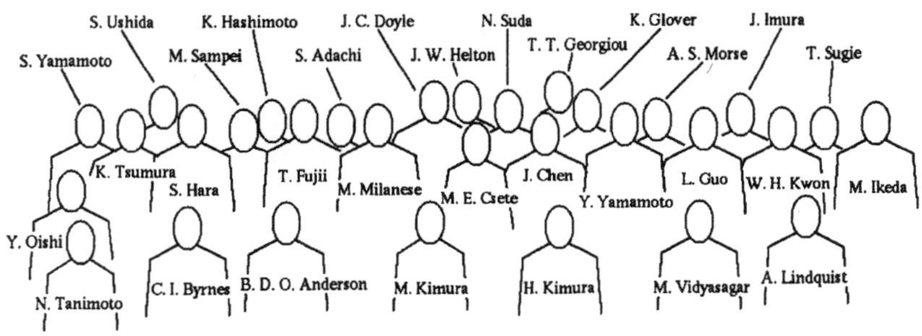

# Contents

# Preface

Hidenori Kimura, renowned system and control theorist, turned 60 years of age in November, 2001. To celebrate this memorable occasion, his friends, collaborators, and former students gathered from all over the world and held a symposium in his honor on November 1 and 2, 2001, at the Sanjo Conference Hall at the University of Tokyo. Reflecting his current research interests, the symposium was entitled "Cybernetics in the 21st Century: Information and Complexity in Control Theory," and it drew nearly 150 attendees. There were twenty-five lectures, on which the present volume is based.

Hidenori Kimura was born on November 3, 1941, in Tokyo, just prior to the outbreak of the Second World War. It is not hard to imagine, then, that his early days, like those of so many of his contemporaries, must have been difficult. Fortunately, the war ended in 1945, and his generation found itself thoroughly occupied with the rebuilding effort and with Japan's uphill journey in the last half-century. He entered the University of Tokyo in 1963, received a B.S. in 1965, an M.S. in 1967, and, in 1970, a Ph.D. degree for his dissertation "A Study of Differential Games."

After obtaining his doctorate, he joined the Department of Control Engineering at Osaka University as a research associate, and in 1973 he was promoted to an associate professor. From 1974 to 1975, he was a Visiting Fellow at the University of Warwick; during this visit, he was exposed to the British focus on the practicality of control engineering. While this influenced him, he envisioned a theory that, though practical, was also more unifying.

He was promoted to Professor at Osaka University in 1987; in 1995, he moved to his current position at the Department of Mathematical Engineering and Information Physics of the University of Tokyo—the very same department from which he had graduated 30 years before. Since 1999, he has been with the Department of Complexity Science and Engineering, in the Graduate School of Frontier Sciences. He is now also affiliated with the Bio-Mimetic Control Research Center of RIKEN as the Laboratory Head.

Throughout his career, Hidenori made key contributions to several important problems in control theory. His early paper on the pole assignment problem by output gain feedback [35], published in the IEEE Transactions on Automatic Control in 1975, continues to be regarded as a fundamental result in the resolution of this important, and in a sense still open, question. From then onward, his investigations rapidly expanded to include algebraic theory of linear systems, design theory, and practical aspects of design methods, the latter motivated in part by his interactions with industries and interest in making control theory more applicable to the real world. Real plants are often exposed to varied uncertainties, parametric or nonparametric, and LQ theory appeared at the time to show only a limited capacity for handling such situations. Those precise limitations catalyzed his deeper studies in the field of robust control, which paralleled the rising fascination with the subject among control theorists in the West.

At just around that time, in 1981, the 8th IFAC World Congress was held in Kyoto, and at the conference, George Zames gave a talk on his new approach to robust control. Zames's lecture was Hidenori's first encounter with $H^\infty$-control, a branch of control theory in which he would ultimately author various influential papers. In fact, his research in $H^\infty$-control culminated in the paper [53]: "Robust stabilizability for a class of transfer functions," published in 1984 in the IEEE Transactions on Automatic Control; the paper received the Society's Outstanding Paper Award (now known as the G. S. Axelby Outstanding Paper Award). Hidenori was also a two-time recipient of the Automatica award for his papers on perfect regulation [48] and application in cold tandem mill [67].

In the second half of the 1980s, Hidenori's efforts revolved around a unified theory of $H^\infty$-control. He pursued a different approach from that presented in the Glover–Doyle and "DGKF" papers. He presented his method, based upon a concept called "conjugation," in his book "Chain-Scattering Approach to $H^\infty$-Control" published by Birkhäuser in 1996.

In 1990, Hidenori was elected a fellow of the IEEE. During the decade that followed, he put much effort into real applications of $H^\infty$-control. He felt that $H^\infty$-control could now accommodate and handle real-world uncertainties to some extent; he believed that this ought to be substantiated by real applications in industry. Today, of course, one can witness many such industrial applications of $H^\infty$-control.

We cannot forget his leadership in making the world of Japanese control and system theory what it is today; it would simply not be the same without him. As with any eminent scientist, Hidenori's role was an almost automatic consequence of his own extraordinary research. His papers stimulated generations of young Japanese researchers: we believe that every Japanese scientist who contributed to this volume was enlightened by his legacy.

In addition to his scholarly contributions, Hidenori's influence on the community overall became even more direct when, during the 9th IFAC World Congress in Budapest in 1984, he expressed frustration with particular aspects of the state of control theory in Japan. This prompted a serious discussion among several young scientists on how to make the community more internationally recognized. This discussion materialized itself as a sequence of closed seminars among selected researchers. Yamamoto still recalls with pleasure the intense, serious, yet friendly discussions shared in those seminars: the average age of the members was under 40, none of them were full professors (some were still research associates), and, due to the lack of open positions in academia, they were frequently quite concerned about the future of the discipline. Each seminar talk often extended for more than three hours, late into the night, and delved into the fundamental motivation for the discussed research topic itself. While the general atmosphere was open and constructive, it was simultaneously very critical, and, as a consequence, it provided the members with an excellent opportunity to test their ideas. Hidenori was always at the core of these seminars.

The outcome was tremendous. It led to a great deal of joint research among members of the group, and the collaboration had significant effects on the community at large. The results were published in major journals and garnered Japanese control theorists considerable international recognition. Moreover, an indirect outgrowth of the aforementioned seminars was the organization of two international conferences in Japan: MTNS'91[1] in 1991 and the 35th IEEE CDC[2] in 1996, both held in Kobe. They were milestones for Japanese system and control theorists; again, while such success depended on the collaboration of many individuals, it was Hidenori's leadership specifically which facilitated and strengthened much of the interaction.

This tradition of cooperative research and desire for innovation has been inherited in different ways by the younger generation; today we are surrounded by prospective researchers whose futures depend on the paths that control and system theory will take. Because of the common apprehension that control theory has reached a plateau, Hidenori acutely understood the need to venture in novel directions. Inspired by his current affiliation with the Department of Complexity Science and Engineering, he has recently begun a new attempt to apply control theory to problems such as motor control in biology. In the spirit of Wiener, his analysis has centered around comprehending and elucidating general mechanisms in biological, information, or other complex systems with control as a guiding principle. Hence the title of this Festschrift Symposium.

The papers in the present volume are generally based upon the presentations given in the Festschrift Symposium; unfortunately, however, we could not include all the papers that were presented there. The contributions are broadly classified into five categories: Signal Processing, Identification, Robust Control, Hybrid, Chaotic and Nonlinear Systems, and Control Applications. The first part consists of four papers on signal processing, by Byrnes and Lindquist, Georgiou, Y. Yamamoto and Nagahara, and Kwon and Han. Hidenori was and still remains very interested in signal processing and its connection to discrete-time control theory. The second part contains three papers on identification, by Milanese and Taragna, Vidyasagar and Karandikar, and Oishi. Hidenori has published many papers in identification, and he has constantly emphasized its importance in practical

[1]Mathematical Theory of Networks and Systems
[2]Conference on Decision and Control

applications. Part Three includes five papers on robust control, by Anderson and Bombois, Chen and Hara, Tsumura, Sugie, and Hara and Iwasaki. It is well known that robust control is the central area in which Hidenori made a number of key contributions. Part Four contains five papers on hybrid, chaotic and nonlinear systems, by Guo, Imura, S. Yamamoto and Ushio, Mita and Nam, and Sampei, et al. They represent a new horizon for control theory, and these papers show us forthcoming challenges in these areas. The final part consists of three papers in control applications, by Tomizuka, Adachi, and Hashimoto, et al. In view of Hidenori's profound leadership in the application of control theory to real systems, this seems to be an appropriate conclusion to the entire volume.

Each of the contributions is in one way or another related to Hidenori's research: the collection represents the state-of-the-art and also points to future directions for the subject as a whole. It is, therefore, our great pleasure to make available to the public the highlights of the Festschrift Symposium. We are deeply indebted to the authors who did such a superb job of writing valuable manuscripts in spite of their tight schedules, and we would like to thank the organizers of the symposium: particularly, Masao Ikeda, Shinji Hara, Koji Tsumura and Shun Ushida, for their smooth operation of the conference. We are very grateful to Ms. Ann Kostant of Birkhäuser for her help and encouragement at every step of the editorial process and for her careful answers to our most detailed questions.

Hidenori has always been our hero. In honor of his sixtieth birthday, we wish him equal or even greater success for the *next* sixty years!

*Koichi Hashimoto*
*Yasuaki Oishi*
*Yutaka Yamamoto*
September 2002

# Awards and Honors of Hidenori Kimura

1972: Outstanding Paper Award, the Society of Instrument and Control Engineers [29]

1983: Outstanding Paper Award, the Society of Instrument and Control Engineers [51]

1984: Automatica Prize Paper Award, IFAC [48]

1985: Outstanding Paper Award, IEEE Control Systems Society [53]

1989: Fellow, the Society of Instrument and Control Engineers

1990: Fellow, IEEE

1990: Automatica Prize Paper Award, IFAC [67]

1990: Outstanding Invention Award, the Science and Technology Agency of Japan [67]

1993: Outstanding Paper Award, the Society of Instrument and Control Engineers [75]

1993: Chubu District Invention Promotion Award, Japan Institute of Invention and Innovation [67]

1994: Outstanding Paper Award, Fluid Power Technology Promotion Foundation [89]

1996: Distinguished Member Award, IEEE Control Systems Society

1997: Outstanding Paper Award, the Society of Instrument and Control Engineers [98]

1998: Outstanding Book Award, the Society of Instrument and Control Engineers [8]

# Technical Publications of Hidenori Kimura

## Books


[1] H. Kimura, *Theory of Dynamical Systems* (in Japanese). Tokyo, Japan: Sangyo Tosho Publishing, 1974.

[2] M. Ito, H. Kimura, and S. Hosoe, *Design Theory of Linear Control Systems* (in Japanese). Tokyo, Japan: The Society of Instrument and Control Engineers, 1978.

[3] H. Kimura, *Digital Signal Processing and Control* (in Japanese). Tokyo, Japan: Shokodo, 1982.

[4] H. Kimura, H. Mayeda, Y. Inouye, Y. Misaka, and R. Takahashi, *Control System Design Using PC9801* (in Japanese). Tokyo, Japan: Nikkan Kogyo Shimbun, 1984.

[5] H. Kimura and S. Kodama, Eds., *Recent Advances in Mathematical Theory of Systems, Control, Networks and Signal Processing I and II: Proceedings of the International Symposium MTNS-91.* Tokyo, Japan: Mita Press, 1992.

[6] H. Kimura, *Fourier and Laplace Analyses* (in Japanese). Tokyo, Japan: Iwanami Shoten, Publishers, 1993.

[7] H. Kimura, T. Fujii, and T. Mori, *Robust Control* (in Japanese). Tokyo, Japan: Corona Publishing, 1993.

[8] H. Kimura, *Chain-Scattering Approach to $H^\infty$-Control.* Systems & Control: Foundations & Applications, Boston, MA: Birkhäuser, 1996.

[9] H. Kimura, T. Mita, S. Shin, and H. Kuzuya, *Design Theory of Control Systems and CAD Tools* (in Japanese). Tokyo, Japan: Corona Publishing, 1998.

[10] H. Kimura, *$H^\infty$ Control* (in Japanese). Tokyo, Japan: Corona Publishing, 2000.

# Book Chapters

[11] H. Kimura and H. Iwase, "On directional interpolation in $H^\infty$," in *Linear Circuits, Systems and Signal Processing: Theory and Application*, C. I. Byrnes, C. F. Martin, and R. E. Saeks, Eds. Amsterdam, The Netherlands: North-Holland, 1988, pp. 551–560.

[12] H. Kimura, "State space approach to the classical interpolation problem and its applications," in *Three Decades of Mathematical System Theory: A Collection of Surveys at the Occasion of the 50th Birthday of Jan C. Willems*, H. Nijmeijer and J. M. Schumacher, Eds. Berlin, Germany: Springer-Verlag, 1989, pp. 241–275.

[13] H. Kimura, "Application of classical interpolation theory," in *Linear Circuits, Systems and Signal Processing: Advanced Theory and Applications*, N. Nagai, Ed. New York, NY: Marcel Dekker, 1990, pp. 61–85.

[14] H. Kimura, "LQG as a design theory," in *Mathematical System Theory: The Influence of R. E. Kalman: A Festschrift in Honor of Professor R. E. Kalman on the Occasion of His 60th Birthday*, A. C. Antoulas, Ed. Berlin, Germany: Springer-Verlag, 1991, pp. 147–157.

[15] H. Kimura, "Generalized chain-scattering approach to $H^\infty$ control problems," in *Control of Uncertain Dynamic Systems: A Collection of Papers Presented at the International Workshop on Robust Control, San Antonio, Texas, March 1991*, S. P. Bhattacharyya and L. H. Keel, Eds. Boca Raton, FL: CRC Press, 1991, pp. 21–38.

[16] H. Kimura, "$(J, J')$-lossless factorization using conjugations of zero and pole extractions," in *Robust Control: Proceedings of a Workshop Held in Tokyo, Japan, June 23–24, 1991*, S. Hosoe, Ed. Berlin, Germany: Springer-Verlag, 1992, pp. 1–8.

[17] X. Xin and H. Kimura, "Chain-scattering approach to non-standard $H^\infty$ control problems," in *Systems and Networks: Mathematical Theory and Applications: Proceedings of the International Symposium MTNS '93 Held in Regensburg, Germany, August 2–6, 1993*, vol. 1: Key Invited Lectures, U. Helmke, R. Mennicken, and J. Saurer, Eds. Berlin, Germany: Akademie Verlag, 1994, pp. 183–208.

[18] S. Yamamoto and H. Kimura, "Robust stabilization for parametric uncertainty with application to magnetic levitation," in *Robust Control Theory*, B. A. Francis and P. P. Khargonekar, Eds. New York, NY: Springer-Verlag, 1995, pp. 129–142.

[19] H. Kimura, "Nonuniqueness, uncertainty, and complexity in modeling," in *Applied and Computational Control, Signals, and Circuits*, vol. 1, B. N. Datta, Ed. Boston, MA: Birkhäuser, 1999, pp. 121–150.

[20] H. Kimura, "Towards a system theory for model set: Chain-scattering approach," in *Dynamical Systems, Control, Coding, Computer Vision: New Trends, Interfaces, and Interplay*, G. Picci and D. S. Gilliam, Eds. Basel, Switzerland: Birkhäuser, 1999, pp. 135–156.

[21] H. Kimura, "Monotonicity of performance with respect to its specification in $H^\infty$ control," in *Open Problems in Mathematical Systems and Control Theory*, V. D. Blondel, E. D. Sontag, M. Vidyasagar, and J. C. Willems, Eds. London, UK: Springer-Verlag, 1999, pp. 131–136.

[22] M. Yanagisawa and H. Kimura, "A control problem for Gaussian states," in *Learning, Control and Hybrid Systems: Festschrift in Honor of Bruce Allen Francis and Mathukumalli Vidyasagar on the Occasion of Their 50th Birthdays*, Y. Yamamoto and S. Hara, Eds. London, UK: Springer-Verlag, 1999, pp. 294–313.

# Journal Articles

[23] H. Kimura, "C-form of switching functions and its applications" (in Japanese), *Transactions of the Society of Instrument and Control Engineers*, vol. 3, no. 4, pp. 316–322, 1967.

[24] H. Kimura, "Linear differential games with terminal payoff," *IEEE Transactions on Automatic Control*, vol. AC-15, no. 1, pp. 58–66, 1970.

[25] H. Kimura, "A game-theoretic approach to control and stabilization of systems with disturbances" (in Japanese), *Transactions of the Society of Instrument and Control Engineers*, vol. 6, no. 4, pp. 366–372, 1970.

[26] S. Arimoto and H. Kimura, "Optimum input test signals for system identification: An information-theoretical approach," *International Journal of Systems Science*, vol. 1, no. 3, pp. 279–290, 1971.

[27] H. Kimura, "Recursive state estimation with nonlinear sensor" (in Japanese), *Transactions of the Society of Instrument and Control Engineers*, vol. 7, no. 4, pp. 324–330, 1971.

[28] H. Kimura, "Differential games of prescribed duration," *Journal of Optimization Theory and Its Applications*, vol. 8, no. 6, pp. 431–440, 1971.

[29] H. Kimura, "Differential game and its applications" (in Japanese), *Journal of the Society of Instrument and Control Engineers*, vol. 10, no. 9, pp. 637–650, 1971.

[30] H. Kimura, "Sequentially best estimators for linear systems with nonlinear noise-free sensors," *International Journal of Control*, vol. 17, no. 1, pp. 145–160, 1973.

[31] H. Kimura, "Quasi-static process and control" (in Japanese), *Transactions of the Society of Instrument and Control Engineers*, vol. 9, no. 6, pp. 647–651, 1973.

[32] H. Kimura, "Synthesis of self-oscillating system generating a prescribed periodic waveform" (in Japanese), *The Transactions of the Institute of Electronics and Communication Engineers of Japan*, vol. 56-A, no. 4, pp. 210–217, 1973.

[33] H. Kimura, "An inverse problem for second order differential systems and waveform design," *SIAM Journal on Applied Mathematics*, vol. 26, no. 4, pp. 730–744, 1974.

[34] H. Kimura and K. Hirase, "A nonlinear conveyer model" (in Japanese), *Transactions of the Society of Instrument and Control Engineers*, vol. 11, no. 1, pp. 120–121, 1975.

[35] H. Kimura, "Pole assignment by gain output feedback," *IEEE Transactions on Automatic Control*, vol. AC-20, no. 4, pp. 509–516, 1975.

[36] H. Kimura, "A further result on the problem of pole assignment by output feedback," *IEEE Transactions on Automatic Control*, vol. AC-22, no. 3, pp. 458–463, 1977.

[37] H. Kimura, "Geometric structure of observers for linear feedback control laws," *IEEE Transactions on Automatic Control*, vol. AC-22, no. 5, pp. 846–855, 1977.

[38] H. Kimura, "On pole assignment by output feedback," *International Journal of Control*, vol. 28, no. 1, pp. 11–22, 1978.

[39] H. Kimura, "Linear function observers of mixed-type," *International Journal of Control*, vol. 28, no. 3, pp. 441–455, 1978.

[40] H. Kimura, "Deadbeat function observers for discrete-time linear systems," *SIAM Journal on Control and Optimization*, vol. 16, no. 6, pp. 880–894, 1978.

[41] A. Sugie, Y. Inoue, and H. Kimura, "Stabilizing control of the coupled inverted pendulums" (in Japanese), *Transactions of the Society of Instrument and Control Engineers*, vol. 14, no. 5, pp. 591–597, 1978.

[42] T. Yamada and H. Kimura, "Multivariable control of cold tandem mill" (in Japanese), *Transactions of the Society of Instrument and Control Engineers*, vol. 15, no. 5, pp. 647–653, 1979.

[43] A. Sugie and H. Kimura, "An algorithm for decentralized pole assignment and its application to the coupled inverted pendulums" (in Japanese), *Transactions of the Society of Instrument and Control Engineers*, vol. 16, no. 1, pp. 70–76, 1980.

[44] H. Kimura, "A new approach to the perfect regulation and the bounded peaking in linear multivariable control systems," *IEEE Transactions on Automatic Control*, vol. AC-26, no. 1, pp. 253–270, 1981.

[45] H. Kimura and Y. Tanaka, "Minimal-time minimal-order deadbeat regulator with internal stability," *IEEE Transactions on Automatic Control*, vol. AC-26, no. 6, pp. 1276–1282, 1981.

[46] Y. Tanaka and H. Kimura, "Minimal time output deadbeat control with internal stability" (in Japanese), *Transactions of the Society of Instrument and Control Engineers*, vol. 17, no. 2, pp. 210–216, 1981.

[47] M. Kamitori and H. Kimura, "Calculation of output response of linear systems by splitting method" (in Japanese), *Transactions of the Society of Instrument and Control Engineers*, vol. 17, no. 4, pp. 524–525, 1981.

[48] H. Kimura, "Perfect and subperfect regulation in linear multivariable control systems," *Automatica*, vol. 18, no. 2, pp. 125–145, 1982.

[49] H. Kimura, "Synthesis of high-gain feedback systems based on root loci assignment" (in Japanese), *Transactions of the Society of Instrument and Control Engineers*, vol. 18, no. 5, pp. 429–435, 1982.

[50] H. Kimura and M. Toboh, "Design of digital controller based on partial model matching" (in Japanese), *Transactions of the Society of Instrument and Control Engineers*, vol. 18, no. 8, pp. 777–783, 1982.

[51] H. Kimura and O. Sugiyama, "Design of robust controller using perfect regulation and perfect observation" (in Japanese), *Transactions of the Society of Instrument and Control Engineers*, vol. 18, no. 10, pp. 955–960, 1982.

[52] H. Kimura, "Optimal $L_2$-approximation with fixed poles," *Systems & Control Letters*, vol. 2, no. 5, pp. 257–261, 1983.

[53] H. Kimura, "Robust stabilizability for a class of transfer functions," *IEEE Transactions on Automatic Control*, vol. AC-29, no. 9, pp. 788–793, 1984.

[54] K. Hashimoto and H. Kimura, "Rounding error analysis of finite settling time control with floating point arithmetic" (in Japanese), *Transactions of the Society of Instrument and Control Engineers*, vol. 20, no. 1, pp. 36–42, 1984.

[55] H. Kimura, "Generalized Schwarz form and lattice-ladder realizations of digital filters," *IEEE Transactions on Circuits and Systems*, vol. CAS-32, no. 11, pp. 1130–1139, 1985.

[56] K. Hashimoto and H. Kimura, "A parallel architecture for recursive least square method" (in Japanese), *The Transactions of the Institute of Electronics and Communication Engineers of Japan*, vol. J68-A, no. 11, pp. 1148–1156, 1985.

[57] H. Kimura, "On interpolation-minimization problem in $H^\infty$," *Control: Theory and Advanced Technology*, vol. 2, no. 1, pp. 1–25, 1986.

[58] M. Vidyasagar and H. Kimura, "Robust controllers for uncertain linear multivariable systems," *Automatica*, vol. 22, no. 1, pp. 85–94, 1986.

[59] K. Hashimoto and H. Kimura, "An adaptive equalizer based on parallel computation scheme" (in Japanese), *The Transactions of the Institute of Electronics and Communication Engineers of Japan*, vol. J69-A, no. 3, pp. 400–407, 1986.

[60] Y. Zhao and H. Kimura, "Dead-beat control with robustness," *International Journal of Control*, vol. 43, no. 5, pp. 1427–1440, 1986.

[61] H. Kimura, "Directional interpolation approach to $H^\infty$-optimization and robust stabilization," *IEEE Transactions on Automatic Control*, vol. AC-32, no. 12, pp. 1085–1093, 1987.

[62] H. Kimura and T. Osada, "Canonical pipelining of lattice filters," *IEEE Transactions on Acoustics, Speech and Signal Processing*, vol. ASSP-35, no. 6, pp. 878–887, 1987.

[63] H. Kimura and Y. Honoki, "A hybrid approach to high resolution two dimensional spectrum estimation," *IEEE Transactions on Acoustics, Speech and Signal Processing*, vol. ASSP-35, no. 7, pp. 1024–1036, 1987.

[64] H. Kimura, "Directional interpolation in the state space," *Systems & Control Letters*, vol. 10, no. 5, pp. 317–324, 1988.

[65] Y. Zhao and H. Kimura, "Multivariable dead-beat control with robustness," *International Journal of Control*, vol. 47, no. 1, pp. 229–255, 1988.

[66] Y. Zhao and H. Kimura, "Two-degree-of-freedom dead-beat control system with robustness," *International Journal of Control*, vol. 48, no. 1, pp. 303–315, 1988.

[67] I. Hoshino, Y. Maekawa, T. Fujimoto, H. Kimura, and H. Kimura, "Observer-based multivariable control of the aluminum cold tandem mill," *Automatica*, vol. 24, no. 6, pp. 741–754, 1988.

[68] H. Kimura, "Conjugation, interpolation and model-matching in $H^\infty$," *International Journal of Control*, vol. 49, no. 1, pp. 269–307, 1989.

[69] K. Hashimoto and H. Kimura, "A new parallel algorithm for inverse dynamics," *International Journal of Robotics Research*, vol. 8, no. 1, pp. 63–76, 1989.

[70] Y. Zhao and H. Kimura, "Two-degree-of-freedom dead-beat control system with robustness: Multivariable case," *International Journal of Control*, vol. 49, no. 2, pp. 667–679, 1989.

[71] R. Kawatani and H. Kimura, "Synthesis of reduced-order $H^\infty$ controllers based on conjugation," *International Journal of Control*, vol. 50, no. 2, pp. 525–541, 1989.

[72] K. Hashimoto, T. Kimoto, K. Ohashi, K. Arai, M. Kagoshima, and H. Kimura, "Cartesian-space nonlinear feedback control with a parallel processing system," *Journal of the Robotics Society of Japan*, vol. 8, no. 1, pp. 48–56, 1990.

[73] K. Hashimoto, K. Ohashi, and H. Kimura, "An implementation of a parallel algorithm for real-time model-based control on a network of microprocessors," *International Journal of Robotics Research*, vol. 9, no. 6, pp. 37–47, 1990.

[74] H. Kimura, T. Oike, A. Miura, K. Akai, and T. Kida, "Robust stability: Degree assignment and its applications to the control of flexible structures," *International Journal of Robust and Nonlinear Control*, vol. 1, no. 3, pp. 153–169, 1991.

[75] R. Kawatani, M. Yamashita, K. Fujimori, and H. Kimura, "Control of automobile active suspension based on $H^\infty$ control theory: State feedback control of a quarter-car model" (in Japanese), *Transactions of the Society of Instrument and Control Engineers*, vol. 27, no. 5, pp. 554–561, 1991.

[76] H. Kimura, Y. Lu, and R. Kawatani, "On the structure of $H^\infty$ control systems and related extensions," *IEEE Transactions on Automatic Control*, vol. AC-36, no. 6, pp. 653–667, 1991.

[77] R. Kawatani, T. Fujii, and H. Kimura, "On the role of disturbance estimation in $H^\infty$ control systems" (in Japanese), *Transactions of the Society of Instrument and Control Engineers*, vol. 27, no. 9, pp. 990–995, 1991.

[78] H. Kimura, "$(J, J')$-lossless factorization based on conjugation," *Systems & Control Letters*, vol. 19, no. 2, pp. 95–109, 1992.

[79] S. Yamamoto, T. Nishida, and H. Kimura, "Quadratic stabilization tracking control and its application to a high speed positioning control system" (in Japanese), *Transactions of the Society of Instrument and Control Engineers*, vol. 29, no. 1, pp. 55–62, 1993.

[80] S. Yamamoto, Q. Shen, and H. Kimura, "Quadratic stabilization for a class of uncertain linear systems and its application to a magnetic levitation system" (in Japanese), *Transactions of the Society of Instrument and Control Engineers*, vol. 29, no. 3, pp. 334–339, 1993.

[81] T. Zhou and H. Kimura, "Time domain identification for robust control," *Systems & Control Letters*, vol. 20, no. 3, pp. 167–178, 1993.

[82] M. Kawabata, K. Hashimoto, and H. Kimura, "$H_\infty$ control of robot with linearized model," *Transactions of the Japan Society of Mechanical Engineers*, ser. C, vol. 59, no. 566, pp. 3151–3156, 1993.

[83] I. Hoshino, M. Kawai, M. Kokubo, T. Matsuura, H. Kimura, and H. Kimura, "Observer-based multivariable flatness control of a cold rolling mill," *Control Engineering Practices*, vol. 1, no. 6, pp. 917–925, 1993.

[84] H. Kimura, "Chain-scattering representation, $J$-lossless factorization and $H^\infty$ control," *Journal of Mathematical Systems, Estimation, and Control*, vol. 4, no. 4, pp. 401–450, 1994.

[85] S. Yamamoto, T. Nishida, T. Kawabata, and H. Kimura, "Robust motion control based on quadratic stabilization and its application to two-inertia systems," *JSME International Journal*, ser. C, vol. 37, no. 3, pp. 559–564, 1994.

[86] T. Zhou and H. Kimura, "Simultaneous identification of nominal model, parametric uncertainty and unstructured uncertainty for robust control," *Automatica*, vol. 30, no. 3, pp. 391–402, 1994.

[87] T. Zhou and H. Kimura, "Robust control of the Sydney benchmark problem with intermittent adaptation," *Automatica*, vol. 30, no. 4, pp. 629–632, 1994.

[88] Y. Zhao and H. Kimura, "Optimization of control input for deadbeat control systems," *International Journal of Control*, vol. 59, no. 4, pp. 1107–1117, 1994.

[89] K. Hayakawa and H. Kimura, "Modeling and decoupling control of an automobile suspension system" (in Japanese), *Transactions of the Society of Instrument and Control Engineers*, vol. 30, no. 5, pp. 568–576, 1994.

[90] X. Xin and H. Kimura, "Singular $(J, J')$-lossless factorization for strictly proper functions," *International Journal of Control*, vol. 59, no. 6, pp. 1383–1400, 1994.

[91] X. Xin and H. Kimura, "$(J, J')$-lossless factorization for descriptor systems," *Linear Algebra and Its Applications*, vols. 205–206, pp. 1289–1318, 1994.

[92] M. Yamashita, K. Fujimori, K. Hayakawa, and H. Kimura, "Application of $H_\infty$ control to active suspension systems," *Automatica*, vol. 30, no. 11, pp. 1717–1729, 1994.

[93] S. Ogura, T. Ebine, K. Hashimoto, and H. Kimura, "Visual feedback control based on $H^\infty$ control theory" (in Japanese), *Transactions of the Society of Instrument and Control Engineers*, vol. 30, no. 12, pp. 1505–1511, 1994.

[94] K. Hashimoto, T. Inoue, and H. Kimura, "Visual servoing: Nonlinear control approach" (in Japanese), *Journal of the Robotics Society of Japan*, vol. 13, no. 2, pp. 263–269, 1995.

[95] T. Zhou and H. Kimura, "Structure of model uncertainty for a weakly corrupted plant," *IEEE Transactions on Automatic Control*, vol. AC-40, no. 4, pp. 639–655, 1995.

[96] H. Kimura, "$J$-lossless conjugation and classical interpolation problems," *Archiv für Elektronik und Übertragungstechnik*, vol. 49, nos. 5–6, pp. 313–319, 1995.

[97] S. Yamamoto and H. Kimura, "On structured singular values of reciprocal matrices," *Systems & Control Letters*, vol. 26, no. 3, pp. 163–166, 1995.

[98] I. Hoshino, T. Abe, H. Kimura, and H. Kimura, "Observer-based roll eccentricity control via roll balancing force" (in Japanese), *Transactions of the Society of Instrument and Control Engineers*, vol. 31, no. 8, pp. 1114–1121, 1995.

[99] H. Oku, S. Yamamoto, and H. Kimura, "Structured singular value of symmetric matrix and its lower bound" (in Japanese), *Transactions of the Society of Instrument and Control Engineers*, vol. 31, no. 11, pp. 1824–1827, 1995.

[100] S. Ushida and H. Kimura, "A counterexample to Mustafa–Glovers' monotonicity conjecture," *Systems & Control Letters*, vol. 28, no. 3, pp. 129–137, 1996.

[101] H. Kimura, "From output feedback pole-assignment to robust stabilization," *European Journal of Control*, vol. 2, no. 3, pp. 160–165, 1996.

[102] S. Yamamoto and H. Kimura, "Quadratic stabilization control via $H^\infty$ controller and tuning up by time-varying gain" (in Japanese), *Transactions of the Society of Instrument and Control Engineers*, vol. 32, no. 4, pp. 486–494, 1996.

[103] K. Hashimoto, T. Ebine, and H. Kimura, "Visual servoing with hand-eye manipulator: Optimal control approach," *IEEE Transactions on Robotics and Automation*, vol. 12, no. 5, pp. 766–774, 1996.

[104] W. Kongprawechnon and H. Kimura, "$J$-lossless conjugation and factorization for discrete-time systems," *International Journal of Control*, vol. 65, no. 5, pp. 867–884, 1996.

[105] S. Yamamoto, T. Ueda, and H. Kimura, "Quadratic stabilization approach to coupled three-inertia benchmark problem" (in Japanese), *Transactions of the Society of Instrument and Control Engineers*, vol. 32, no. 7, pp. 1027–1034, 1996.

[106] L. Baramov and H. Kimura, "Nonlinear local $J$-lossless conjugation and factorization," *International Journal of Robust and Nonlinear Control*, vol. 6, no. 8, pp. 869–893, 1996.

[107] P. H. Lee, H. Kimura, and Y. C. Soh, "On the lossless and $J$-lossless embedding theorems in $H^\infty$," *Systems & Control Letters*, vol. 29, no. 1, pp. 1–7, 1996.

[108] S. Yamamoto, S. Ushida, and H. Kimura, "Quadratic stabilization by $H^\infty$ state feedback control with adjustable parameter" (in Japanese), *Transactions of the Institute of Systems, Control and Information Engineers*, vol. 10, no. 1, pp. 11–17, 1997.

[109] P. H. Lee, H. Kimura, and Y. C. Soh, "$J$-lossless conjugations and $J$-lossless factorizations of exponentially stable linear time-varying systems," *International Journal of Control*, vol. 66, no. 2, pp. 199–212, 1997.

[110] L. Baramov and H. Kimura, "Nonlinear coprime factorizations and parametrization of a class of stabilizing controllers," *International Journal of Control*, vol. 66, no. 3, pp. 413–434, 1997.

[111] P. H. Lee, H. Kimura, and Y. C. Soh, "Co-$(J'', J)$-lossless co-conjugations and co-$(J'', J)$-lossless factorizations of linear time-varying systems," *International Journal of Control*, vol. 68, no. 4, pp. 731–752, 1997.

[112] N. Nishiuma, H. Futatsugi, and H. Kimura, "Estimation of air mass into cylinder with extended Kalman filter for the control of injected SI engine" (in Japanese), *Transactions of the Society of Instrument and Control Engineers*, vol. 33, no. 5, pp. 397–402, 1997.

[113] L. Baramov and H. Kimura, "Nonlinear $L_2$-gain suboptimal control," *Automatica*, vol. 33, no. 7, pp. 1247–1262, 1997.

[114] P. H. Lee, H. Kimura, and Y. C. Soh, "Characterizations of $J$-lossless time-varying systems via the chain scattering approach," *International Journal of Robust and Nonlinear Control*, vol. 8, no. 1, pp. 39–48, 1998.

[115] P. H. Lee, H. Kimura, and Y. C. Soh, "Chain scattering approach to $H^\infty$ control for time-varying systems," *International Journal of Control*, vol. 71, no. 2, pp. 195–218, 1998.

[116] H. Itoh, D. Kyono, Y. Oishi, and H. Kimura, "Identification of a model set and its performance analysis based on diameters" (in Japanese), *Transactions of the Society of Instrument and Control Engineers*, vol. 34, no. 8, pp. 1005–1012, 1998.

[117] M. Yanagisawa and H. Kimura, "Control for Gaussian states," *The Transactions of the Institute of Electronics, Information and Communication Engineers*, vol. J81-A, no. 12, pp. 1715–1721, 1998.

[118] K. Hayakawa, K. Matsumoto, M. Yamashita, Y. Suzuki, K. Fujimori, and H. Kimura, "Robust $H^\infty$-output feedback control of decoupled automobile active suspension systems," *IEEE Transactions on Automatic Control*, vol. AC-44, no. 2, pp. 392–396, 1999.

[119] H. Oku and H. Kimura, "Recursive algorithms of the state-space subspace system identification" (in Japanese), *Transactions of the Society of Instrument and Control Engineers*, vol. 35, no. 6, pp. 800–805, 1999.

[120] H. Oku and H. Kimura, "A recursive 4SID from the input-output point of view," *Asian Journal of Control*, vol. 1, no. 4, pp. 258–269, 1999.

[121] N. Nishiuma and H. Kimura, "Gain-scheduling control of uncertain systems with time-varying delay" (in Japanese), *Transactions of the Society of Instrument and Control Engineers*, vol. 35, no. 12, pp. 1559–1566, 1999.

[122] W. Kongprawechnon, S. Ushida, and H. Kimura, "Structure of discrete-time $H^\infty$ controller," *ScienceAsia*, vol. 26, no. 1, pp. 49–55, 2000.

[123] I. Shinohara, S. Kikuchi, Y. Oishi, and H. Kimura, "An algorithm to compute the admissible parameter region of a single-layered perceptron" (in Japanese), *Transactions of the Society of Instrument and Control Engineers*, vol. 36, no. 7, pp. 622–624, 2000.

[124] S. Ushida, T. Ohba, H. Kimura, and Y. Oishi, "On the closed-loop structure of $H^\infty$ control systems," *Transactions of the Society of Instrument and Control Engineers*, vol. 36, no. 12, pp. 1093–1100, 2000.

[125] K. Doya, H. Kimura, and A. Miyamura, "Motor control: Neural models and systems theory," *International Journal of Applied Mathematics and Computer Science*, vol. 11, no. 1, pp. 77–104, 2001.

[126] I. Jikuya and H. Kimura, "Analysis of inclusion and equivalence between model sets in the chain-scattering representation" (in Japanese), *Transactions of the Society of Instrument and Control Engineers*, vol. 37, no. 2, pp. 178–180, 2001.

[127] Y. Oishi and H. Kimura, "A learning-theoretic approach to model-set identification," *Asian Journal of Control*, vol. 3, no. 1, pp. 10–20, 2001.

[128] Q. Zheng and H. Kimura, "Just-in-time modeling for function prediction and its applications," *Asian Journal of Control*, vol. 3, no. 1, pp. 35–44, 2001.

[129] A. Miyamura and H. Kimura, "Stability of feedback error learning scheme," *Systems & Control Letters*, vol. 45, no. 4, pp. 303–316, 2002.

[130] H. Oku and H. Kimura, "Recursive 4SID algorithms using gradient type subspace tracking," *Automatica*, vol. 38, no. 6, pp. 1035–1043, 2002.

# Part I

# Signal Processing

# A Convex Optimization Approach to Generalized Moment Problems

**Christopher I. Byrnes and Anders Lindquist**

*To Hidenori Kimura on the occasion of his 60th birthday*

ABSTRACT In this paper we present a universal solution to the generalized moment problem, with a nonclassical complexity constraint. We show that this solution can be obtained by minimizing a strictly convex nonlinear functional. This optimization problem is derived in two different ways. We first derive this intrinsically, in a geometric way, by path integration of a one-form which defines the generalized moment problem. It is observed that this one-form is closed and defined on a convex set, and thus exact with, perhaps surprisingly, a strictly convex primitive function. We also derive this convex functional as the dual problem of a problem to maximize a cross entropy functional. In particular, these approaches give a constructive parameterization of all solutions to the Nevanlinna–Pick interpolation problem, with possible higher-order interpolation at certain points in the complex plane, with a degree constraint as well as all solutions to the rational covariance extension problem — two areas which have been advanced by the work of Hidenori Kimura. Illustrations of these results in system identification and probability are also mentioned.

**Keywords:** Moment problems, Convex optimization, Nevanlinna–Pick interpolation, Covariance extension, Systems identification, Kullback–Leibler distance

## 1 Introduction

Hidenori Kimura has been at the forefront of many major developments in the theory of systems and control. Among the many things for which he is widely recognized, two of his contributions have influenced the research results we present in this paper, which we dedicate to him on the occasion of his 60th birthday. The first is his pioneering work on the rational covariance extension problem, and the second is his profound work on robust stabilization and control.

In the rational covariance extension problem, one seeks to parameterize and compute all shaping filters, of a bounded degree, which shape white noise into a process with a given window of covariance lags. In [31], Kimura was able to give a neat parameterization of rational filters of a bounded degree and with a given window of Laurent coefficients, in terms of Szegö polynomials of the first and second kind. This parameterization was independently discovered by Georgiou [23]. From the Kimura–Georgiou parameterization one can see that the space of shaping filters with a fixed window of covariance lags is a smooth manifold [3]. Georgiou [23] used degree theory for functions on manifolds to give a very basic existence result and conjectured a refinement of his result that would give a complete parameterization of all solutions, of no more than a given degree, to the rational covariance extension problem. In [6] Georgiou's conjecture was proved by first noting that the Kimura–Georgiou parameterization defined the leaves of a foliation of a space of positive real rational functions having a bounded degree. A second observation used in [6] was that the fast filtering algorithm of [40, 41], viewed as a dynamical system on the same space of positive real functions, also defines a foliation of this space, with its leaves being the stable manifolds through various equilibria [4, 5]. The proof of Georgiou's conjecture was obtained as a corollary of a theorem about the geometry of these two foliations, including the fact that leaves of one intersect leaves of the other transversely.

The paper [32] on robust stabilization of plants with a fixed number of unstable poles and Nyquist plot in a neighborhood of the Nyquist plot of a nominal plant was one of the key contributions which ushered in the era of $H^\infty$ control. In this paper, Kimura gave necessary and sufficient conditions for the solution of this problem in terms of the classical Nevanlinna–Pick interpolation problem, a methodology he would continue to use and develop for robust control [33, 34, 35, 36, 37]. We also refer the reader to the books [21, 53] and the references therein. Nevanlinna–Pick interpolation for bounded-real rational functions is now one of the tools commonly used in robust control. Indeed, it is widely known that the sensitivity function of a controlled plant must take on certain prescribed values at the unstable poles and zeroes of the plant to be internally stabilized and that the bounded-real condition gives a bound on the $H^\infty$ norm of the sensitivity function. Again it is desirable that the interpolant, being the sensitivity function, has an a priori bounded degree. This leads naturally to the Nevanlinna–Pick problem with degree constraint. A complete parameterization of the class of such interpolants was conjectured by Georgiou in [24] and recently settled in [25] for interpolation at distinct points in the complex plane. This can again be enhanced using the geometry of foliations on a space of rational, positive real functions, with the foliation defined by covariance windows being replaced by a foliation whose leaves are defined by fixing the interpolation values [8].

The question of actually finding, or computing solutions to either problem can be solved in the context of nonlinear convex optimization. In [7], we presented a convex optimization approach for determining an arbitrary solution to the rational covariance extension problem with degree constraint for interpolation at distinct points in the complex plane. In this way, one obtains both an algorithm for solving the covariance extension problem and a constructive proof of Georgiou's conjecture. Similarly, in [9] a generalized entropy criterion is developed for solving the rational Nevanlinna–Pick problem with degree constraints. In both problems, the primal problem of maximizing this entropy gain has a very well-behaved dual problem in a finite-dimensional space and gives algorithms for solving both problems with the degree constraints.

At this point, one should ask whether there is a unified point of view from which one can also see why there must be a strictly convex functional whose minimization solves a given problem. While it is true that the general rational Nevanlinna–Pick interpolation problem, allowing for higher-order interpolation at points in the extended complex plane, does include the rational covariance extension problem, there is a more compelling generalization of these problems. It is classical mathematical fact that both problems can be formulated as special cases of the generalized moment problem, dating back to Chebychev and Markov and formulated in terms of convex functional analysis by Carathéodory, Toeplitz, and others. We refer to the classic book [38] for a neat exposition of these topics. In fact, we shall adopt their notation and refer to positive real functions as Carathéodory functions.

In this paper we present a universal solution to the generalized moment problem, with a nonclassical complexity constraint, obtained by minimizing a strictly convex nonlinear functional. This optimization problem is derived in two different ways. We first answer the question of why, intrinsically, there should always be an equivalent convex optimization problem. We settle this question in a geometric way by path integration of a one-form which defines the generalized moment problem. This exposition follows the original calculation in [16] where it is observed that this one-form is closed and defined on a convex set, and thus exact. Since its integral is therefore path-independent, it is intrinsic and is, perhaps surprisingly, a strictly convex functional. In Section 5 we give a new derivation of this convex functional as the dual problem of a problem to maximize a cross entropy functional. In particular, these approaches give a constructive parameterization of all solutions to the Nevanlinna–Pick interpolation problem, with possible higher-order interpolation at certain points in the complex plane, with a degree constraint.

# 2 The rational covariance extension problem as a trigonometric moment problem with a complexity constraint

A function of a complex variable that is analytic in the open unit disc and maps points there into the open right half-plane is called a *Carathéodory function* in this paper. The *covariance extension problem* is to find an infinite extension $c_{n+1}, c_{n+2}, \dots$ of a given positive sequence $n+1$ real numbers $c_0, c_1, \dots, c_n$ such that

$$f(z) = \frac{1}{2}c_0 + c_1 z + c_2 z^2 + c_3 z^3 + \cdots$$

is a rational Carathéodory function of degree at most $n$. Here, by positivity of a sequence we mean positive definiteness of the Toeplitz matrix

$$\begin{bmatrix} c_0 & c_1 & \cdots & c_n \\ c_1 & c_0 & \cdots & c_{n-1} \\ \vdots & \vdots & \ddots & \vdots \\ c_n & c_{n-1} & \cdots & c_0 \end{bmatrix}.$$

A function $f$ that is analytic in the disc is a Carathéodory function if and only if

$$\Phi(\theta) := \text{Re}\{f(e^{i\theta})\}$$

is positive for all $\theta \in [-\pi, \pi]$. Note that

$$\Phi(\theta) = c_0 + 2\sum_{k=0}^{\infty} c_k \cos k\theta.$$

In particular,

$$\frac{1}{2\pi}\int_{-\pi}^{\pi} \Phi(\theta) \cos k\theta \, d\theta = c_k \tag{2.1}$$

for $k = 0, 1, 2, \dots$.

As pointed out in [20], the rational covariance extension problem is a trigonometric moment problem: Given a positive sequence $c_0, c_1, \dots, c_n$, find all positive $\Phi$ such that (2.1) holds for $k = 0, 1, \dots, n$. However, the rational covariance extension problem involves additional constraints. Indeed, for $f$ to be rational of degree at most $n$ it is necessary and sufficient that

$$\Phi(\theta) = \frac{P(\theta)}{Q(\theta)}, \tag{2.2}$$

where $P$ and $Q$ are positive trigonometric polynomials in the span of $\{\cos k\theta\}_{k=0}^{n}$. This is the complexity constraint.

# 3 The generalized moment problem with complexity constraint

There is a vast literature on the generalized moment problem (see, e.g., [1, 2, 27, 38, 49]), in part because so many problems and theorems in pure and applied mathematics, physics and engineering can be formulated as moment problems. The classical problem can be formulated in the following way.

Consider a sequence of continuous functions

$$\alpha_0, \alpha_1, \dots, \alpha_n,$$

defined on some interval $\mathcal{I}$ of the real line. Suppose that $\alpha_0, \dots, \alpha_{r-1}$ are real functions and $\alpha_r, \dots, \alpha_n$ are complex-valued functions whose real and imaginary parts, taken together with $\alpha_0, \dots, \alpha_{r-1}$ are linearly independent over $\mathbb{R}$. This is no restriction since a purely imaginary moment condition can always be reduced to a real one. We also assume that $\alpha_0 = 1$. Let $\mathfrak{P}$ be the real vector space that is the sum of the real span of $\alpha_0, \dots, \alpha_{r-1}$ and the complex span of $\alpha_r, \dots, \alpha_n$. Hence, the real dimension of $\mathfrak{P}$ is $2n - r + 2$. Moreover, let $\mathfrak{P}_+$ be the subset of all functions in $\mathfrak{P}$ that have a positive real part on $\mathcal{I}$. Classically, one can replace the assumption that $\alpha_0 = 1$ by simply assuming that $\mathfrak{P}_+$ is nonempty, in which case $\mathfrak{P}_+$ is an open convex subset of $\mathfrak{P}$.

The moment problem is then to find all monotone, nondecreasing functions $\mu$ of bounded variation such that

$$\int_{\mathcal{I}} \alpha_k(t) d\mu(t) = c_k, \quad k = 0, 1, \dots, n, \tag{3.1}$$

where $c_0, c_1, \dots, c_n$ is a given sequence of complex numbers. This problem has a solution if and only if the sequence $c_0, c_1, \dots, c_n$ is *positive* in the sense that

$$\langle c, q \rangle := \mathrm{Re} \sum_{k=0}^{n} q_k c_k > 0 \tag{3.2}$$

for all $(q_0, q_1, \dots, q_n) \in \mathbb{R}^r \times \mathbb{C}^{n-r+1}$ such that

$$q := \sum_{k=0}^{n} q_k \alpha_k \in \mathfrak{P}_+. \tag{3.3}$$

Clearly, this is a necessary condition since

$$\langle c, q \rangle = \int_{\mathcal{I}} \left[ \mathrm{Re} \sum_{k=0}^{n} q_k c_k \right] d\mu > 0,$$

whenever (3.3) holds, but it can be shown [38] that it is also sufficient. Denote by $\mathfrak{C}_+$ the space of positive sequences. Since $\mathfrak{P}_+$ is nonempty, so is $\mathfrak{C}_+$.

Next, we introduce the nonclassical *complexity constraint*

$$\frac{d\mu}{dt} = \frac{\Psi(t)}{Q(t)}, \quad Q = \mathrm{Re}\{q\}, q \in \mathfrak{P}_+, \tag{3.4}$$

where $\Psi$ is a given positive function in $L^1(\mathcal{I})$ such that

$$\left| \int_{\mathcal{I}} \alpha_k \frac{\Psi}{Q}\, dt \right| < \infty, \quad k = 0, 1, \ldots, n, \tag{3.5}$$

for all $Q = \text{Re}\{q\}, q \in \mathfrak{P}_+$. Let us call the class of such functions $\mathcal{G}_+$. If $\mathcal{I}$ is a finite interval, (3.5) holds for all $\Psi$ in the space $L^1_+(\mathcal{I})$ of positive functions in $L^1(\mathcal{I})$ so then $\mathcal{G}_+ = L^1_+(\mathcal{I})$.

To motivate this problem formulation, let us give a couple of examples, the first of which we have already encountered, for real data, in Section 2.

**Example 3.1.** A function of a complex variable that is analytic in the open unit disc and maps points there into the open left half-plane is called a *Carathéodory function.* Given $n+1$ complex numbers $c_0, c_1, \ldots, c_n$, find an infinite extension $c_{n+1}, c_{n+2}, \ldots$ such that

$$f(z) = \frac{1}{2} c_0 + c_1 z + c_2 z^2 + c_3 z^3 + \cdots$$

is a rational Carathéodory function of degree at most $n$. This *covariance extension problem* is a trigonometric moment problem with degree constraint and can be formulated as above by taking $\Phi = 2\text{Re}\{f\}, \mathcal{I} = [-\pi, \pi]$ and

$$\alpha_k(\theta) = e^{ik\theta}, \quad k = 0, 1, \ldots, n.$$

Moreover, the degree constraint is equivalent to (3.4) if we take

$$\Psi = \text{Re}\{\psi\}, \quad \psi \in \mathfrak{P}_+. \tag{3.6}$$

Finally, $c_0, c_1, \ldots, c_n$ is positive if and only if the Toeplitz matrix

$$\begin{bmatrix} c_0 & c_1 & \cdots & c_n \\ \bar{c}_1 & c_0 & \cdots & c_{n-1} \\ \vdots & \vdots & \ddots & \vdots \\ \bar{c}_n & \bar{c}_{n-1} & \cdots & c_0 \end{bmatrix}$$

is positive definite. □

**Example 3.2.** Given $n+1$ distinct points $z_0, z_1, \ldots, z_n$ in the open unit disc, consider the problem to determine the rational Carathéodory functions $f$ of degree at most $n$ satisfying the interpolation condition

$$f(z_k) = c_k, \quad k = 0, 1, \ldots, n, \tag{3.7}$$

where $c_0, c_1, \ldots, c_n$ are prescribed values in the open right half of the complex plane with $c_0$ real. If the points $z_0, z_1, \ldots, z_n$ are not distinct, the interpolation conditions are modified in the following way. If $z_k = z_{k+1} = \cdots = z_{k+m-1}$, the corresponding interpolation conditions are replaced by

$$f(z_k) = c_k, \quad f'(z_k) = c_{k+1}, \quad \ldots, \quad \frac{1}{(m-1)!} f^{(m-1)}(z_k) = c_{k+m-1}. \tag{3.8}$$

This Nevanlinna–Pick interpolation problem differs from the classical one in that a degree constraint on the interpolant $f$ has been introduced, a restriction motivated by applications [9, 10]. In fact, many problems in systems and control can be reduced to Nevanlinna–Pick interpolation [19, 21, 28, 32, 53, 50], and, as the interpolant generally can be interpreted as a transfer function, the bound on the degree is a natural complexity constraint.

To reformulate this interpolation problem as a generalized moment problem of the type described above, we note that, by the Herglotz Theorem, any rational Carathéodory function of degree at most $n$ such that $f(0)$ is real can be represented as

$$f(z) = \frac{1}{2\pi} \int_{-\pi}^{\pi} \frac{e^{i\theta} + z}{e^{i\theta} - z} \Phi(e^{i\theta}) d\theta \tag{3.9}$$

in terms of a rational spectral density $\Phi$ of degree at most $2n$, which is given by

$$\Phi(e^{i\theta}) = \text{Re}\{f(e^{i\theta})\}. \tag{3.10}$$

Differentiating (3.9), we obtain

$$\frac{1}{\nu!} f^{(\nu)}(z) = \frac{1}{2\pi} \int_{-\pi}^{\pi} \frac{2e^{i\theta}}{(e^{i\theta} - z)^{\nu+1}} \Phi(e^{i\theta}) d\theta, \quad \nu = 1, 2, \ldots.$$

Consequently, the interpolation conditions can be written

$$\int_{-\pi}^{\pi} \alpha_k(\theta) \Phi(e^{i\theta}) d\theta = c_k, \quad k = 0, 1, \ldots, n, \tag{3.11}$$

where

$$\alpha_k(\theta) = \frac{1}{2\pi} \frac{e^{i\theta} + z_k}{e^{i\theta} - z_k}$$

for single interpolation points and

$$\begin{aligned} \alpha_k(\theta) &= \frac{1}{2\pi} \frac{e^{i\theta} + z_k}{e^{i\theta} - z_k}, \\ \alpha_{k+1}(\theta) &= \frac{1}{2\pi} \frac{2e^{i\theta}}{(e^{i\theta} - z_k)^2}, \\ &\vdots \\ \alpha_{k+m-1}(\theta) &= \frac{1}{2\pi} \frac{2e^{i\theta}}{(e^{i\theta} - z_k)^m}, \end{aligned}$$

whenever $z_k = z_{k+1} = \cdots = z_{k+m-1}$. Taking $\mathcal{I} = [-\pi, \pi]$ and $d\mu = \Phi d\theta$, (3.11) are precisely of the moment conditions (3.1). Moreover the degree constraint is equivalent to (3.4) if we take $\Psi$ in the class (3.6).

If the interpolation points $\{z_k\}_0^n$ are distinct, the sequence $c_0, c_1, \dots, c_n$ is positive if and only if the Pick matrix

$$\left[\frac{c_k + \bar{c}_\ell}{1 - z_k \bar{z}_\ell}\right]_{k,\ell=0}^n$$

is positive definite. If $z_k = 0$, $k = 0, 1, \dots, n$, we have the trigonometric moment problem with degree constraint, described in Example 3.1. □

**Example 3.3.** A popular method in systems identification amounts to estimating the first $n+1$ coefficients in an orthogonal basis function expansion

$$G(z) = \frac{1}{2} c_0 f_0(z) + \sum_{k=1}^{\infty} c_k f_k(z)$$

of a transfer function $G(z)$ [51, 52]. The functions $f_0, f_1, f_2, \dots$ are orthonormal on the unit circle, i.e.,

$$\frac{1}{2\pi} \int_{-\pi}^{\pi} f_j(e^{i\theta})^* f_k(e^{i\theta}) d\theta = \delta_{jk}.$$

A general class of such functions is given by

$$f_k(z) = \frac{\sqrt{1 - |\xi_k|^2}}{z - \xi_k} \prod_{j=0}^{k-1} \frac{1 - \xi_j^* z}{z - \xi_j},$$

where $\xi_0, \xi_1, \xi_2, \dots$ are poles to be selected by the user [29].

Given the estimated coefficients $c_0, c_1, \dots, c_n$, the usual problem considered in the literature [47] is to find a rational function $G$ of smallest degree which match these coefficients. Here, however, we consider the corresponding problem where $G$ is a Carathéodory function of degree at most $n$, a problem that has remained open. It can be reformulated as a generalized moment problem with complexity constraint by observing that

$$c_k = \frac{1}{2\pi} \int_{-\pi}^{\pi} f_k(e^{i\theta}) \Phi(e^{i\theta})\, d\theta, \quad k = 0, 1, \dots, n,$$

where $\Phi(e^{i\theta}) = 2\mathrm{Re}\{G(e^{i\theta})\}$. The degree constraint is enforced by choosing $\Psi$ as in (3.6). □

There is one and only one solution to the generalized moment problem with complexity constraint formulated above, and it is determined by the unique minimum of a strictly convex functional. In fact, the following theorem is a version of a result proved in [16], generalizing similar results in analytic interpolation theory [7, 9, 10, 11, 15].

**Theorem 3.4.** *Let $\mathfrak{P}$ be spanned by $C^2$ functions $\alpha_0, \alpha_1, \dots, \alpha_n$, whose nonzero real and imaginary parts form a linearly independent set over $\mathbb{R}$. Moreover, let $c \in \mathfrak{C}_+$. Then, for any $\Psi \in \mathcal{G}_+$, there is one and only one $q \in \mathfrak{P}_+$ solving the generalized moment problem*

$$\int_{\mathcal{I}} \alpha_k(t) \frac{\Psi(t)}{Q(t)}\, dt = c_k, \quad k = 0, 1, \dots, n, \tag{3.12}$$

*where $Q = \text{Re}\{q\}$. In fact, $c$ and $\Psi$ determine $Q$ as the unique minimum in $\mathfrak{P}_+$ of the strictly convex functional*

$$\mathbb{J}_\Psi(q) = \langle c, q \rangle - \int_{\mathcal{I}} \Psi \log Q dt. \tag{3.13}$$

Note that $\mathbb{J}_\Psi(q)$ is finite for all $q \in \mathfrak{P}_+$. In fact, by Jensen's inequality,

$$-\log \int_{\mathcal{I}} \frac{\Psi}{Q} dt \leq \int_{\mathcal{I}} \Psi \log Q dt \leq \log \int_{\mathcal{I}} Q\Psi dt,$$

where both bounds are finite by (3.5). (To see this, for the lower bound take $k = 0$; for the upper bound first take $Q = 1$ in (3.5) and then form the appropriate linear combination.)

In Section 4 we shall outline a geometric proof of Theorem 3.4, in which the functional (3.13) is derived by integrating a 1-form along a path. It is the proof presented in [16]. Another proof is based on duality theory and follows the same pattern as in [7, 9, 10, 11]. In fact, the problem to maximize $\mathbb{J}_\Psi$ over $\mathfrak{P}_+$ is the dual problem in the sense of mathematical programming of the constrained optimization problem in the following theorem.

**Theorem 3.5.** *Let $\mathfrak{P}$ be spanned by $C^2$ functions $\alpha_0, \alpha_1, \dots, \alpha_n$, whose nonzero real and imaginary parts form a linearly independent set over $\mathbb{R}$. For any choice of $\Psi \in \mathcal{G}_+$, the constrained optimization problem to minimize the functional*

$$\mathbb{I}_\Psi(\Phi) = \int_{\mathcal{I}} \Psi(t) \log \frac{\Psi(t)}{\Phi(t)} dt \tag{3.14}$$

*over $L^1_+(\mathcal{I})$ subject to the constraints*

$$\int_{\mathcal{I}} \alpha_k(t)\Phi(t) dt = c_k, \quad k = 0, 1, \dots, n, \tag{3.15}$$

*has a unique solution, and it has the form*

$$\Phi = \frac{\Psi}{Q}, \quad Q = \text{Re}\{q\},$$

*where $q \in \mathfrak{P}_+$ is the unique minimum of (3.13).*

The proof of this theorem can be found in Section 5.

**Example 3.6.** Consider the problem of determining a probability density $\Phi$ on $\mathcal{I} := (-\infty, \infty)$ of the form

$$\Phi(t) = \frac{\Psi(t)}{Q(t)}, \quad Q(t) = q_n t^{2n} + q_{n-1} t^{2n-1} + \cdots + q_0 \tag{3.16}$$

and satisfying the moment conditions

$$\int_{-\infty}^{\infty} t^k \Phi(t) dt = c_k, \quad k = 0, 1, \ldots, 2n, \tag{3.17}$$

where, of course, $c_0 = 1$. Here $\Psi$ could be an a priori selected probability distribution. This is a moment problem of the type described above with

$$\alpha_k(t) = t^k, \quad k = 0, 1, \ldots, 2n.$$

This is a Hamburger moment problem with complexity constraint. The sequence $c_0, c_1, \ldots, c_{2n}$ is positive if and only if the Hankel matrix

$$\begin{bmatrix} c_0 & c_1 & \cdots & c_n \\ c_1 & c_2 & \cdots & c_{n+1} \\ \vdots & \vdots & \ddots & \vdots \\ c_n & c_{n+1} & \cdots & c_{2n} \end{bmatrix}$$

is positive definite. Remarkably, the function

$$\mathbb{I}(\Phi, \Psi) = \mathbb{I}_\Psi(\Phi)$$

is the cross-entropy [26], gain of information [46], directed divergency [39], or the Kullback–Leibler distance between $\Psi$ and $\Phi$. Then the optimization problem of Theorem 3.5 is equivalent to minimizing $\mathbb{I}(\Phi, \Psi)$ subject to the moment conditions (3.1). This gives an interesting interpretation, further pursued in [12], to the present problem: Given an a priori probability density $\Psi$, we want to find another probability density $\Phi$ that has prescribed moments up to order $2n$ and that minimizes the Kullback–Leibler distance to $\Psi$. Similar optimization problems have been considered in the statistical literature, where, however, minimization is generally with respect to $\Psi$; see, e.g., [17, 18]. □

# 4 A geometric derivation of the convex optimization problem

Following [16], we prove Theorem 3.4 by first constructing the dual functional from the moment equations (3.12). To this end, for any $\Psi \in \mathcal{G}_+$, we define the map

$$F^\Psi : \mathfrak{P}_+ \to \mathfrak{C}_+$$

componentwise via

$$F_k^{\Psi}(q) = \int_{\mathcal{I}} \alpha_k \frac{\Psi}{Q}\, dt.$$

We want to prove that the map $F^{\Psi}$ is a homeomorphism. For this we use the following global inverse function theorem proven in [16].

**Lemma 4.1.** *Suppose $M$ and $N$ are $n$-dimensional, topological manifolds and that $N$ is connected. Consider a continuous map $f : M \to N$. Then, $f$ is a homeomorphism if and only if $f$ is injective and proper. In this case, $M$ is connected.*

Recall that $f$ is said to be *proper* if the inverse image $f^{-1}(K)$ is compact for all compact $K \in M$. That $F^{\Psi}$ is proper follows from the following lemma, also proven in [16].

**Lemma 4.2.** *Suppose $\mathfrak{P}$ is a vector space consisting of $C^2$-smooth functions. Then $F^{\Psi} : \mathfrak{P}_+ \to \mathfrak{C}_+$ is proper.*

The domain $\mathfrak{P}_+$ is an open convex subset of $\mathbb{R}^{2n-r+2}$, and clearly the same holds for $\mathfrak{C}_+$. Therefore, it will follow from Lemma 4.1 that $F^{\Psi}$ is a homeomorphism if we can show that $F^{\Psi}$ is injective. We proceed to proving precisely this by constructing the functional (3.13).

Since $\mathfrak{P}_+$ is an open, convex subset of $\mathbb{R}^{2n-r+2}$, it is diffeomorphic to $\mathbb{R}^{2n-r+2}$ and hence Euclidean. Parameterizing $q \in \mathfrak{P}_+$ via $q = \sum_{k=0}^{n} q_k \alpha_k$, we construct the 1-form

$$\omega = \mathrm{Re}\left\{ \sum_{k=0}^{n} [c_k - F_k(q)]\, dq_k \right\},$$

on $\mathfrak{P}_+$, which can be written as

$$\begin{aligned} \omega &= \mathrm{Re}\left\{ \sum_{k=0}^{n} c_k dq_k - \int_{\mathcal{I}} \sum_{k=0}^{n} \alpha_k \frac{\Psi}{Q} dq_k dt \right\} \\ &= \mathrm{Re} \sum_{k=0}^{n} c_k dq_k - \int_{\mathcal{I}} \frac{\Psi}{Q}\, dQ\, dt. \end{aligned}$$

Taking the exterior derivative on $\mathfrak{P}_+$, we obtain

$$d\omega = \int_{\mathcal{I}} \frac{\Psi}{Q^2}\, dQ \wedge dQ\, dt = 0,$$

establishing that the 1-form $\omega$ is closed.

Therefore, by the Poincaré Lemma, there exists a smooth function $\mathbb{J}_{\Psi}$ such that $\omega = d\mathbb{J}_{\Psi}$, and hence

$$\mathbb{J}_{\Psi} = \int \omega = \int \left( \mathrm{Re} \sum_{k=0}^{n} c_k dq_k - \int_{\mathcal{I}} \frac{\Psi}{Q}\, dQ\, dt \right),$$

with the integral being independent of the path between two endpoints. Computing the path integral

$$\int_{q_0}^{q_1} \omega = \left[ \langle c, q \rangle - \int_{\mathcal{I}} \Psi \log Q \, dt \right]_{q_0}^{q_1},$$

we obtain, modulo a constant of integration,

$$\mathbb{J}_\Psi(q) = \langle c, q \rangle - \int_{\mathcal{I}} \Psi \log Q \, dt. \tag{4.1}$$

Since

$$\frac{\partial \mathbb{J}_\Psi}{\partial q_k} = c_k - \int_{\mathcal{I}} \alpha_k \frac{\Psi}{Q} \, dt, \quad k = 0, 1, \dots, n, \tag{4.2}$$

the functional (4.1) has a critical point precisely at the solution of the generalized moment problem. To see this on the second factor of $\mathbb{R}^r \times \mathbb{C}^{n-r+1}$, we decompose the exterior differential as the sum $d = \partial + \bar{\partial}$, where $\bar{\partial}$ is the Cauchy–Riemann differential. Since $\mathbb{J}_\Psi$ is real, to say that $d\mathbb{J}_\Psi = 0$ is to say that $\partial \mathbb{J}_\Psi = 0$ or, equivalently, that $\bar{\partial} \mathbb{J}_\Psi = 0$. Finally, by inspection we see that $\partial \mathbb{J}_\Psi = 0$ is the set of defining equations of the generalized moment problem.

Now, it is easy to see that $\mathbb{J}_\Psi$ is strictly convex. Hence there is at most one critical point, which must be a minimum. Consequently, $F^\Psi$ is injective, as claimed.

# 5 Duality theory

Next we turn to the constraint optimization problem of Theorem 3.5, reformulated as a maximization problem: Maximize the cross entropy

$$\mathbb{I}_\Psi(\Phi) = \int_{\mathcal{I}} \Psi \log \frac{\Phi}{\Psi} \, dt \tag{5.1}$$

over $L^1_+(\mathcal{I})$ subject to the constraints

$$\int_{\mathcal{I}} \alpha_k(t) \Phi(t) dt = c_k, \quad k = 0, 1, \dots, n. \tag{5.2}$$

Note that $\mathbb{I}_\Psi(\Phi)$, here redefined to be the negative of that in Theorem 3.5, is bounded from above. To see this, we may without restriction assume that $\int_{\mathcal{I}} \Psi dt = 1$. Then, Jensen's inequality [48, p. 61] yields $\mathbb{I}_\Psi(\Phi) \le \log c_0$.

To solve this optimization problem, we form the Lagrangian

$$L(\Phi, q) = \mathbb{I}_\Psi(\Phi) + \operatorname{Re} \sum_{k=0}^{n} q_k \left[ c_k - \int_{\mathcal{I}} \alpha_k \Phi dt \right],$$

where $(q_0, q_1, \dots, q_n) \in \mathbb{R}^r \times \mathbb{C}^{n-r+1}$ are Lagrange multipliers, complex for the complex moment conditions and real for the real ones. Setting

$$Q = \mathrm{Re} \sum_{k=0}^{n} q_k \alpha_k,$$

we obtain

$$L(\Phi, q) = \int_{\mathcal{I}} \Psi \log \frac{\Phi}{\Psi}\, dt + \langle c, q \rangle - \int_{\mathcal{I}} Q\Phi dt.$$

Clearly, the dual function

$$\psi(q) = \sup_{\Phi \in L^1_+(\mathcal{I})} L(\Phi, q)$$

takes finite values only if $q \in \mathfrak{P}_+$. For such $q$, form the directional derivative

$$\delta L(\Phi, q; h) = \lim_{\epsilon \to 0} \frac{L(\Phi + \epsilon h, q) - L(\Phi, q)}{\epsilon}$$

in any direction $h \in L^1(\mathcal{I})$ to obtain

$$\delta L(\Phi, q; h) = \int_{\mathcal{I}} \left[ \frac{\Psi}{\Phi} - Q \right] h dt,$$

provided the limit exists. Consequently, for each such $\Phi$,

$$\delta L(\Phi, q; h) = 0, \quad \text{for all } h \in L^1(\mathcal{I})$$

if and only if

$$\Phi = \frac{\Psi}{Q},$$

which inserted into the dual functional yields

$$\psi(q) = \langle c, q \rangle - \int_{\mathcal{I}} \Psi \log Q\, dt - \int_{\mathcal{I}} \Psi\, dt.$$

Since the last term is constant, the dual problem to minimize $\psi(q)$ over $\mathfrak{P}_+$ is equivalent to the optimization problem

$$\min_{q \in \mathfrak{P}_+} \mathbb{J}_\Psi(q)$$

of Theorem 3.4.

In Section 4 we proved that this optimization problem has a unique solution at some $\hat{q} \in \mathfrak{P}_+$ satisfying the moment conditions (5.2). Then, setting $\hat{Q} = \mathrm{Re}\{\hat{q}\}$,

$$\hat{\Phi} := \frac{\Psi}{\hat{Q}} \in L^1_+(\mathcal{I}). \tag{5.3}$$

Since the function $\Phi \to L(\Phi, \hat{q})$ is strictly concave and

$$\delta L(\hat{\Phi}, \hat{q}; h) = \int_{\mathcal{I}} \left[ \frac{\Psi}{\hat{\Phi}} - \hat{Q} \right] h \, dt = 0, \quad \text{for all } h \in L^1(\mathcal{I}),$$

we have

$$L(\Phi, \hat{q}) \leq L(\hat{\Phi}, \hat{q}) \quad \text{for all } \Phi \in L^1_+(\mathcal{I}), \tag{5.4}$$

with equality if and only if $\Phi = \hat{\Phi}$.

However, $L(\Phi, \hat{q}) = \mathbb{I}_\Psi(\Phi)$ for all $\Phi$ satisfying the moment conditions (5.2). In particular, since (3.12) holds for $Q = \hat{Q}$, $L(\hat{\Phi}, \hat{q}) = \mathbb{I}_\Psi(\hat{\Phi})$. Consequently, (5.4) implies that

$$\mathbb{I}_\Psi(\Phi) \leq \mathbb{I}_\Psi(\hat{\Phi})$$

for all $\Phi \in L^1_+(\mathcal{I})$ satisfying the moment conditions (5.2) with equality if and only if $\Phi = \hat{\Phi}$. Hence, $\mathbb{I}_\Psi$ has a unique maximum on the space of all $\Phi \in L^1_+(\mathcal{I})$ satisfying the constraints (5.2), and it is given by (5.3). This proves Theorem 3.5.

# 6 Conclusions

It is well known that many important problems in control theory and statistical signal processing can be formulated as moment problems. However, modern engineering applications require complexity constraints, thus significantly altering the mathematical problem so that classical theory does not apply. These applications also require complete classes of solutions, rather than just one single solution, and complete smooth parameterizations of these classes.

In this paper we presented a universal solution to the generalized moment problem, with a nonclassical complexity constraint, obtained by minimizing a strictly convex nonlinear functional. This optimization problem was derived in two different ways, first geometrically by path integration of a one-form, and second via duality theory of mathematical programming.

We have thus provided a unified framework that generalizes previous work on interpolation of the Carathéodory and of the Nevanlinna–Pick type [7, 9, 10, 15]. We refer to [9, 13, 14] for applications to signal processing and to [42, 43, 44, 45] for applications to robust control. Algorithms using homotopy continuation methods, based on our convex optimization approach, have been developed for Carathéodory extension in [22] and for Nevanlinna–Pick interpolation in [43, 45].

The results presented in this paper have interesting interpretations also in probability and statistics, where moment problems are prevalent. Indeed, the complexity constraint may, for example, represent a priori information about a probability distribution to be estimated from moments.

# Acknowledgement

This research was supported in part by grants from AFOSR, the Swedish Research Council, the Göran Gustafsson Foundation, and Southwestern Bell.

**References**

[1] N. I. Akhiezer, *The Classical Moment Problem and Some Related Questions in Analysis.* New York: Hafner Publishing, 1965.

[2] N. I. Ahiezer and M. Krein, *Some Questions in the Theory of Moments*, Providence, Rhode Island: American Mathematical Society, 1962.

[3] C. I. Byrnes and A. Lindquist, "On the geometry of the Kimura–Georgiou parameterization of modelling filter," *Int. J. Control*, vol. 50, pp. 99–105, 1989.

[4] C. I. Byrnes, A. Lindquist, and T. McGregor, "Predictability and unpredictability in Kalman filtering," *IEEE Trans. Automatic Control*, vol. 36, pp. 563–579, 1991.

[5] C. I. Byrnes, A. Lindquist, and Y. Zhou, "On the nonlinear dynamics of fast filtering algorithms," *SIAM J. Control and Optimization*, vol. 32, pp. 744–789, 1994.

[6] C. I. Byrnes, A. Lindquist, S. V. Gusev and A. S. Matveev, "A complete parametrization of all positive rational extensions of a covariance sequence," *IEEE Trans. Automatic Control*, vol. AC-40, pp. 1841–1857, 1995.

[7] C. I. Byrnes, S. V. Gusev, and A. Lindquist, "A convex optimization approach to the rational covariance extension problem," *SIAM J. Control and Optimization*, vol. 37, pp. 211–229, 1999.

[8] C. I. Byrnes and A. Lindquist, "On the duality between filtering and Nevanlinna–Pick interpolation," *SIAM J. Control and Optimization*, vol. 39, pp. 757–775, 2000.

[9] C. I. Byrnes, T. T. Georgiou and A. Lindquist, "A generalized entropy criterion for Nevanlinna–Pick interpolation with degree constraint," *IEEE Trans. Automatic Control*, vol. AC-46, pp. 822–839, 2001.

[10] C. I. Byrnes, T. T. Georgiou, and A. Lindquist, "A new approach to spectral estimation: A tunable high-resolution spectral estimator," *IEEE Trans. Signal Processing*, vol. SP-49, pp. 3189–3205, Nov. 2000.

[11] C. I. Byrnes, T. T. Georgiou, and A. Lindquist, "Generalized interpolation in $H^\infty$: Solutions of bounded complexity," preprint.

[12] C. I. Byrnes, T. T. Georgiou, and A. Lindquist, forthcoming paper.

[13] C. I. Byrnes, P. Enqvist, and A. Lindquist, "Cepstral coefficients, covariance lags and pole-zero models for finite data strings," *IEEE Trans. Signal Processing*, vol. SP-50, pp. 677–693, 2001.

[14] C. I. Byrnes, P. Enqvist, and A. Lindquist, "Identifiability and well-posedness of shaping-filter parameterizations: A global analysis approach," *SIAM J. Control and Optimization*, vol. 41, pp. 23–59, 2002.

[15] C. I. Byrnes, S. V. Gusev, and A. Lindquist, "From finite covariance windows to modeling filters: A convex optimization approach," *SIAM Review*, vol. 43, pp. 645–675, Dec. 2001.

[16] C. I. Byrnes and A. Lindquist, "Interior point solutions of variational problems and global inverse function theorems," submitted for publication.

[17] I. Csiszár, "I-divergence geometry of probability distributions and minimization problems," *Ann. Probab.*, vol. 3, pp. 146–158, 1975.

[18] I. Csiszár and F. Matúš, "Information projections revisited," preprint, 2001.

[19] Ph. Delsarte, Y. Genin, and Y. Kamp, "On the role of the Nevanlinna–Pick problem in circuits and system theory," *Circuit Theory and Applications*, vol. 9, pp. 177–187, 1981.

[20] Ph. Delsarte, Y. Genin, Y. Kamp, and P. van Dooren, "Speech modelling and the trigonometric moment problem," *Philips J. Res.*, vol. 37, pp. 277–292, 1982.

[21] J. C. Doyle, B. A. Frances, and A. R. Tannenbaum, *Feedback Control Theory*. New York: Macmillan Publ. Co., 1992.

[22] P. Enqvist, "Spectral estimation by geometric, topological and optimization Methods," Ph.D. thesis, Optimization and Systems Theory, KTH, Stockholm, Sweden, 2001.

[23] T. T. Georgiou, "Realization of power spectra from partial covariance sequences," *IEEE Trans. Acoustics, Speech, and Signal Processing*, vol. ASSP-35, pp. 438–449, 1987.

[24] T. T. Georgiou, "A topological approach to Nevanlinna–Pick interpolation," *SIAM J. Math. Analysis*, vol. 18, pp. 1248–1260, 1987.

[25] T. T. Georgiou, "The interpolation problem with a degree constraint," *IEEE Trans. Automatic Control*, vol. 44, no. 3, pp. 631–635, 1999.

[26] I. J. Good, "Maximum entropy for hypothesis formulation, especially for multidimensional contingency tables," *Annals Math. Stat.*, vol. 34, pp. 911–934, 1963.

[27] U. Grenander and G. Szegö, "Toeplitz forms and their applications," Univ. California Press, 1958.

[28] J. W. Helton, "Non-Euclidean analysis and electronics," *Bull. Amer. Math. Soc.*, vol. 7, pp. 1–64, 1982.

[29] P. S. C. Heuberger, T. J. de Hoog, P. M. J. Van den Hof, Z. Szabó, and J. Bokar, "Minimal partial realization from orthonormal basis function expansions," in *Proc. 40th IEEE Conf. Decision and Control*, Orlando, Florida, USA, December 2001, pp. 3673–3678.

[30] R. E. Kalman, "Realization of covariance sequences," in *Proc. Toeplitz Memorial Conference (1981)*, Tel Aviv, Israel, 1981.

[31] H. Kimura, "Positive partial realization of covariance sequences," in *Modelling, Identification and Robust Control*, C. I. Byrnes and A. Lindquist, Eds. North-Holland, 1987, pp. 499–513.

[32] H. Kimura, "Robust stabilizability for a class of transfer functions," *IEEE Trans. Automatic Control*, vol. AC-29, pp. 788–793, 1984.

[33] H. Kimura, "State space approach to the classical interpolation problem and its applications," in *Three Decades of Mathematical System Theory*, Lecture Notes in Control and Inform. Sci., vol. 135, Berlin: Springer, 1989, pp. 243–275.

[34] H. Kimura and H. Iwase, "On directional interpolation in $H^\infty$," in *Linear circuits, systems and signal processing: theory and application* (Phoenix, AZ, 1987), Amsterdam: North-Holland, 1988, pp. 551–560.

[35] H. Kimura, "Conjugation, interpolation and model-matching in $H^\infty$," *Int. J. Control*, vol. 49, pp. 269–307, 1989.

[36] H. Kimura, "Directional interpolation in the state space," *Systems Control Lett.*, vol. 10, pp. 317–324, 1988.

[37] H. Kimura, "Directional interpolation approach to $H^\infty$-optimization and robust stabilization," *IEEE Trans. Automatic Control*, vol. AC-32, pp. 1085–1093, 1987.

[38] M. G. Krein and A. A. Nudelman, *The Markov Moment Problem and Extremal Problems*, Providence, Rhode Island: American Mathematical Society, 1977.

[39] S. Kullback, *Information Theory and Statistics*, New York: John Wiley, 1959.

[40] A. Lindquist, "A new algorithm for optimal filtering of discrete-time stationary processes," *SIAM J. Control*, vol. 12, pp. 736–746, 1974.

[41] A. Lindquist, "Some reduced-order non-Riccati equations for linear least-squares estimation: the stationary, single-output case," *Int. J. Control*, vol. 24, pp. 821–842, 1976.

[42] R. Nagamune and A. Lindquist, "Sensitivity shaping in feedback control and analytic interpolation theory," in *Optimal Control and Partial Differential Equations*, J. L. Medaldi et al. Eds. Amsterdam: IOS Press, 2001, pp. 404–413.

[43] R. Nagamune, "A robust solver using a continuation method for Nevanlinna–Pick interpolation with degree constraint," *IEEE Trans. Automatic Control*, to appear.

[44] R. Nagamune, "Closed-loop shaping based on Nevanlinna–Pick interpolation with a degree bound," submitted to *IEEE Trans. Automatic Control.*

[45] R. Nagamune, "Robust control with complexity constraint," Ph.D. thesis, Optimization and Systems Theory, Royal Institute of Technology, Stockholm, Sweden, 2002.

[46] A. Renyi, *Probability Theory.* North-Holland, 1970.

[47] Z. Szabó, P. Heuberger, J. Bokar, and P. Van den Hof, "Extended Ho–Kalman algorithm for systems represented in generalized orthonormal bases," *Automatica*, vol. 36, pp. 1809–1818, 2000.

[48] W. Rudin, *Real and Complex Analysis.* McGraw-Hill, 1966.

[49] J. A. Shohat and J. D. Tamarkin, *The Problem of Moments*, Mathematical Surveys I. New York: American Mathematical Society, 1943.

[50] A. Tannenbaum, "Feedback stabilization of linear dynamical plants with uncertainty in the gain factor," *Int. J. Control*, vol. 32, pp. 1–16, 1980.

[51] B. Wahlberg, "Systems identification using Laguerre models," *IEEE Trans. Automatic Control*, vol. AC-36, pp. 551–562, 1991.

[52] B. Wahlberg, "System identification using Kautz models," *IEEE Trans. Automatic Control*, vol. AC-39, pp. 1276–1282, 1994.

[53] K. Zhou, *Essentials of Robust Control.* Prentice-Hall, 1998.

Christopher I. Byrnes
Department of Systems Science and Mathematics
Washington University
St. Louis, MO 63130, USA
chrisbyrnes@seas.wustl.edu

Anders Lindquist
Department of Mathematics
Division of Optimization and Systems Theory
Royal Institute of Technology
100 44 Stockholm, Sweden
alq@math.kth.se

# Toeplitz Covariance Matrices and the von Neumann Relative Entropy

**Tryphon T. Georgiou**

*Dedicated to Professor Hidenori Kimura on the occasion of his 60th birthday*

ABSTRACT Sample covariances based on time-series data fail to be Toeplitz. The purpose of the paper is to suggest the von Neumann relative-entropy as a distance measure for approximating a positive-definite sample covariance by one having the Toeplitz structure. This leads to a convex optimization problem whose solution retains the property of being positive-definite.

**Keywords:** Approximation, Structured matrices, Positivity, von Neumann entropy

## 1 Motivation

A basic problem of signal analysis is to estimate the power spectrum of a stationary zero-mean stochastic process $\{\mathbf{y}_k : k \in \mathbb{Z}\}$, from a finite observation record $\{y_k : k = 0, 1, \ldots, N\}$. The paradigm which underlies most of the modern nonlinear methods begins with an estimate of a matrix covariance

$$
\begin{aligned}
T &:= \begin{bmatrix} c_0 & c_1^* & \cdots & c_n^* \\ c_1 & c_0 & \cdots & c_{n-1}^* \\ \vdots & \vdots & \ddots & \vdots \\ c_n & c_{n-1} & \cdots & c_0 \end{bmatrix} \\
&:= E\left\{ \begin{bmatrix} \mathbf{y}_k \\ \mathbf{y}_{k+1} \\ \vdots \\ \mathbf{y}_{k+n} \end{bmatrix} \begin{bmatrix} \mathbf{y}_k^* & \mathbf{y}_{k+1}^* & \cdots & \mathbf{y}_{k+n}^* \end{bmatrix} \right\}
\end{aligned}
$$

which is built out of a finite number of covariance lags

$$c_\ell = E\{\mathbf{y}_k \mathbf{y}_{k+\ell}^*\},\ \ell = 0, 1, \ldots, n,$$

with $n << N$. The covariance matrix has a Toeplitz structure.

Fundamentally, assuming ergodicity of the process, there are two alternatives for estimating such second-order statistics from data. Either estimate the $c_k$'s independently, e.g., via sample averages

$$\hat{c}_k = \frac{1}{N-k}\sum_{\ell=0}^{N-k} y_k y_{k+\ell}^*,$$

or estimate the Toeplitz covariance, e.g., again via sample averaging

$$\hat{T} = \frac{1}{N-n}\sum_{\ell=0}^{N-n} \begin{bmatrix} y_k \\ y_{k+1} \\ \vdots \\ y_{k+n} \end{bmatrix} \begin{bmatrix} y_k^* & y_{k+1}^* & \cdots & y_{k+n}^* \end{bmatrix},$$

(backward averaging, or both). However, there is a caveat. The Toeplitz matrix of sample-covariances

$$\begin{bmatrix} \hat{c}_0 & \hat{c}_1 & \dots & \hat{c}_n \\ \hat{c}_1^* & \hat{c}_0 & \dots & \hat{c}_{n-1} \\ \vdots & \vdots & \ddots & \vdots \\ \hat{c}_n^* & \hat{c}_{n-1}^* & \dots & \hat{c}_0 \end{bmatrix}$$

may fail to be positive-definite whereas $\hat{T}$, which is positive definite, fails to be a Toeplitz matrix. Either way such estimates are problematic.

A vast literature has grown around this issue and several algorithms have been devised to remedy the situation. Most importantly, Burg's algorithm which has spurred much of the research to this day. The idea of the algorithm is to compute recursively certain other parameters, the reflection coefficients (otherwise known as Schur, Szegö parameters) in a nonlinear way that forces the estimates to be admissible, i.e., corresponding to covariance lags for which $T$ is automatically positive-definite. Following Burg several *ad hoc* alternatives that guarantee positivity have been devised (see [8]) and, to this day, the subject is an active area of research. The motivation in subsequent literature was to understand and remedy problems of "line-splitting," "line-shifting," etc., that arise.

Herein, we suggest an alternative to the traditional thinking. We begin with an estimate $\hat{T}$ of the covariance matrix which is positive-definite and seek a Toeplitz matrix $T$ which is closest to $\hat{T}$ in a suitable sense.

## 2 Approximation by a Toeplitz matrix

A "least-squares" approach to approximating a positive-definite matrix $\hat{T}$ by one which is Toeplitz (i.e., by minimizing their distance in the Frobenius norm), does not necessarily yield a positive-definite approximant. If the criterion for measuring distances does not "respect" positivity, then imposing such a property as an added constraint is computationally challenging. To this end we explore a certain distance measure which has been introduced in the context of quantum mechanics for comparing states of quantum systems. Such states are represented by positive operators, and consequently, the measure is pertinent in our context as well.

## 2.1 The von Neumann relative entropy

The notion of quantum entropy was introduced by von Neumann to quantify information and uncertainty in a quantum system. In general, the states of such a system are described by non-negative operators $T$ on a Hilbert space with trace equal to 1, and the quantum entropy is defined (see [10]) by

$$S(T) := -\text{trace}\,(T\log T) = -\sum_i \lambda_i \log(\lambda_i),$$

where $\lambda_i$ $(i = 1, 2, \dots)$ denote the eigenvalues of $T$ (when the spectrum is discrete). Then, a quantum version of relative entropy

$$\mathbb{S}(\hat{T}\|T) \;:=\; \text{trace}\left(\hat{T}\left(\log\hat{T} - \log T\right)\right), \tag{2.1}$$

where $\hat{T}$ and $T$ are again trace 1 positive operators, quantifies distance between relevant quantum systems. Note that when $\hat{T}$ and $T$ commute and $\hat{\lambda}_i$ and $\lambda_i$ $(i = 1, 2, \dots, n)$ denote the corresponding eigenvalues (via correspondence of invariant subspaces),

$$\mathbb{S}(\hat{T}\|T) \;=\; \sum_i \left(\hat{\lambda}_i\left(\log\hat{\lambda}_i - \log\lambda_i\right)\right)$$

reduces to the classical notion of relative entropy [3, 9].

While $\mathbb{S}(\cdot\|\cdot)$ is not a metric it generates a useful topology. In particular, it is non-negative, jointly convex in its arguments, and it is equal to zero only when the two arguments are identical (see [10, Theorem 11.7]). The requirement that the trace be 1 is not essential. In particular, if we require that the two arguments have equal trace, then $\mathbb{S}$ remains non-negative. In fact, for the purposes of this paper, requiring that $\hat{T}$ and $T$ have equal trace is a natural assumption.

So we let $\hat{T}$ be a positive-definite $n \times n$ sample covariance matrix and consider approximating $\hat{T}$ by a positive-definite Toeplitz matrix with the same trace. The set of positive-definite matrices is convex and the above restrictions on $T$, namely that it is Toeplitz and that it has a fixed trace, are linear constraints. Thus, with trace$(\hat{T}) = \tau$,

$$\mathcal{T}(\tau) := \{T \;:\; T > 0,\, T \text{ is Toeplitz, and trace}(T) = \tau\}$$

is a convex and bounded set. On the other hand, $\mathbb{S}(\hat{T}\|\cdot)$ is strictly convex over positive-definite matrices ([10, Section 11.3]). It is also clear that $\mathbb{S}(\hat{T}\|\cdot)$ grows unbounded at the boundary of $\mathcal{T}$. Hence, the following statement is now evident.

**Proposition 2.1.** *If $\hat{T} > 0$ and $\tau = \text{trace}(\hat{T})$, then*

$$\{\mathbb{S}(\hat{T}\|T) \; : \; T \in \mathcal{T}(\tau)\}$$

*has a minimum, attained at a unique minimizer $T > 0$.*

Analogous conclusions can be drawn in the case where the structure of $T$ is dictated by more general linear constraints as in [4, 5, 6].

## 2.2 An example

We now demonstrate by way of example the possibility that the optimal least-squares Toeplitz approximant of a positive definite matrix may not be positive-definite. At the same time we give the Toeplitz minimizer of the von Neumann relative entropy and display the shape of the relative entropy surface as a function of the entries of its argument (which is an arbitrary positive-definite Toeplitz matrix).

Consider the following $3 \times 3$ positive-definite matrix

$$\hat{T} = \frac{1}{3}\begin{bmatrix} 1.1 & 0.9 & 1.05 \\ 0.9 & 0.8 & 0.9 \\ 1.05 & 0.9 & 1.1 \end{bmatrix}.$$

Positivity can be readily verified. In fact, the eigenvalues of $\hat{T}$ are

$$\{0.0114, 0.0167, 0.9719\}.$$

By inspection, the closest Toeplitz approximant of $\hat{T}$ in a least-squares sense is

$$T_2 = \frac{1}{3}\begin{bmatrix} 1 & 0.9 & 1.05 \\ 0.9 & 1 & 0.9 \\ 1.05 & 0.9 & 1 \end{bmatrix}.$$

Since $\det(T_2) < 0$, $T_2$ fails to be positive-definite.

On the other hand, the minimizer of the von Neumann relative entropy $\mathbb{S}(\hat{T}\|T)$ is

$$T_{\text{vN}} = \frac{1}{3}\begin{bmatrix} 1 & 0.942 & .957 \\ .942 & 1 & .942 \\ .957 & .942 & 1 \end{bmatrix}.$$

This was obtained by minimizing the convex function $\mathbb{S}(\hat{T}\|T)$ with

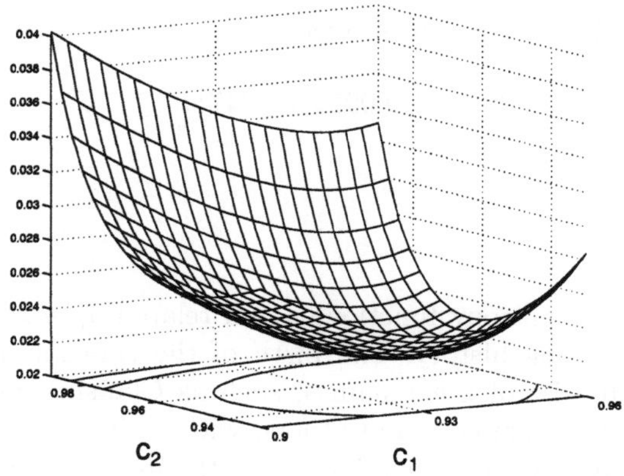


FIGURE 2.1. $\mathbb{S}(\hat{T}\|T)$ as a function of $(c_1, c_2)$ near the minimum

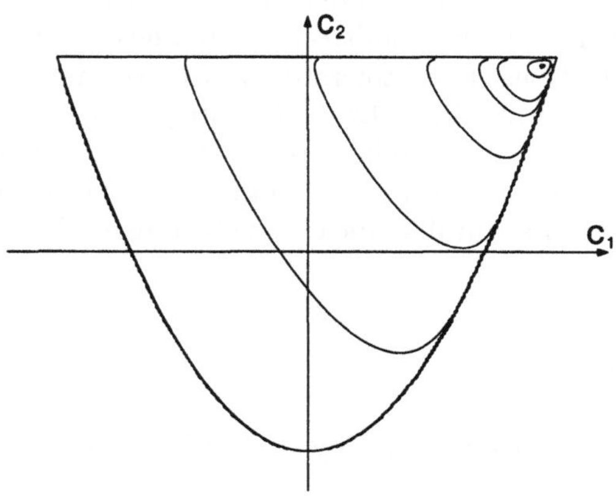


FIGURE 2.2. Level curves of $\mathbb{S}(\hat{T}\|T)$ on the $(c_1, c_2)$ plane

$$T = \frac{1}{3}\begin{bmatrix} 1 & c_1 & c_2 \\ c_1 & 1 & c_1 \\ c_2 & c_1 & 1 \end{bmatrix}$$

over the parameter space $(c_1, c_2)$ where positive-definiteness is ensured. Note that we conveniently normalized all matrices to have trace 1. The surface $\mathbb{S}(\hat{T}\|T)$, as a function of $(c_1, c_2)$, is drawn in Figure 2.1 in a neighborhood of the minimum. Figure 2.2 shows contours of equal value of $\mathbb{S}(\hat{T}\|T)$ on the $(c_1, c_2)$ plane. Note that the region of the plane which corresponds to a positive Toeplitz approximant is specified by

$$\begin{aligned} -1 < \ & c_1 < 1, \\ 2c_1^2 - 1 < \ & c_2 < 1. \end{aligned}$$

The minimal value of $\mathbb{S}(\hat{T}\|T)$ is 0.0215, and for comparison, we evaluate the eigenvalues of the minimizing $T$ which are:

$$\{0.0143, 0.0210, 0.9647\}.$$

# 3 Remarks

Extremal solutions of entropy functionals and related approximation theoretic ideas have a long history. The birth of the concept of entropy is connected to a long list of great names, such as Clausius, Gibbs, Boltzmann, von Neumann, Shannon, and others, with E. T. Jaynes one of the most persuasive advocates for the relevance of the maximum entropy principle [3]. The use of entropy functionals in signal analysis originates in the work of J.P. Burg (see [8]).

The present work has been motivated by recent collaborative work with A. Lindquist [7], which in turn built on [1, 2]. These references focus on spectral density functions, their consistency with second-order statistics, and the minimization of suitable entropy-like integrals as a rule for selecting suitable spectra. Herein, in a similar spirit, we consider minimization of the von Neumann distance to a given sample covariance for selecting a covariance matrix with a suitable linear structure (e.g., Toeplitz).

# Acknowledgement

This research was supported in part by the NSF and AFOSR.

## References

[1] C. Byrnes, T. T. Georgiou, and A. Lindquist, "A generalized entropy criterion for Nevanlinna–Pick interpolation: A convex optimization approach to certain problems in systems and control," *IEEE Trans. on Automatic Control*, vol. 45, no. 6, pp. 822–839, June 2001.

[2] C. I. Byrnes, S. V. Gusev, and A. Lindquist, "A convex optimization approach to the rational covariance extension problem," *SIAM J. Control and Opt.*, vol. 37, pp. 211–229, 1999.

[3] T. M. Cover and J. A. Thomas, *Elements of Information Theory*. Wiley, 1991.

[4] T. T. Georgiou, "Spectral estimation by selective harmonic amplification," *IEEE Trans. on Automatic Control*, vol. 46, no. 1, pp. 29–42, January 2001.

[5] T. T. Georgiou, "The structure of state covariances and its relation to the power spectrum of the input," *IEEE Trans. on Automatic Control*, vol. 47, no. 7, pp. 1056–1066, July 2002.

[6] T. T. Georgiou, "Spectral analysis based on the state covariance: the maximum entropy spectrum and linear fractional parameterization," *IEEE Trans. on Automatic Control*, to appear.

[7] T. T. Georgiou and A. Lindquist, "Kullback–Leibler approximation of spectral density functions," preprint, May 2002.

[8] S. Haykin, *Nonlinear Methods of Spectral Analysis.* New York: Springer-Verlag, New York, 247 pages, 1979.

[9] S. Kullback, *Information Theory and Statistics*, 2nd ed. New York: Dover Books, 1968 (1st ed. New York: John Wiley, 1959).

[10] M. A. Nielsen and I. L. Chuang, *Quantum Computation and Quantum Information.* Cambridge University Press, 2000.

Tryphon T. Georgiou
Dept. of Electrical and Computer Engineering
University of Minnesota
MN 55455, USA
tryphon@ece.umn.edu

# Digital Filter Design via Sampled-Data Control Theory


## Yutaka Yamamoto and Masaaki Nagahara


*Dedicated to Professor Hidenori Kimura on the occasion of his 60th birthday*


ABSTRACT This paper describes a new filter design method based on modern sampled-data control theory. Unlike the conventional design methods executed in the discrete-time domain, the method has the advantage of optimizing an analog performance with built-in intersample behavior. Such a design method is seen to be effective when the original analog signals are not ideally band-limited. A design example is given to illustrate the method.



**Keywords:** Digital signal processing, Sampled-data control theory, Analog performance, $H^\infty$ control


## 1 Preface

I[1] met Kimura-sensei perhaps for the first time at the Joint Automatic Control Conference, held in Nara, in the fall of 1979. Or could it be 1980? The memory is now quite vague, but I still remember that we walked around the neighborhood of the conference site, and he told me many things about science, engineering, and about himself, how he got the job in Osaka University, etc. It was not much after I returned to Japan, after four years of my Ph. D. study in Florida. I had some troubles adjusting myself to the Japanese society; I was young, iconoclastic, and frustrated. He told me many things about science, academia, etc., and was encouraging; I felt much better afterwards.

Since then, I have had the pleasure of many occasions of interaction with him both personally and scientifically.

Of particular importance was perhaps the organization of a small research group to enhance the research activities in control. Approximately 10 people gathered together to have a symposium every half year, the themes of which ranged according to the interest of the speakers. All members were young, enthusiastic, and not so well established in the Japanese

---

[1]In this section, "I" refers to the first author, Yutaka Yamamoto.

academic hierarchy. The promotion system of this society was not quite impartial at that time, and each of us was perhaps quite frustrated. Each of us volunteered to give a fairly long seminar talk, often lasting more than three hours with an extended discussion section after that. The discussions were critical, straight and constructive. Such an atmosphere perhaps never before existed in the Japanese control research environment, and was highly stimulating to all of us. The outcome was enormous: The seminars were an excellent occasion to test and check our own ideas. Several joint research projects took off out of this group, and they had much impact on the Japanese control scene. I recall with pleasure that my collaboration with Shinji Hara started with his proposal to jointly study the stabilizability of repetitive control systems in 1984 at one of these meetings.

During this activity Kimura-sensei was most critical, and active in all discussions. It would be an exaggeration that he never made mistakes in such discussions. I still recall that I presented the function space model for sampled-data control systems (now called *lifting*) in one of these meetings in 1989, but his reaction to this new model was not so positive. In spite of this, his enthusiasm in all discussions was contagious and gave a driving force to everyone. The topic in the subsequent sections grew partly out of this effort in sampled-data control theory. It in particular intends to explore the potential of sampled-data control theory in digital signal processing.

# 2 Signal reconstruction problem

Digital signals are all around us: CD's, MP3 formats, digital images, communications, just to name a few. They are processed by digital filters, which have advantages that are not possible with analog filters, e.g., the linear phase property.

Today, their designs are performed mostly in the discrete-time domain. To capture the continuous-time performance, the notion of aliasing is utilized, and deviation from the ideal filter has to be discussed. Many design specifications such as smooth pass-band transmission, high attenuation level in the stop-band, etc., have to be taken into account, and there are a large variety of design methods and guiding measures to cope with them [9, 17, 10].

We should however note that, in many applications, the performance we wish to optimize is still in the analog domain: speech/audio is one example; visual images are another. While in some cases we may be concerned only with precise transmission or storing of digitized data, there are many other cases where we are more concerned with the overall analog performance. From this point of view, the design methodology in the discrete-time domain may be regarded only as an approximation of designs based on the analog performance, and it should be certainly desirable to develop a direct design method that gives an optimal analog domain performance.

To see the situation in the simplest case, let us consider the following problem: *Given an analog signal $f(t)$, sample it with sampling period $h$ to obtain a discrete-time signal $f(kh)$, $k = 0, 1, 2, \ldots$. How can we recover the original analog signal $f(t)$ from the sampled-data $\{f(kh)\}_{k=0}^{\infty}$?* Without any condition on the intersample behavior, the problem is clearly ill-posed. We must assume some a priori knowledge.

The celebrated Shannon–Nyquist–Whittaker[2] sampling theorem answers the question by assuming the ideal band-limited hypothesis as follows:

**Theorem 2.1 (Shannon and others).** *Let $f(t)$ be a continuous-time signal ideally band-limited to the range $(-\pi/h, \pi/h)$, that is, its Fourier transform $\hat{f}(\omega)$ is zero outside this interval. Then $f(t)$ can uniquely be recovered from its sampled values $f(nh)$, $n = 0, \pm 1, \pm 2, \ldots$ via the formula*

$$f(t) = \sum_{n=-\infty}^{\infty} f(nh) \frac{\sin \pi(t/h - n)}{\pi(t/h - n)}. \tag{2.1}$$

While it is commonly believed that this theorem is the key to the current digital signal processing, we should also note the following:

- Formula (2.1) is not causal, i.e., one needs infinitely many future values (as well as the past values) $\{f(kh)\}_{kh>t}$ to reconstruct $f(t)$.
- Formula (2.1) is not compatible with the more familiar zero-order hold employed in digital signal processing.
- Ideally band-limited signals are entire (i.e., analytic over the whole complex plane). Thus it is not expected to occur in real applications. Furthermore, no real filter can realize the ideal band-limiting characteristic.

In other words, Theorem 2.1 describes only an idealized situation. To apply this to a real application we take it only as an approximation of the real situation. That is, the true signal may be not fully band-limited, but only approximately. Formula (2.1) cannot be exactly realized, and it must be approximated by those with more realistic hold devices, etc.

To this end, a standard approach is to

- take the sampling rate sufficiently fast, e.g., greater than twice the so-called Nyquist rate $\pi/h$,

---

[2]While the thoerem is largely attributed to Claude Shannon, it is difficult to determine who was indeed the first to find this theorem. The history apparently dates back to a mathematical result by Whittaker; Nyquist also contributed significantly to the theorem (both much earlier than Shannon). A detailed discussion can be found in [16]. I am indebted to Steve Morse for pointing out the contribution of Nyquist.

- try to approximate the ideal band-limiting filter with a digital filter; and
- an $H^2$ performance index is often employed in such an approximation.

Due to a variety of design considerations and the nonexistence of a unified optimization method, there are a large number of design techniques to accommodate such a design policy. As a result, one has to combine various design specifications and techniques, and the procedure can often be quite intricate. Furthermore, the effect of hold devices in the analog domain cannot be incorporated in the design methods as described above [9, 4, 10].

In contrast, we here propose to use the sampled-data $H^\infty$ control theory. The major advantage here is that we can incorporate sample and hold devices into the design framework, and optimize an analog performance index. This is in marked contrast with the current design methods as described above. Furthermore, the main objective in such a filter design is to shape the frequency characteristic, for which the $H^\infty$ norm criterion is more appropriate than that by the $H^2$ norm.

The modern sampled-data control theory thus provides an optimal platform for designing digital filters. Chen and Francis [3] formulated and solved a discrete-time multirate filter bank design problem with an $H^\infty$ norm criterion. Khargonekar and Yamamoto [7] formulated a single-rate signal reconstruction problem in the sampled-data setting. Subsequent work by Yamamoto et al. [12, 13] extends this to a multirate setting. The sample-rate conversion problem is solved by [5, 8].

This paper describes the idea of an optimal low-pass filter design for upsampled signals (i.e., so-called interpolators [10]) in the sampled-data setting following [15]. The objective is to reconstruct the original signal. As discussed above, this problem is usually dealt with under the assumption that the original signal is band-limited; in practice, however, no signals are entirely band-limited. Instead, we often have some a priori knowledge of an approximate frequency characteristic that the original signals obey, and it is these original analog signals we wish to reconstruct optimally.

We thus formulate a digital signal reconstruction problem under the following assumptions:

- the original analog signal is subject to a certain frequency characteristic, but not fully band-limited;
- the digital signal can be upsampled to employ a faster hold device;
- the overall analog $H^\infty$ performance must be optimized.

This may also be regarded as an optimal D/A converter design. We will see that performance improvement is possible over a conventional low-pass filter. Even though we do not explicitly place the constraint on the linear-phase property, it is interesting to note that the obtained filter is very close

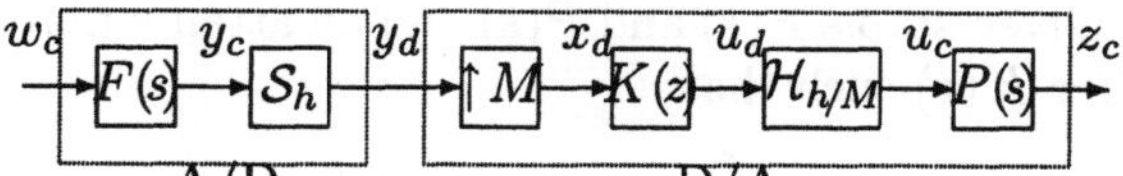


FIGURE 3.1. Multirate signal reconstruction

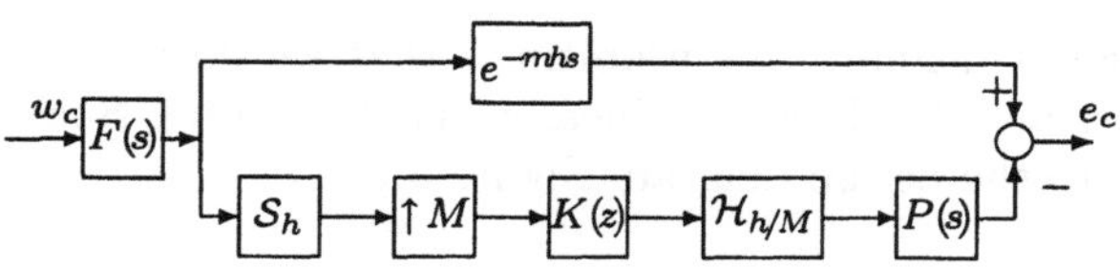


FIGURE 3.2. Signal reconstruction error system

to linear phase up to the Nyquist frequency, due to the $H^\infty$ performance requirement. In the sequel, we will briefly review the design procedure following [15] etc.

# 3 Problem formulation

Consider the block diagram Figure 3.1. The incoming signal $w_c$ first goes through an anti-aliasing filter $F(s)$ and the filtered signal $y_c$ becomes nearly (but not entirely) band-limited. $F(s)$ governs the frequency-domain characteristic of the analog signal $y_c$. This signal is then sampled by $\mathcal{S}_h$ to become the discrete-time signal $y_d$ with sampling period $h$. This signal is usually stored or transmitted with some media (e.g., CD) or a channel.

To restore $y_c$ we first upsample the discrete-time signal $y_d$ by factor $M$:

$$\uparrow M : y_d \mapsto x_d : x_d[k] = \begin{cases} y_d[l], \; k = Ml, \; l = 0, 1, \dots, \\ 0, \qquad\qquad \text{otherwise.} \end{cases}$$

The signal then becomes another discrete-time signal $x_d$ with sampling period $h/M$. The discrete-time signal $x_d$ is then processed by a digital filter $K(z)$, becomes a continuous-time signal $u_c$ by going through the 0-order hold $\mathcal{H}_{h/M}$ (that works in sampling period $h/M$), and then becomes the final signal by passing through an analog filter $P(s)$. An advantage here is that one can use a fast hold device $\mathcal{H}_{h/M}$ thereby making more precise signal restoration possible. The objective here is to design the digital filter $K(z)$ for given $F(s)$, $M$ and $P(s)$.

Figure 3.2 shows the block diagram for the error system for the design. The delay in the upper portion of the diagram corresponds to the fact that we allow a certain amount of time delay for signal reconstruction. Let $T_{ew}$ denote the input/output operator from $w_c$ to $e_c(t) := z_c(t) - u_c(t - mh)$. Our design objective is as follows:

**Problem 3.1.** Given stable $F(s)$ and $P(s)$ and an attenuation level $\gamma > 0$, find a digital filter $K(z)$ such that

$$\|T_{ew}\|_\infty := \sup_{w_c \in L^2[0,\infty)} \frac{\|T_{ew}w_c\|_2}{\|w_c\|_2} < \gamma, \tag{3.1}$$

where $\|\cdot\|_2$ denotes the $L^2$ norm. □

An advantage of defining this problem is that we do not need any extra design constraint (such as the linear phase property). Such issues are incorporated into the design diagram Figure 3.2.

# 4 Reduction to a finite-dimensional problem

Problem 3.1 is not finite-dimensional since it involves a continuous-time delay. Another difficulty is that it contains the upsampler $\uparrow M$, so that it makes the overall system (periodically) time-varying.

Following the method of [7, 12], however, we can reduce this problem to a finite-dimensional single-rate problem. The idea is to invoke lifting both in continuous-time and discrete-time (blocking):

**Theorem 4.1.** *There exist (finite-dimensional) discrete-time systems $G_{d11}(z)$, $G_{d12}(z)$ and $G_{d21}(z)$ such that (3.1) is equivalent to*

$$\|z^{-m}G_{d11}(z) - G_{d12}(z)\widetilde{K}(z)G_{d21}(z)\|_\infty < \gamma, \tag{4.1}$$

*where $\widetilde{K}(z)$ is the blocking system of $K(z)$.*

**Proof.** We first reduce the problem to a single-rate problem. Define the blocking operator $\mathbf{L}_M$ and its inverse $\mathbf{L}_M^{-1}$ by

$$\begin{aligned}\mathbf{L}_M &:= (\downarrow M)\left[1\ z\ \cdots\ z^{M-1}\right]^T,\\ \mathbf{L}_M^{-1} &:= \left[1\ z^{-1}\ \cdots\ z^{-M+1}\right](\uparrow M),\end{aligned}$$

where $\downarrow M$ denotes the downsampler

$$\downarrow M : x_d \mapsto y_d : y_d[k] = x_d[Mk].$$

Then $K(z)(\uparrow M)$ can be rewritten as

$$\begin{aligned}K(z)(\uparrow M) &= \mathbf{L}_M^{-1}\widetilde{K}(z),\\ \widetilde{K}(z) &:= \mathbf{L}_M K(z)\mathbf{L}_M^{-1}\left[1\ 0\ \cdots\ 0\right]^T.\end{aligned}$$

$\widetilde{K}(z)$ is an LTI, single-input/$M$-output system that satisfies

$$K(z) = \left[1\ z^{-1}\ \cdots\ z^{-M+1}\right]\widetilde{K}(z^M).$$

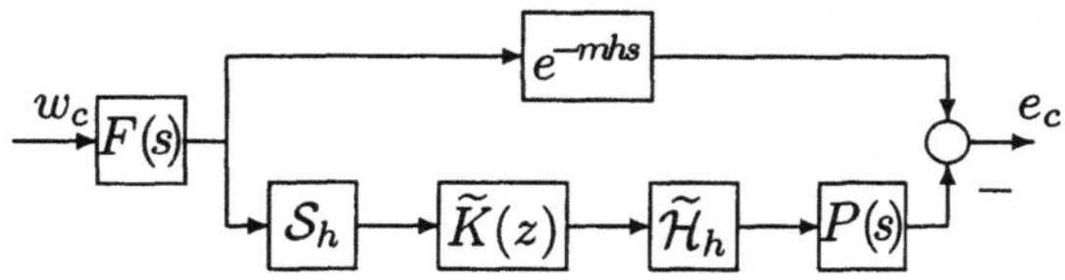


FIGURE 4.1. Reduced single-rate problem

Using the generalized hold $\widetilde{\mathcal{H}}_h$ defined by

$$\widetilde{\mathcal{H}}_h \ : \ l^2 \ni v \mapsto u \in L^2, \ \ u(kh+\theta) = \mathbf{H}(\theta)v[k]$$
$$\theta \in [0,h), \ \ k = 0,1,2,\dots,$$
$$\mathbf{H}(\theta) := \begin{cases} \begin{bmatrix} 1\,0\,0\dots\,0 \end{bmatrix}, & \theta \in [0, h/M), \\ \begin{bmatrix} 0\,1\,0\dots\,0 \end{bmatrix}, & \theta \in [h/M, 2h/M), \\ \dots & \\ \begin{bmatrix} 0\,0\dots\,0\,1 \end{bmatrix}, & \theta \in [(M-1)h/M, h), \end{cases}$$

we obtain the identity

$$\mathcal{H}_{h/M}\mathbf{L}_M^{-1} = \widetilde{\mathcal{H}}_h.$$

This yields

$$\mathcal{H}_{h/M}K(z)(\uparrow M)\mathcal{S}_h = \widetilde{\mathcal{H}}_h\widetilde{K}(z)\mathcal{S}_h.$$

Hence Figure 3.2 is equivalent to Figure 4.1. We can then invoke the technique of [7] to reduce this to a finite-dimensional design problem (4.1). □

# 5 Approximation via fast sample/hold

While the procedure above reduces Problem 3.1 to a finite-dimensional $H^\infty$ problem, it is in general not numerically suitable for actual computation; the formulas are quite involved, and not so numerically tractable. It is often more convenient to resort to an approximation method. We employ the fast sample/hold approximation [2, 12]. This method approximates continuous-time inputs and outputs via a sampler and hold that operate in the period $h/M$. The convergence of such an approximation is guaranteed in [14]; see also [11] for a more elaborate study.

By this method, our design problem (3.1) is approximated as

$$\|z^{-m}G_{dN11}(z) + G_{dN12}(z)\widetilde{K}(z)G_{dN21}(z)\|_\infty < \gamma,$$

where $G_{dN11}$, $G_{dN12}$ and $G_{dN21}$ are discrete-time systems obtained by the technique in [2, 12]. Once the problem has been reduced to this form, it can be solved by a control design toolbox such as those given by MATLAB (see, e.g., [1]).

# 6 A design example

We present a design example for

$$F(s) = \frac{1}{(0.70223s + 1)(7.0223s + 1)}, \quad P(s) = 1$$

with $h = 1$, $m = 2$ and upsampling factor $M = 2$. (In commercial CD players, $M$ is usually $8 \sim 32$.) For the CD format, the above frequency response corresponds to the decay of −20dB/dec beyond 1kHz, and −40dB/dec beyond 10kHz. In many such applications, we often have such a frequency characteristic for original analog signals. An approximate suboptimal design is executed by the method in the previous section. For comparison, we also show the gain characteristic of the Johnston filter[6] of order 31, which is often used in commercial applications.

The computed (via MATLAB [1]) (sub) optimal filter $K_{SD}(z)$ is IIR of order 7 with the filter coefficients (**b**: numerator, **a**: denominator) of the transfer function:

$$\mathbf{b} = \begin{bmatrix} -0.053136 \\ -0.081530 \\ 0.252018 \\ 0.809684 \\ 0.917880 \\ 0.512434 \\ 0.134899 \\ 0.010296 \end{bmatrix}, \quad \mathbf{a} = \begin{bmatrix} 1.000000 \\ 0 \\ 0.255479 \\ 0 \\ 0 \\ 0 \\ 0 \\ 0 \end{bmatrix}.$$

The impulse response of this filter becomes almost zero after $12 \sim 14$ steps, so that it can be well approximated by an FIR filter of 14 taps.

Figure 6.1 shows the gain characteristics of these filters. The Johnston filter shows a very sharp decay beyond the cutoff frequency ($\pi$ [rad/sec]) and the sampled-data design shows a rather slow decay. If the original signals are ideally band-limited, then a sharp decay such as that of the Johnston filter would be desirable. But if the original signal is not fully band-limited, but obeys the frequency characteristic as specified by $F(s)$, such a sharp cut in the transition band may not be optimal.

In fact, the reconstruction error characteristic in Figure 6.2 shows that the performance of $K_{SD}(z)$ is comparable to that of the 32 tap Johnston filter. Furthermore, we have made sure (frequency response plot omitted) that the sampled-data design gives 6-10 dB improvement over the Johnston filter when we increase the upsampling factor to $M = 4$.

Furthermore, in the time-domain, the sampled-data design shows a clear advantage over the Johnston filter. Let us see the time responses against rectangular waves in Figures 6.3, 6.4:

While the Johnston filter exhibits a very typical Gibbs phenomenon, the one by $K_{SD}(z)$ has much less peak around the edge. We also note that

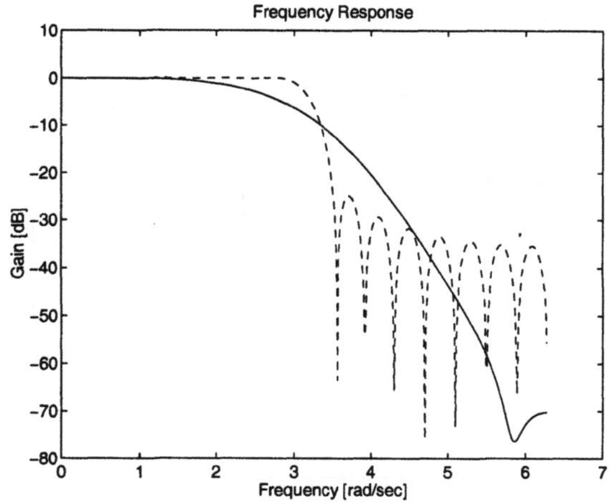


FIGURE 6.1. Frequency response of filter: $K_{SD}(z)$ (solid) and $K_J(z)$ (dash)

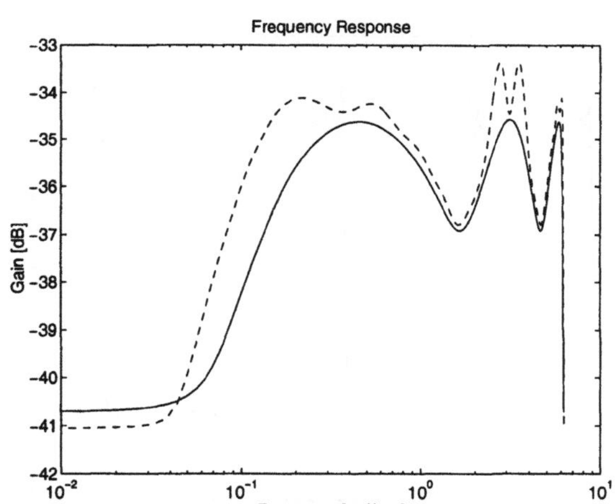


FIGURE 6.2. Frequency response of error system $T_{ew}$ : sampled-data $H^\infty$ synthesis (solid) and Johnston filter (dash)

$K_{SD}(z)$ is of nearly linear phase, as shown in Figure 6.5. The Gibbs phenomenon in the conventional design is simply an outcome of the very sharp cut beyond the cut-off frequency, since it corresponds to the truncation of a Fourier series with finitely many terms.

When we employ the upsampling factor $M = 4$, the improvement in distortion becomes much greater. Figure 6.6 shows the error magnitude frequency response for the case $M = 4$. The designed filter in the present method has an advantage up to 12dB compared to the Johnston filter.

The obtained filters are actually implemented using a digital signal pro-

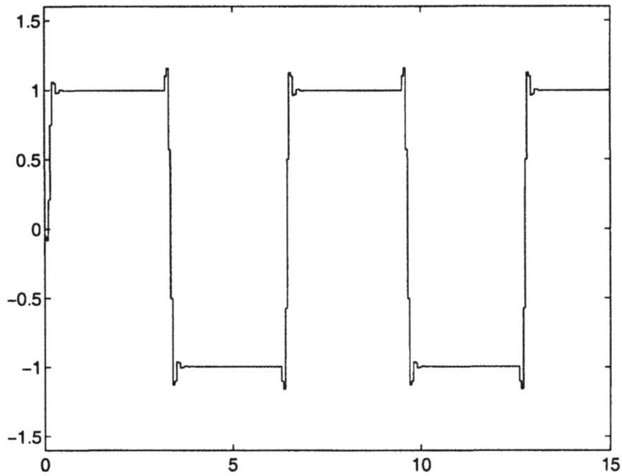


FIGURE 6.3. Time response (sampled-data design)

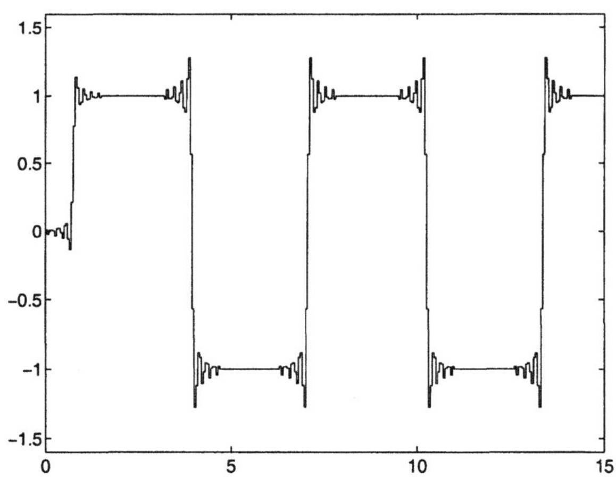


FIGURE 6.4. Time response (Johnston filter)

cessor, and showed a clear advantage over the conventionally designed DA converters.

# 7 Concluding remarks

We have presented a method based on sampled-data control theory for designing a digital filter in a multirate signal reconstruction problem. An advantage here is that an analog optimal performance can be obtained, and

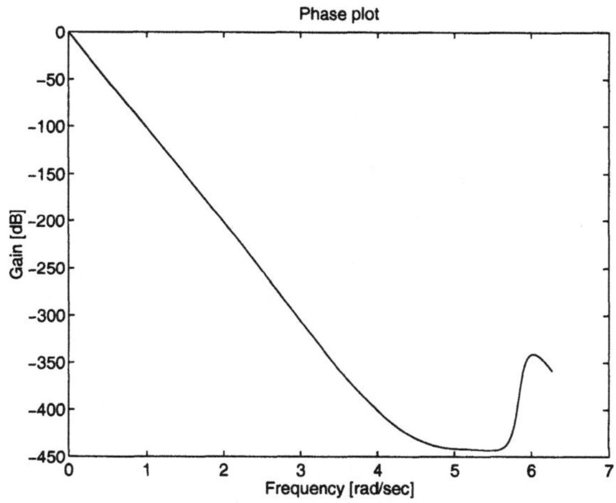


FIGURE 6.5. Phase plot of $K_{SD}$

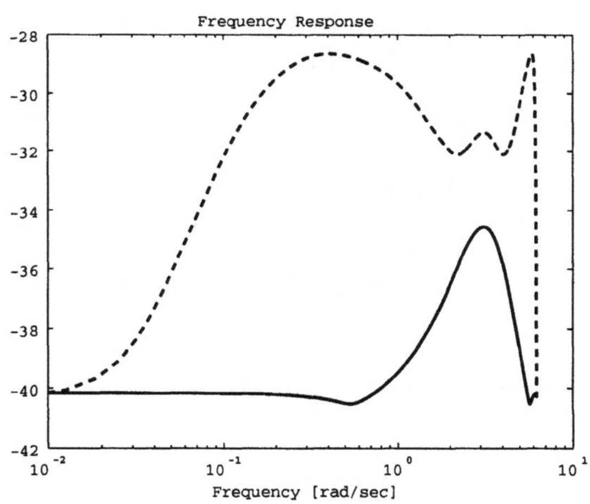


FIGURE 6.6. Error frequency response for $M = 4$

this can be advantageous in audio signal reconstruction. Another advantage is that once we fix the performance index and weighting functions, the design can be done in solving one optimization problem, using MATLAB routines (e.g., [1]), unlike those in more conventional digital filter design techniques where the design procedure usually takes several different steps. For other related developments, see [5, 8, 12, 13, 11], etc.

## References


[1] G. Balas, J. C. Doyle, K. Glover, A. Packard, and R. Smith, *μ-Analysis and Synthesis Toobox.* MUSYN and The MathWorks, 1994.

[2] T. Chen and B. A. Francis, *Optimal Sampled-Data Control Systems.* New York: Springer, 1995.

[3] T. Chen and B. A. Francis, "Design of multirate filter banks by $\mathcal{H}_\infty$ optimization," *IEEE Trans. Signal Processing*, vol. SP-43, pp. 2822–2830, 1995.

[4] N. J. Fliege, *Multirate Digital Signal Processing.* New York: John Wiley, 1994.

[5] H. Ishii, Y. Yamamoto, and B. A. Francis, "Sample-rate conversion via sampled-data $H^\infty$ control," in *Proc. 38th IEEE Conf. Decision and Control*, pp. 3440–3445, 1999.

[6] J. D. Johnston, 'A filter family designed for use in quadrature mirror filter banks," in *Proc. ICASSP*, pp. 291–294, 1980.

[7] P. P. Khargonekar and Y. Yamamoto, "Delayed signal reconstruction using sampled-data control," in *Proc. 35th IEEE Conf. Decision and Control*, pp. 1259–1263, 1996.

[8] M. Nagahara and Y. Yamamoto, "A new design for sample-rate converters," in *Proc. 39th IEEE Conf. Decision and Control*, Sydney, 2000.

[9] T. W. Parks and C. S. Burrus, *Digital Filter Design.* New York: John-Wiley, 1987.

[10] P. P. Vidyanathan, *Multirate Systems and Filter Banks.* Englewood Cliffs: Prentice Hall, 1993.

[11] Y. Yamamoto, B. D. O. Anderson and M. Nagahara, "Approximating sampled-data systems with applications to digital redesign," *Proc. 41st IEEE CDC*, Las Vegas, 2002.

[12] Y. Yamamoto, H. Fujioka, and P. P. Khargonekar, "Signal reconstruction via sampled-data control with multirate filter banks," in *Proc. 36th IEEE Conf. Decision and Control*, pp. 3395–3400, 1997.

[13] Y. Yamamoto and P. P. Khargonekar, "From sampled-data control to signal processing," in *Learning, Control and Hybrid Systems*, Springer Lecture Notes in Control and Information Sciences, vol. 241, 1998, pp. 108–126.

[14] Y. Yamamoto, A. G. Madievski, and B. D. O. Anderson, "Approximation of frequency response for sampled-data control systems," *Automatica*, vol. 35, pp. 729-734, 1999.

[15] Y. Yamamoto, M. Nagahara, and H. Fujioka, "Multirate signal reconstruction and filter design via sampled-data $H^\infty$ control," in *Proc. MTNS 2000*, Perpignan, France, 2000.

[16] A. I. Zayed, *Advances in Shannon's Sampling Theory.* Boca Raton: CRC Press, 1993.

[17] G. Zelniker and F. J. Taylor, *Advanced Digital Signal Processing: Theory and Applications.* Marcel Dekker, 1994.

Yutaka Yamamoto
Department of Applied Analysis and
Complex Dynamical Systems
Graduate School of Informatics
Kyoto University
Kyoto 606-8501, Japan
yy@i.kyoto-u.ac.jp

Masaaki Nagahara
Department of Applied Analysis and
Complex Dynamical Systems
Graduate School of Informatics
Kyoto University
Kyoto 606-8501, Japan
nagahara@acs.i.kyoto-u.ac.jp

# Time-Domain FIR Filters for Stochastic and Deterministic Systems

**Wook Hyun Kwon and Soo Hee Han**

*Dedicated with great respect and admiration, to Prof. Hidenori Kimura on the occasion of his 60th birthday in appreciation of our long-standing friendship and collaboration*

ABSTRACT In this paper, several FIR filters based on the time-domain are proposed for stochastic and deterministic systems. For state space signal models, FIR filters with linear structure are required to be independent of the initial state and to minimize the given optimal criteria subject to being unbiased for stochastic systems and quasi-deadbeat for deterministic systems. The concept of FIR filter design is shown to be applicable to parameter estimation problems.

**Keywords:** Time domain, Finite Impulse Response (FIR), Initial state independence, Unbiased, Quasi-deadbeat

## 1 Introduction

Signals are often represented without models, but in some cases can be represented by certain models, such as state space models and input-output models. In this paper, mostly, state space signal models are discussed. Filters are used to estimate the current state from the measured output data.

Because the initial state is also a state, it is logical to assume that the initial information cannot be measured, or is unknown. In fact, if the initial information is known, the future state information can be determined when the noise information is given. Therefore, the filter must be obtained without the initial state information. This means that initial state independence (ISI) is a natural assumption for filters. When the filter has ISI, then the gain matrices of the filter must also be independent of the initial state information. In the case of stochastic systems, the initial state information is usually given by the mean and covariance.

Filters can be either Infinite Impulse Response (IIR) or Finite Impulse Response (FIR). In frequency domain design, the FIR structure has been much investigated and is known to have several advantages for signal processing. FIR filters are easier to design for frequency shaping although the order may be large. The computation load is higher for FIR filters. These disadvantages may be alleviated by a high-speed computing signal processor. FIR filters may introduce a linear phase shift, which means there is no distortion with respect to the original signal. An FIR filter has built-in bounded-input bounded-output (BIBO) stability because of its FIR structure. The FIR filters are known to be more robust with respect to parameter changes and round-off errors. These desirable properties, derived in the frequency domain, may suggest that even in the time domain the FIR filter may have some desirable properties. However, FIR filters for state space models in the time domain have been investigated less than the many existing IIR filters. In the case of FIR filters, only limited error criteria, such as minimum variance, maximum likelihood, and least squares, have been investigated for special state space signal models. Therefore, FIR filters for general state space models with several error criteria often seen in IIR filters need to be investigated with initial state independence.

For filter design, another property may be required in addition to the above properties. Usually the property that the filter is unbiased is used for stochastic systems. It is reasonable that no matter what the real state, the estimation should, on average, yield the true value of the state. A similar concept, for deterministic systems, is the deadbeat property applied to the filter design. Because systems with disturbances behave around nominal systems, obtained by removing the disturbances, it is desirable that the filters without disturbances be required to give exactly the same state as the nominal systems. In this paper, this is called the quasi-deadbeat property. This paper concerns the quasi-deadbeat property and optimality simultaneously so that the proposed filter keeps the quasi-deadbeat property without degrading performance.

The FIR filter design concept is applied to parameter estimations. Because the generic linear models for parameter estimation problems are simpler than state models, the design concepts proposed in the filter can more easily be applied to parameter estimation.

This paper is organized as follows. In Section 2, some widely used criteria are introduced. In Sections 3 and 4, the FIR filters are proposed for the stochastic and deterministic systems. Each section is composed of two subsections: one for continuous-time systems and the other for discrete-time systems. In Section 5, it is shown that the FIR filter design concepts can be extended to parameter estimation problems. Finally, conclusions are presented in Section 6.

## 2 Performance criterion

Filtering problems are optimization problems with optimal criteria. The optimal criteria should properly reflect the design objective. Widely used optimal criteria for stochastic systems are minimum variance, minimax variance, and least square as follows:

$$\min_{H(\cdot)} E\Big\{ \big[x(t)-\hat{x}(t)\big]^T \big[x(t)-\hat{x}(t)\big] \Big\}, \tag{2.1}$$

$$\min_{H(\cdot)} \max_{Q\ R} E\Big\{ \big[x(t)-\hat{x}(t)\big]^T \big[x(t)-\hat{x}(t)\big] \Big\}, \tag{2.2}$$

$$\min_{H(\cdot)} \int_{t_0}^{t} \Big\{ \big[y(\tau)-L(\hat{x}(t))\big]^T \big[y(\tau)-L(\hat{x}(t))\big] \Big\} d\tau. \tag{2.3}$$

Historically the criterion (2.1), which is a very general optimal criterion, has been extensively exploited. The famous Kalman filter is derived from this criterion. In the case with unknown noise, i.e., mean and covariance in some range, the criterion (2.2) can give reasonable results that can cope with the worst case. More especially, the least square (2.3) can be used for both stochastic and deterministic systems. Minimum variance FIR filters for stochastic systems are introduced in Section 3.

For deterministic systems, the following criteria are often used for filter design:

$$\inf_{H(\cdot)} \sup_{w(\cdot)\neq 0} \frac{\int_{t_0}^{t} [x(\tau)-\hat{x}(\tau)]^T [x(\tau)-\hat{x}(\tau)] d\tau}{\int_{t_0}^{t} w^T(\tau) w(\tau) d\tau}, \tag{2.4}$$

$$\min_{H(\cdot)} \gamma^{-2} \int_{t_0}^{t} [x(\tau)-\hat{x}(\tau)]^T [x(\tau)-\hat{x}(\tau)] d\tau - \int_{t_0}^{t} w^T(\tau) w(\tau) d\tau. \tag{2.5}$$

Because the criterion (2.4) describes the worst case that may happen in systems it has been considered proper for engineering specifications. Therefore, many results for various problems exist. However, the exact solution to the criterion (2.4) is difficult to obtain. In this paper, a new $L_2$-$E_2$ criterion considering the worst case at the recent time is proposed as

$$\min_{H(\cdot)} \max_{w(\cdot)\neq 0} \frac{[x(t)-\hat{x}(t)]^T [x(t)-\hat{x}(t)]}{\int_{t-T}^{t} w^T(\tau) w(\tau) d\tau}, \tag{2.6}$$

and the exact solution to this criterion is obtained. The FIR filter based on the above $L_2$-$E_2$ criterion is introduced in Section 4.

# 3 Minimum variance FIR filters for stochastic systems

## 3.1 Continuous-time systems

FIR filters for continuous-time systems have not been extensively investigated because the entailed functional optimization problem is difficult to solve explicitly. Specifically, system noise makes the design of the filter more difficult. For general state space models with systems and measurement noises, the recursive limited memory filter was derived using the scattering description [2]. The horizon initial state was obtained from the limited memory smoother for all horizons. The filter has problems, such as stability, computational burden, and a complicated algorithm. For general state space models with systems and measurement noises, the optimal FIR filter was developed using a Fredholm integral equation [27]. The horizon initial state information is assumed to be known, which is not reasonable. Additionally the filter coefficients are dependent on the initial state information. The optimal criterion used in these studies is

$$\min_{H(\cdot)} E\Big\{ \big[x(t) - \hat{x}(t|t)\big]^T \big[x(t) - \hat{x}(t|t)\big] \Big\}. \tag{3.1}$$

All the above FIR filters did not deal with external control inputs and therefore could not be utilized in feedback control problems. Therefore, it is necessary to introduce a simple alternative to obtain FIR filters for general state space models.

Let us consider a linear continuous-time state space model with control input:

$$\dot{x}(t) = Ax(t) + Bu(t) + Gw(t), \tag{3.2}$$

$$y(t) = Cx(t) + v(t) \tag{3.3}$$

where $x(t) \in \Re^n$, $u(t) \in \Re^l$, and $y(t) \in \Re^q$ are the state, the input, and the measurement, respectively. At the initial time $t_0$ of the system, the state $x(t_0)$ is a random variable with a mean $\bar{x}(t_0)$ and a covariance $P(t_0)$. The system noise $w(t) \in \Re^p$ and the measurement noise $v(t) \in \Re^q$ are zero-mean white Gaussian and mutually uncorrelated. These noises are uncorrelated with the initial state $x(t_0)$. The covariances of $w(t)$ and $v(t)$ are denoted by $Q$ and $R$, respectively.

For general state models with controls, the FIR filter without *a priori* initial state information was introduced by a modification of the Kalman filter in [26], where the transformation method was introduced to avoid singularity. However, this work includes problems, such as the difficulty of handling infinite covariance of the initial state information and uncertain efficiency. To avoid the infinite covariance, the filter can be obtained from the Kalman filter with initial state estimation [15]. Because this filter is required to calculate the initial state estimation each time, the calculation load can be heavy. Although this filter provides an alternative to the weird infinite covariance, optimality remains unsolved. Therefore, to overcome these problems, FIR filters with *a priori* built-in unbiasedness regardless of the initial state information were proposed and called receding horizon unbiased FIR filters (RHUFF) in [16]. The RHUFF is obtained by directly minimizing a performance index of minimum variance subject to the unbiasedness condition. The RHUFF for an observable system can be expressed as

$$\hat{x}(t|t) = \int_{t-T}^{t} H(t-\sigma)y(\sigma)d\sigma + \int_{t-T}^{t} L(t-\sigma)u(\sigma)d\sigma \tag{3.4}$$

where $H(t-\sigma)$ and $L(t-\sigma)$ are as follows:

$$\begin{aligned} H(t-\sigma) &= \left[\int_{t-T}^{t} W^T(\tau-t+T)C^T R^{-1} C e^{A(\tau-t)} d\tau\right]^{-1} \\ & \quad W^T(\sigma-t+T)C^T R^{-1} \end{aligned} \tag{3.5}$$

and

$$L(t-\sigma) = \int_{t-T}^{\sigma} H(t-\tau)Ce^{A(\tau-\sigma)}Bd\tau. \tag{3.6}$$

Note that

$$\mathbf{H} = \begin{bmatrix} -A^T & \frac{1}{2}C^T R^{-1} C \\ 2GQG^T & A \end{bmatrix},$$

and

$$e^{\mathbf{H}(\sigma-t+T)} = \begin{bmatrix} X(\sigma-t+T) & Y(\sigma-t+T) \\ Z(\sigma-t+T) & W(\sigma-t+T) \end{bmatrix}. \tag{3.7}$$

The recursive form corresponding to the above batch form can be obtained as

$$\begin{aligned} \hat{x}(t|t) &= S^{-1}(T)\hat{\eta}(t|t), \\ \frac{dS(\tau)}{d\tau} &= -S(\tau)A - A^T S(\tau) - S(\tau)GQG^T S(\tau) + C^T R^{-1} C, \\ \frac{\partial\hat{\eta}(\tau+t-T|t)}{\partial\tau} &= -[A^T + S(\tau)GQG^T]\hat{\eta}(\tau+t-T|t) + C^T R^{-1} y(\tau+t-T) \\ &+ S(\tau)Bu(\tau+t-T) \end{aligned}$$

where $0 \le \tau \le T$, $S(0) = 0$, and $\hat{\eta}(t-T|t) = 0$.

For stochastic FIR filters, the minimax variance optimal criterion can be applied as

$$\min_{H(\cdot)L(\cdot)} \max_{Q\ R} E\Big\{ [x(t) - \hat{x}(t|t)]^T [x(t) - \hat{x}(t|t)] \Big\}. \tag{3.8}$$

Note that the noise covariance $Q$ and $R$ are positive definite and bounded by $Q_0$ and $R_0$, respectively. The optimal solution to the optimal criterion (3.8) is the same as the receding horizon unbiased FIR filter with $Q$ and $R$ replaced by $Q_0$ and $R_0$, respectively.

## 3.2 Discrete-time systems

In discrete-time and stochastic systems, limited memory filters [21] and finite memory filters [47] were derived based on a maximum likelihood optimal criterion. These FIR filters are designed only for limited state spaces without system noises. Therefore, when the actual systems have system noises, they may show poor performance.

Consider a general linear discrete-time state space model:

$$x_{k+1} = Ax_k + Bu_k + Gw_k, \tag{3.9}$$
$$y_k = Cx_k + v_k \tag{3.10}$$

where $x_k \in \Re^n$ is the state, $u_k \in \Re^l$ and $y_k \in \Re^q$ are the input and measurement, respectively. At the initial time $k_0$ of the system, the state $x_{k_0}$ is a random variable with a mean $\bar{x}_{k_0}$ and a covariance $P_{k_0}$. The system noise $w_k \in \Re^p$ and the measurement noise $v_k \in \Re^q$ are zero-mean white Gaussian and mutually uncorrelated. The covariances of $w_k$ and $v_k$ are denoted by $Q$ and $R$, respectively, and are assumed to be positive definite matrices. These noises are uncorrelated with the initial state $x_{k_0}$.

In a similar way to continuous systems, there are FIR filters without *a priori* initial state information derived by a modification of the Kalman filter in [25], the FIR filter from the Kalman filter with initial state estimation [15], and the FIR filter with *a priori* built-in unbiasedness regardless of the initial state information [24].

For discrete-time systems, the state space model can be changed to the following linear model:

$$Y_{k-1} = \bar{C}_N x_k + \bar{B}_N U_{k-1} + \bar{G}_N W_{k-1} + V_{k-1} \tag{3.11}$$

where

$$Y_{k-1} \triangleq [y_{k_N}^T \; y_{k_N+1}^T \; \cdots \; y_{k-1}^T]^T, \tag{3.12}$$
$$U_{k-1} \triangleq [u_{k_N}^T \; u_{k_N+1}^T \; \cdots \; u_{k-1}^T]^T, \tag{3.13}$$
$$W_{k-1} \triangleq [w_{k_N}^T \; w_{k_N+1}^T \; \cdots \; w_{k-1}^T]^T, \tag{3.14}$$
$$V_{k-1} \triangleq [v_{k_N}^T \; v_{k_N+1}^T \; \cdots \; v_{k-1}^T]^T, \tag{3.15}$$

and $\bar{C}_N$, $\bar{B}_N$, $\bar{G}_N$ are obtained from

$$\bar{C}_i \triangleq \begin{bmatrix} CA^{-i} \\ CA^{-i+1} \\ CA^{-i+2} \\ \vdots \\ CA^{-1} \end{bmatrix} = \begin{bmatrix} \bar{C}_{i-1} \\ C \end{bmatrix} A^{-1}, \tag{3.16}$$

$$\bar{B}_i \triangleq - \begin{bmatrix} CA^{-1}B & CA^{-2}B & \cdots & CA^{-i}B \\ 0 & CA^{-1}B & \cdots & CA^{-i+1}B \\ 0 & 0 & \cdots & CA^{-i+2}B \\ \vdots & \vdots & \vdots & \vdots \\ 0 & 0 & \cdots & CA^{-1}B \end{bmatrix} \tag{3.17}$$

$$= \begin{bmatrix} \bar{B}_{i-1} & -\bar{C}_{i-1}A^{-1}B \\ 0 & -CA^{-1}B \end{bmatrix}, \tag{3.18}$$

$$\bar{G}_i \triangleq - \begin{bmatrix} CA^{-1}G & CA^{-2}G & \cdots & CA^{-i}G \\ 0 & CA^{-1}G & \cdots & CA^{-i+1}G \\ 0 & 0 & \cdots & CA^{-i+2}G \\ \vdots & \vdots & \vdots & \vdots \\ 0 & 0 & \cdots & CA^{-1}G \end{bmatrix} \tag{3.19}$$

$$= \begin{bmatrix} \bar{G}_{i-1} & -\bar{C}_{i-1}A^{-1}G \\ 0 & -CA^{-1}G \end{bmatrix}, \tag{3.20}$$

$$1 \le i \le N,$$

$$\begin{aligned} \Xi_i &\triangleq \bar{G}_i\bar{G}_i^T + \bar{D}_i^T\bar{D}_i \\ &= \begin{bmatrix} \bar{G}_{i-1}\bar{G}_{i-1}^T + \bar{D}_{i-1}^T\bar{D}_{i-1} & 0 \\ 0 & I \end{bmatrix} \\ &\quad + \begin{bmatrix} \bar{C}_{i-1} \\ C \end{bmatrix} A^{-1}GG^TA^{-T} \begin{bmatrix} \bar{C}_{i-1} \\ C \end{bmatrix}^T \\ &= \begin{bmatrix} \Xi_{i-1} & 0 \\ 0 & I \end{bmatrix} + \begin{bmatrix} \bar{C}_{i-1} \\ C \end{bmatrix} A^{-1}GG^TA^{-T} \begin{bmatrix} \bar{C}_{i-1} \\ C \end{bmatrix}^T . \end{aligned} \tag{3.21}$$

Using the above notation, the RHUFF $\hat{x}_{k|k}$ with the optimal gain matrix $H_B$ can be represented as

$$\hat{x}_{k|k} = H_B(Y_{k-1} - \bar{B}_N U_{k-1}) \tag{3.22}$$

with the optimal gain matrix $H_B$ determined by

$$H_B = (\bar{C}_N^T \Xi_N^{-1} \bar{C}_N)^{-1} \bar{C}_N^T \Xi_N^{-1} \tag{3.23}$$

where $Y_{k-1}$, $U_{k-1}$, $\bar{C}_N$, $\bar{B}_N$, and $\Xi_N$ are given by (3.12)-(3.18) and (3.21), respectively.

The RHUFF $\hat{x}_{k|k}$ (3.22) can be represented in an iterative form :

$$\begin{aligned} \Omega_{i+1} &= [I + A^{-T}(\Omega_i + C^T R^{-1} C)A^{-1}GQG^T]^{-1} \\ &\quad A^{-T}(\Omega_i + C^T R^{-1} C)A^{-1} \end{aligned} \tag{3.24}$$

$$\begin{aligned} \hat{\eta}_{k_N+i+1|k} &= [I + A^{-T}(\Omega_i + C^T R^{-1} C)A^{-1}GQG^T]^{-1} \\ &\quad A^{-T}[\hat{\eta}_{k_N+i|k} + C^T R^{-1} y_{k_N+i} \\ &+ (\Omega_i + C^T R^{-1} C)A^{-1} B u_{k_N+i}] \end{aligned} \tag{3.25}$$

with $\Omega_0 = 0$ and $\eta_0 = 0$.

# 4 $L_2$-$E_2$ FIR filters for deterministic systems

In reality, the probabilistic information is difficult to obtain. Therefore, the stochastic operations, such as expectation and variance, may not be applied in many cases because it is difficult to know the probability density functions required in the design. In this section, instead of stochastic systems, deterministic systems are considered to deal with systems that do not assume any statistical quantities. Similarly to Section 3, FIR filters for deterministic systems are proposed for continuous-time systems and discrete-time systems.

## 4.1 Continuous-time systems

Consider a linear continuous-time state space model with control input:

$$\dot{x}(t) = Ax(t) + Bu(t) + Gw(t), \tag{4.1}$$

$$y(t) = Cx(t) + Dw(t) \tag{4.2}$$

where $x(t) \in \Re^n$, $u(t) \in \Re^l$, $y(t) \in \Re^q$, and $w(t) \in \Re^p$ are the state, the input, the measurement, and the disturbance, respectively. $GD^T = O$ and $DD^T = I$ are satisfied to decouple the system disturbance and the measurement disturbance.

For deterministic systems, we introduce the quasi-deadbeat condition. Because systems with disturbances behave around nominal systems, obtained by removing the disturbances, it is desirable that the filters without disturbances be required to give exactly the same state as the nominal systems. This is called a quasi-deadbeat property. The quasi-deadbeat condition means that

$$\hat{x}(t|t) = x(t) \tag{4.3}$$

when $w(\cdot) = v(\cdot) = 0$ on $[t-T, t]$. For the systems with disturbances, the filter minimizes the optimal criterion by considering the worst case. This "quasi-deadbeat" property plays the role of a constraint for the optimization problem with the given performance criterion.

Because conventional deadbeat filters consider only the deadbeat property, in many cases the construction of the filter is an issue. These deadbeat filters can be sensitive to unknown variance, i.e., numerical error, temporary uncertainty, etc. because the performance is not considered. Therefore, the deadbeat property together with performance is desirable in filter design.

A new optimal criterion, called $L_2$-$E_2$ criterion, for deterministic systems such as

$$\min_{H(\cdot)L(\cdot)} \max_{w(\cdot)\neq 0} \frac{[x(t)-\hat{x}(t|t)]^T[x(t)-\hat{x}(t|t)]}{\int_{t-T}^{t} w^T(\tau)w(\tau)d\tau} \tag{4.4}$$

will be introduced. Note that the numerator of (4.4) considers only the current estimation error. Because the current estimation error is important, it is reasonable to take only the current estimation error in the numerator of the cost function instead of the accumulated estimation error, including previous estimation errors. The solution to the above $L_2$-$E_2$ criterion with the quasi-deadbeat condition is

$$\hat{x}(t|t) = \int_{t-T}^{t} H(t-\sigma)y(\sigma)d\sigma + \int_{t-T}^{t} L(t-\sigma)u(\sigma)d\sigma \tag{4.5}$$

where $H(t-\sigma)$ and $L(t-\sigma)$ are as follows:

$$\begin{aligned} H(t-\sigma) &= \left[\int_{t-T}^{t} W^T(\tau-t+T)C^TCe^{A(\tau-t)}d\tau\right]^{-1} \\ &\quad W^T(\sigma-t+T)C^T \end{aligned} \tag{4.6}$$

and

$$L(t-\sigma) = \int_{t-T}^{\sigma} H(t-\tau)Ce^{A(\tau-\sigma)}Bd\tau \tag{4.7}$$

where

$$\mathbf{H} = \begin{bmatrix} -A^T & \frac{1}{2}C^TC \\ 2GG^T & A \end{bmatrix}$$

and

$$e^{\mathbf{H}(\sigma-t+T)} = \begin{bmatrix} X(\sigma-t+T) & Y(\sigma-t+T) \\ Z(\sigma-t+T) & W(\sigma-t+T) \end{bmatrix}. \tag{4.8}$$

The recursive form corresponding to the above batch form is

$$\begin{aligned}
\hat{x}(t|t) &= S^{-1}(T)\hat{\eta}(t|t), \\
\frac{dS(\tau)}{d\tau} &= -S(\tau)A - A^T S(\tau) - S(\tau)GG^T S(\tau) + C^T C, \\
\frac{\partial\hat{\eta}(\tau + t - T|t)}{\partial\tau} &= -[A^T + S(\tau)GG^T]\hat{\eta}(\tau + t - T|t) \\
&+ C^T y(\tau + t - T) + S(\tau)Bu(\tau + t - T)
\end{aligned}$$

where $0 \le \tau \le T$, $S(0) = 0$, and $\hat{\eta}(t - T|t) = 0$. The above filter is called the $L_2$-$E_2$ FIR filter [17].

As in the optimal criterion (2.4), we can check the $H_\infty$ performance for FIR filters. An upper bound of the $H_\infty$ cost can be obtained and minimized to determine the filter coefficients. It is surprising that the solution to this filter problem is the same as for the $L_2$-$E_2$ FIR filters. Therefore, an $L_2$-$E_2$ FIR filter guarantees the $H_\infty$ performance while optimizing the $L_2$-$E_2$ criterion.

### 4.2 Discrete-time systems

In the case of discrete state space models, least square FIR filters corresponding to

$$\min_H \left[Y_{k-1} - \bar{C}_N x_k\right]^T \left[Y_{k-1} - \bar{C}_N x_k\right] \tag{4.9}$$

were suggested for both a batch form and a recursive form in [31]. However, this filter is only suitable for limited state models with no system and measurement disturbances.

Deadbeat filters for discrete-time systems are developed to make the estimation error exactly zero in finite time. However, the filters of this type may have poor performance in some environments. Similar to continuous-time systems, the quasi-deadbeat property with performance index is proposed.

Consider a linear discrete-time state space signal model:

$$x_{k+1} = Ax_k + Bu_k + Gw_k, \tag{4.10}$$

$$y_k = Cx_k + Dw_k \tag{4.11}$$

where $x_k \in \Re^n$ is the state, and $u_k \in \Re^l$, $y_k \in \Re^q$, and $w_k \in \Re^p$ are the input, measurement, and disturbance, respectively. $GD^T = 0$ and $DD^T = I$ are satisfied to decouple the system disturbance and the measurement disturbance.

For discrete-time systems, the state space model can be changed to the following linear models:

$$Y_{k-1} = \bar{C}_N x_k + \bar{B}_N U_{k-1} + \bar{G}_N W_{k-1} + \bar{D}_N W_{k-1} \tag{4.12}$$

where

$$\bar{D}_i \triangleq \left[\mathrm{diag}(\overbrace{D\ D\ \cdots\ D}^{i})\right] = \left[\mathrm{diag}(\bar{D}_{i-1}, D)\right] \quad 1 \le i \le N. \tag{4.13}$$

and $\bar{C}_N$, $\bar{B}_N$, $\bar{G}_N$ are defined in equations (3.11)-(3.20).

For discrete-time systems, the quasi-deadbeat condition is

$$\hat{x}_k = x_k \tag{4.14}$$

The objective now is to obtain the optimum gain matrix $H_B$, subject to the quasi-deadbeat constraint, (4.14), in such a way that the worst case gain between the estimation error of the estimate $\hat{x}_{k|k}$ and disturbance has the minimum value

$$H_B = \min_{H} \max_{w_i} \frac{[x_k - \hat{x}_{k|k}]^T [x_k - \hat{x}_{k|k}]}{\sum_{i=1}^{N} w_{k-i}^T w_{k-i}}. \tag{4.15}$$

If it is assumed that $\{A, C\}$ of the given system is observable, the $L_2$-$E_2$ FIR filter is

$$\hat{x}_{k|k} = (\bar{C}_N^T \Xi_N^{-1} \bar{C}_N)^{-1} \bar{C}_N^T \Xi_N^{-1} (Y_{k-1} - \bar{B}_N U_{k-1}). \tag{4.16}$$

The $L_2$-$E_2$ FIR filter $\hat{x}_{k|k}$ in batch form is represented in an iterative form, for computational ease.

$$\Omega_{i+1} = [I + A^{-T}(\Omega_i + C^T C)A^{-1} G G^T]^{-1} A^{-T}(\Omega_i + C^T C)A^{-1}, \tag{4.17}$$

$$\begin{aligned} \hat{\eta}_{k_N+i+1|k} &= [I + A^{-T}(\Omega_i + C^T C)A^{-1} G G^T]^{-1} A^{-T} \\ &\quad [\hat{\eta}_{k_N+i|k} + C^T y_{k_N+i} + (\Omega_i + C^T C)A^{-1} B u_{k_N+i}] \end{aligned} \tag{4.18}$$

with $\Omega_0 = 0$ and $\eta_0 = 0$. It is surprising to observe that the $L_2$-$E_2$ FIR filter $\hat{x}_{k|k}$ with an iterative form is the same as the RHUFF with unknown horizon initial state in [25] where the covariances of the system noise and the measurement noise are taken as unit matrices.

# 5 Extension to parameter estimation

The concepts of FIR filter design can be extended to the FIR parameter estimation for the linear generic models

$$z_k = h_k \theta + v_k. \tag{5.1}$$

$L_2$-$E_2$ FIR estimation, and minimum variance FIR estimation can be obtained as reported in [28]. In this section, only the $L_2$-$E_2$ FIR estimation is introduced. We introduce a new optimal criterion

$$\min_F \max_{v_i} \frac{[\theta - \hat{\theta}_{k|k}]^T[\theta - \hat{\theta}_{k|k}]}{\sum_{i=k-N+1}^{k} v_i^T v_i} \tag{5.2}$$

where $\theta_{k-N+i|k}$ is the estimated value of $\theta$ at time $k-N+i$ on the horizon $[k-N, k-1]$. The concept of the quasi-deadbeat property still holds for parameter estimation. The quasi-deadbeat condition is $\hat{\theta}_{k|k} = \theta$ with $v_i = 0$ on $[k-N, k]$. As in the case of filter problems, the FIR estimation problem does not require initial parameter information. The algorithm of the $L_2$-$E_2$ FIR estimation is

$$\begin{aligned}
\hat{\theta}_{k|k} &= P_{k|k}^{-1}\eta_{k|k}, \\
P_{k_N+i+1|k} &= P_{k_N+i|k} + h_{k_N+i}h_{k_N+i}, \\
\eta_{k_N+i+1|k} &= \eta_{k_N+i|k} + h_{k_N+i+1}z_{k_N+i+1}
\end{aligned}$$

with $0 \le i \le N-1$, $\hat{\eta}_{k_N|k} = 0$, $P_{k_N|k} = 0$, and $k_N = k - N$.

# 6 Conclusion

In this paper, FIR filters are introduced for stochastic and deterministic systems.

The conditions for unbiasedness and a quasi-deadbeat property are built into the corresponding filter during its design for stochastic and deterministic systems, respectively. These conditions contribute to the removal of the initial state dependency. Minimum variance FIR filters are proposed. For deterministic systems, a new $L_2$-$E_2$ criterion with the quasi-deadbeat property is proposed. The proposed optimal criterion uses a gain between the disturbance on the horizon and the current estimation error, not including the accumulated estimation error. It is surprising that this $L_2$-$E_2$ FIR filter has a closed form solution, while the $H_\infty$ IIR filter only provides a suboptimal solution.

The design concept of FIR filters based on unbiasedness and a quasi-deadbeat property will provide a new paradigm for filter design. Various FIR filters for state space models in this paper can be widely used instead of IIR filters. The FIR filter design concepts can be applied to other areas such as parameter estimation.

We made the following list to include the references not cited in the present paper hoping that the list can act as a bibiliography in this field.

## References

[1] E.-W. Bai and Y. Ye, "Constrained logarithmic least squares in parameter estimation," *IEEE Transactions on Automatic Control*, vol. 44, pp. 182–186, 1999.

[2] A. M. Bruckstein and T. Kailath, "Recursive limited memory filtering and scattering theory," *IEEE Transactions on Information Theory*, vol. 31, pp. 440–443, 1985.

[3] F. Carravetta and G. Mavelli, "Minimax quadratic filtering of uncertain linear stochastic systems with partial fourth-order information," *IEEE Transactions on Automatic Control*, vol. 44, no. 6, 1999.

[4] B.-S. Chen and W.-S. Hou, "Deconvolution filter design for fractal signal transmission systems: A multiscale Kalman filter bank approach," *IEEE Transactions on Signal Processing*, vol. 45, no. 5, pp. 1359–1364, 1997.

[5] Y.-M. Cheng, B.-S. Chen, and L.-M. Chen, "Minimax deconvolution design of multirate systems with channel noises: A unified approach," *IEEE Transactions on Signal Processing*, vol. 47, no. 11, 1999.

[6] L. Dai, "Filtering and LQG problems for discrete-time stochastic singular systems," *IEEE Transactions on Automatic Control*, vol. 34, no. 10, pp. 1105–1108, 1989.

[7] L. Danyang and L. Xuanhuang, "Optimal state estimation without the requirement of a priori statistics information of the initial state," *IEEE Transactions on Automatic Control*, vol. 39, no. 10, pp. 2087–2091, 1994.

[8] J. C. Darragh and D. P. Looze, "Noncausal minimax linear state estimation for systems with uncertain second-order statistics," *IEEE Transactions on Automatic Control*, vol. 29, no. 6, 1984.

[9] C. E. de Souza, R. M. Palhares, and P. L. D. Peres, "Robust $H_\infty$ filtering for uncertain linear systems with multiple time-varying state delays," *IEEE Transactions on Signal Processing*, vol. 49, no. 3, pp. 569–576, 2001.

[10] D. Z. Feng, Z. Bao, and X. D. Zhang, "Modified RLS algorithm for unbiased estimation of FIR system with input and output noise," *Electronics Letters*, vol. 36, no. 3, pp. 273–274, 2000.

[11] M. Fu, C. E. de Souza, and Z. Q. Luo, "Finite horizon robust Kalman filter design," in *Proceedings of the 38th IEEE Conference on Decision and Control*, 1999, pp. 4555–4560.

[12] J. C. Geromel, "Optimal linear filtering under parameter uncertainty," *IEEE Transactions on Signal Processing*, vol. 47, no. 1, pp. 168–175, 1999.

[13] J. C. Geromel and M. C. de Oliveira, "$H_2$ and $H_\infty$ robust filtering for convex bounded uncertain systems," *IEEE Transactions on Automatic Control*, vol. 46, no. 1, pp. 100–107, 2001.

[14] S. Gollamudi and Y.-F. Huang, "Adaptive minimax filtering via recursive optimal quadratic approximations," *Circuits and Systems*, vol. 3, pp. 142–145, 1999.

[15] S. H. Han, P. S. Kim, and W. H. Kwon, "Receding horizon state estimation with estimated horizon initial state and its application to aircraft engine systems," in *Proceedings of the 8th IEEE International Conference on Control Applications*, Hawai'i, HW, 1999.

[16] S. H. Han, W. H. Kwon, and P. S. Kim, "Receding horizon unbiased FIR filters for continuous-time state space models without a priori initial state information," *IEEE Transactions on Automatic Control*, vol. 46, no. 5, pp. 766–770, 2001.

[17] S. H. Han, W. H. Kwon, and P. S. Kim, "New deadbeat minimax filters for deterministic state space models without the requirement of a priori initial state information," *IEEE Transactions on Automatic Control*, to appear, 2002.

[18] B. Hassibi and T. Kaliath, "$H_\infty$ bounds for least-squares estimators," *IEEE Transactions on Automatic Control*, vol. 46, no. 2, pp. 309–314, 2001.

[19] G. A. Hirchoren and C. E. D'Attellis, "Estimation of fractional Brownian motion with multiresolution Kalman filter banks," *IEEE Transactions on Signal Processing*, vol. 47, no. 5, pp. 1431–1434, 1999.

[20] L. Hong, G. Cheng, and C. K. Chui, "A filter-bank-based Kalman filtering technique for wavelet estimation and decomposition of random signals," *IEEE Transactions on Circuits and Systems II: Analog and Digital Signal Processing*, vol. 45, no. 2, pp. 237–241, 1998.

[21] A. H. Jazwinski, "Limited memory optimal filtering," *IEEE Transactions on Automatic Control*, vol. 13, pp. 558–563, 1968.

[22] A. J. Krener, "Kalman-Bucy and minimax filtering," *IEEE Transactions on Automatic Control*, vol. 25, no. 2, 1980.

[23] O. K. Kwon, C. E. de Souza, and H. S. Ryu, "Robust $H_\infty$ filter for discrete-time uncertain systems," in *Proceedings of the 35th IEEE Conference on Decision and Control*, 1996, pp. 4819–4824.

[24] W. H. Kwon, P. S. Kim, and S. H. Han, "A receding horizon unbiased FIR filters for discrete-time state space models," *Automatica*, vol. 38, no. 3, pp. 545–551, 2002.

[25] W. H. Kwon, P. S. Kim, and P. Park, "A receding horizon Kalman FIR filter for discrete time-invariant systems," *IEEE Transactions on Automatic Control*, vol. 44, no. 9, pp. 1787–1791, 1999.

[26] W. H. Kwon, P. S. Kim, and P. Park, "A receding horizon Kalman FIR filter for linear continuous-time systems," *IEEE Transactions on Automatic Control*, vol. 44, no. 11, pp. 2115–2120, 1999.

[27] W. H. Kwon and O. K. Kwon, "FIR filters and recursive forms for continuous time-invariant state-space models," *IEEE Transactions on Automatic Control*, vol. 32, pp. 352–356, 1987.

[28] W. H. Kwon, K. H. Lee, and S. H. Han, "Quasi-deadbeat $H_\infty$ parameter estimation," in *Proceedings of the Asian Control Conference*, 2002.

[29] H. Li and M. Fu, "A linear matrix inequality approach to robust $H_\infty$ filtering," *IEEE Transactions on Signal Processing*, vol. 45, no. 9, pp. 2338–2350, 1997.

[30] X. R. Li and C. He, "Optimal initialization of linear recursive filters," in *Proceedings of the 37th IEEE Conference on Decision and Control*, 1998, pp. 2335–2340.

[31] K. V. Ling and K. W. Lim, "Receding horizon recursive state estimation," *IEEE Transactions on Automatic Control*, vol. 44, no. 9, pp. 1750–1753, 1999.

[32] D. Liu, "Minimum-time filter for discrete linear time-varying system," in *Proceedings of the 31st IEEE Conference on Decision and Control*, 1992, pp. 2222–2223.

[33] D. P. Looze, H. V. Poor, K. S. Vastola, and J. C. Darragh, "Minimax control of linear stochastic systems with noise uncertainty," *IEEE Transactions on Automatic Control*, vol. 28, no. 9, 1983.

[34] M. S. Mahmoud, N. F. Al-Muthairi, and S. Bingulac, "Robust Kalman filtering for continuous time-lag systems," *Systems and Control Letters*, vol. 38, pp. 309–319, 1999.

[35] M. S. Mahmoud, L. Xie, and Y. C. Soh, "Robust Kalman filtering for discrete state-delay systems," *IEE Proceedings on Control Theory and Applications*, vol. 147, no. 6, pp. 613–618, 2000.

[36] N. C. Martins and M. Athans, "Time-invariant Kalman filtering, a minimax approach," in *Proceedings of the 37th IEEE Conference on Decision and Control*, vol. 3, pp. 2896–2901, 1998.

[37] A. V. Medvedev and T. Toivonen, "Feedforward time-delay structures in state estimation: Finite memory smoothing and continuous deadbeat observers," in *Proceedings of IEE Conference on Control Theory and Applications*, 1994, pp. 121–129.

[38] R. Nikoukhah, S. L. Campbell, and F. Delebecque, "Kalman filtering for general discrete-time linear systems," *IEEE Transactions on Automatic Control*, vol. 44, no. 10, 1999.

[39] R. M. Palhares, C. E. de Souza, and P. L. D. Peres, "Robust $H_\infty$ filter design for uncertain discrete-time state-delayed systems: An LMI approach," in *Proceedings of the 38th IEEE Conference on Decision and Control*, 1999, pp. 2347–2352.

[40] I. R. Petersen and D. C. McFarlane, "Optimal guaranteed cost control and filtering for uncertain linear systems," *IEEE Transactions on Automatic Control*, vol. 39, no. 9, pp. 1971–1977, 1994.

[41] A. P. Petrovic and A. J. Zejak, "Minimax approach to envelope constrained filter design," *Electronics Letters*, vol. 34, no. 25, 1998.

[42] Y. A. Phillis, "Estimation and control of systems with unknown covariance and multiplicative noise," *IEEE Transactions on Automatic Control*, vol. 34, no. 10, 1989.

[43] V. Poor and D. P. Looze, "Minimax state estimation for linear stochastic systems with noise uncertainty," *IEEE Transactions on Automatic Control*, vol. 26, no. 4, pp. 902–906, 1981.

[44] R. G. Reynolds, "Robust estimation with unknown noise statistics," *IEEE Transactions on Automatic Control*, vol. 35, no. 9, 1990.

[45] B. Sayyarrodsari, J. P. How, B. Hassibi, and A. Carrier, "An LMI formulation for the estimation-based approach to the design of adaptive filters," in *Proceedings of the 37th IEEE Conference on Decision and Control*, 1998.

[46] B. Sayyarrodsari, J. P. How, B. Hassibi, and A. Carrier, "Estimation-based synthesis of $H_\infty$-optimal adaptive FIR filters for filtered-LMS problems," *IEEE Transactions on Signal Processing*, vol. 49, no. 1, pp. 164–178, 2001.

[47] F. C. Schweppe, *Uncertain Dynamic Systems.* Englewood Cliffs, NJ: Prentice-Hall, 1973.

[48] U. Shaked and C. E. de Souza, "Robust minimum variance filtering," *IEEE Transactions on Signal Processing*, vol. 43, no. 11, pp. 2474–2483, 1995.

[49] T. Soderstrom, W. X. Zheng, and P. Stoica, "Comments on 'On a least-squares-based algorithm for identification of stochastic linear systems,'" *IEEE Transactions on Signal Processing*, vol. 47, no. 5, pp. 1395–1396, 1999.

[50] Y. S. Suh and J. W. Choi, "Continuous-time deadbeat fault detection and isolation filter design," in *Proceedings of the 38th IEEE Conference on Decision and Control*, 1999, pp. 3130–3131.

[51] Y. Theodor and U. Shaked, "Robust discrete-time minimum-variance filtering," *IEEE Transactions on Signal Processing*, vol. 44, no. 2, pp. 181–189, 1996.

[52] Y. Theodor, U. Shaked, and C. E. de Souza, "A game theory approach to robust discrete-time $H_\infty$-estimation," *IEEE Transactions on Signal Processing*, vol. 42, no. 6, pp. 1486–1495, 1994.

[53] M. E. Valcher, "State observers for discrete-time linear systems with unknown inputs," *IEEE Transactions on Automatic Control*, vol. 44, pp. 397–401, 1999.

[54] S. Verdú and H. V. Poor, "Minimax linear observers and regulators for stochastic systems with uncertain second-order statistics," *IEEE Transactions on Automatic Control*, vol. 29, no. 6, pp. 499–511, 1984.

[55] Z. Wang, B. Huang, and H. Unbehauen, "Robust $H_\infty$ observer design of linear time-delay systems with parametric uncertainty," *Systems and Control Letters*, vol. 42, pp. 303–312, 2001.

[56] Z. Wang, J. Zhu, and H. Unbehauen, "Robust filter design with time-varying parameter uncertainty and error variance constraints," *International Journal of Control*, vol. 72, no. 1, pp. 30–38, 1999.

[57] J. T. Watson, Jr. and K. M. Grigoriadis, "Optimal unbiased filtering via linear matrix inequalities," *Systems and Control Letters*, vol. 35, pp. 111–118, 1998.

[58] L. Xie, Y. C. Soh, and C. E. de Souza, "Robust Kalman filtering for uncertain discrete-time systems," *IEEE Transactions on Automatic Control*, vol. 39, no. 6, pp. 1310–1314, 1994.

[59] G.-H. Yang and J. L. Wang, "Robust nonfragile Kalman filtering for uncertain linear systems with estimator gain uncertainty," *IEEE Transactions on Automatic Control*, vol. 46, no. 2, pp. 343–348, 2001.

[60] Y. Zhang, C. Wen, and Y. C. Soh, "Unbiased LMS filtering in the presence of white measurement noise with unknown power," *IEEE Transactions on Circuits and Systems II: Analog and Digital Signal Processing*, vol. 47, pp. 968–972, 2000.

[61] W.-X. Zheng, "Estimation of the parameters of autoregressive signals from colored noise-corrupted measurements," *IEEE Signal Processing Letters*, vol. 7, no. 7, pp. 201–204, 2000.

[62] W. X. Zheng, "A fast convergent algorithm for identification of noisy autoregressive signals," in *Proceedings of the 2000 IEEE International Symposium on Circuits and Systems*, Geneva, Switzerland, 2000, pp. 497–500.

[63] Y. Zhu, "Efficient recursive state estimator for dynamic systems without knowledge of noise covariances," *IEEE Transactions on Aerospace and Electronic Systems*, vol. 35, pp. 102–114, 1999.

Wook Hyun Kwon
Control Information Systems Lab.
School of Electrical Engr. and Computer Science
Seoul National University
Seoul, 151-742, Korea
whkwon@cisl.snu.ac.kr

Soo Hee Han
Control Information Systems Lab.
School of Electrical Engr. and Computer Science
Seoul National University
Seoul, 151-742, Korea
hsh@cisl.snu.ac.kr

# Part II

# Identification

# Learning $H_\infty$ Model Sets from Data: The Set Membership Approach


## Mario Milanese and Michele Taragna



ABSTRACT The paper investigates the problem of identifying time-invariant, discrete-time, exponentially stable, possibly infinite dimensional, linear systems from noisy experimental data. The aim is to deliver not a single model, but a set of models whose size in $H_\infty$ norm measures the uncertainty in the identification. The noise assumptions can account for information on its maximal magnitude and deterministic uncorrelation properties.

The paper overviews recent results of the authors, focusing on the optimality properties for finite data and on the tradeoff between optimality and complexity of the identified model set.

A method is given for solving the consistency/prior validation problem, requiring the solution of one linear programming problem. An algorithm is presented for evaluating convergent and computationally efficient inner and outer approximations of the value set for a given frequency, i.e., the set of responses at that frequency of all systems not falsified by data. Such approximations provide a method for computing the identification error of any identified model set within any desired accuracy, while in the literature only bounds are available, which are often quite conservative. A method for evaluating an optimal model set at any given number of frequencies is presented. This optimal model set may prove to be too complex and simpler model sets may be looked for, at the expense of identification accuracy degradation. This is measured by the optimality level of the identified model set, i.e., the ratio between the achieved identification error and the minimal one. By suitably approximating the optimal model set, a $\sqrt{2}$-optimal algorithm is derived, called "nearly optimal", thus improving over the 2-optimality of "almost optimal" algorithms available in the literature. Using these results, reduced-order model sets with nominal models in $\mathcal{R}H_\infty$ are derived which tightly include the optimal model set, and their optimality levels are evaluated. Since the optimality level of reduced-order model sets can be made near to $\sqrt{2}$ by increasing their order, the model order selection can be performed by suitably trading off between model set complexity and optimality level degradation.



**Keywords:** Set Membership identification, Identification for control, Model sets, Model quality, Finite samples, Complex systems

# 1 Dedicated to Professor Hidenori Kimura on the occasion of his 60th birthday

Motivated by the important advances of robust control and, more recently, of identification for control, a theory of model set, in contrast to the classical theory of single models, is now widely recognized as a fundamental paradigm to account for the inherent uncertainty in modeling any real system to be studied.

Hidenori Kimura has made seminal contributions to the development of such a theory, such as chain scattering representation of model sets, $H_\infty$ control of model sets, $H_\infty$ identification of model sets, approximation of model sets.

This paper overviews recent results developed by the authors on $H_\infty$ identification and approximation of model sets. These results have been largely inspired by Kimura's work and by stimulating personal interaction with him at so many meetings and during a very pleasant visiting period of the first author at the University of Tokyo.

# 2 Introduction

A quite general problem appearing in many scientific and technical fields is to make some kind of inference on a dynamical system $S^o$, starting from some general information on it and from a finite number of actual measurements corrupted by noise $e$. Typical examples of inference are smoothing, filtering, prediction, control design, decision making, fault detection, diagnosis, etc. The usual approach is to identify a model $M$ of $S^o$ and to make the inference on $M$. Due to the unavoidable discrepancies between the identified model $M$ and the actual system $S^o$, it is of paramount importance to evaluate the inference error, i.e., the error in making inference on $M$ instead of $S^o$.

In particular, let us consider the control design problem. Typically, the system to be controlled is not completely known and a control law has to be designed, able to drive the plant to achieve, if possible, given performance specifications. The classical approach consists in building a mathematical model of the plant, on the basis of the available information (priors and measurements), and then designing a control that meets the desired performance specifications for the identified model. However, this way it is not taken into account that any identified model is only an approximation of the real system. Indeed, the performances actually achievable on the plant may be very poor, according to the size of the modeling error, and even the closed loop stability may be missed. In order to face these problems, robustness had become in past years a central issue in system and control theory, and the attention of researchers has been changing focus from the study of a single model $M$ to the investigation of a set of models $\mathcal{M}$. Such a set, often indicated as an *uncertainty model set* or a model set for short, has to be suitably constructed to describe the inherent uncertainty about the system under consideration and to be used for analysis and design purposes. Typically, model sets are described by a set of perturbations of a "nominal" model, bounded in some norm. Thus, in the control problem, the inference error corresponds to the difference between the performances predicted on the model and the actually achievable ones.

Since data are finite and corrupted by noise, providing only limited information on the actual plant, no bounded model set can be derived if no prior assumption on system $S^o$ and noise $e$ is made. In particular, some information on $S^o$ is required, e.g., by assuming that it belongs to some subset $K$ of dynamical systems. Different theories have been developed, according to different assumptions on the set $K$ and the noise $e$.

Classical statistical identification theory gives deep and extensive results for the case that:

- $K = \{M(p), \; p \in \Re^q\}$ : set of parametric models;
- $e$: stochastic noise with known probability density function, possibly filtered by a parametric noise model.

Since it is assumed that $S^o = M(p^o)$ for some "true" parameter $p^o$, only parametric uncertainty is considered, and parametric uncertainty model set $\mathcal{M} = \{M(p) : p \in \Pi\}$ can be obtained, where $\Pi \subset \Re^q$ represents the uncertainty domain provided by the parameter estimation.

As a matter of fact, in most practical applications $S^o \notin \{M(p), \; p \in \Re^q\}$, thus implying that only approximated models can be estimated and that the effects of unmodeled dynamics have to be accounted for. This appears to be a formidable problem. Some results are available in statistical identification and learning literature, essentially related to asymptotic analysis, see, e.g., [28, 39, 16, 15], while the available finite sample results appear to be quite conservative, see, e.g., [40]. Indeed, evaluation of the identification accuracy with a finite number of samples is of great importance. Set Membership Identification Theory allows one to deal with these aspects (unmodeled dynamics and finite samples), assuming that $K$ is a non-finitely parameterized subset of dynamical systems, and considering that the noise $e$ belongs to some bounded set, accounting for magnitude information and possible deterministic uncorrelation properties. The delivered model sets are described by a parametric (possibly low order) "nominal" model and a dynamic uncertainty, see, e.g., [25, 26, 35, 28, 17, 18, 6] and the references therein.

The case considered in the present paper, where the modeling error is measured by the $H_\infty$ norm, is certainly the most investigated one in the literature. Indeed, $H_\infty$ identification methods deliver uncertainty model sets in a suitable form to be used by well-established robust design techniques, based on $H_\infty$ or $\mu$ optimization methods. The literature on $H_\infty$ identification is now very extensive, and the interested reader may find deep presentations of the state of the art and extensive lists of references in the monographs [32, 3]. Some of the most relevant contributions related to the aspects investigated in this paper, i.e., finite-sample optimality properties and uncertainty evaluation tightness, are now briefly discussed.

The first seminal papers (see, e.g., [12, 31, 10]) proposed different two-stage algorithms guaranteeing convergence, in the sense that the $H_\infty$ identification error converges to zero as the number of data grows to infinity and noise level goes to zero. However, these algorithms lead to high order identified models, typically not less than the number of data, and do not enjoy any specific optimality property. Moreover, bounds on the identification error are obtained, whose conservativeness degree is unknown and may be actually quite loose.

Subsequently, interpolatory algorithms have been introduced (see, e.g., [11, 5, 4, 30, 9]), consisting of two steps: first solve a consistency problem, called in our paper "validation of prior assumptions", which determines whether or not there exist systems and noise sequences satisfying the priors and reproducing the measured data; second, pick one of such systems as the identified model, if the answer to the first step is positive. These algorithms have been developed for measurements both in frequency and time domain, while two-stage algorithms require frequency response data. Interpolatory algorithms are convergent and 2-optimal, i.e., their identification errors overbound the minimal one within a factor of 2. For this reason, they are often indicated as "almost optimal". However, this almost optimality property is not fully exploited, since the derived bounds on the identification error may be much larger than twice the minimal possible one, as shown in the numerical examples reported in [20, 21]. Moreover, models of order equal to the number of data are obtained, and computational problems arise for number of measurements greater than few decades, both for the validation step and the construction of the interpolatory model.

Optimality properties have been investigated in [8, 41]. In [8], tight frequency shaped uncertainty model sets are derived, but the noise is supposed power bounded, and this appears to be a less realistic characterization of noise than the one considered in the present paper. In [41], a characterization is given for optimal model sets with a nominal model of prescribed order, but the proposed solution holds only for "sufficiently small" noise size (actually very small) and is computationally intractable for more than about ten or twenty measurements.

The present paper is also related to the new topic of model set reduction, which only recently has been investigated. In [2], a result is obtained for the reduction of model sets using a special norm the authors introduced. In [1], an Integral Quadratic Constraints description of uncertainty is used. In these two papers, the reduced model set is not guaranteed to contain the original set, an important property in view of the use of the reduced model set for robust control design. In [41, 13] this inclusion problem is considered. The procedure derived in [41] requires the solution of a sequence of mixed complex-real $\mu$-synthesis problems, whose computational complexity appears to be very high. In [13], the problem is formulated for model sets described by homographic transformations and using the Hausdorff distance as a measure of difference among model sets. A computationally tractable method for model set reduction is obtained for the case of single-input single-output, additively perturbed plants.

In this paper, $H_\infty$ identification of linear, time-invariant, discrete-time, BIBO stable, possibly infinite-dimensional, SISO systems is considered. A bound on the system impulse response decay rate is assumed to be known. This assumption is slightly different (actually weaker) from the most usual one in literature, related to a bound on the transfer function magnitude. Experimental information consists of measurements in the time and/or the frequency domain corrupted by an output additive noise. The noise assumptions can account for information on its maximal magnitude and possible uncorrelation properties in a deterministic sense.

The distinctive features of the results here presented, in relation to the above discussed literature, can be summarized as follows. A method is given for solving the consistency/prior validation problem, requiring the solution of one linear programming problem. An algorithm is presented for evaluating convergent and computationally efficient inner and outer approximations of the value set for a given frequency, i.e., the set of responses at that frequency of all systems not falsified by data. Such approximations provide a method for computing the identification error of any identified model set within any desired accuracy, while in the literature only bounds are available, which are often quite conservative, as shown in the examples reported in [20, 21]. A method for evaluating an optimal model set at any given number of frequencies is also obtained. By suitably approximating this optimal model set, a $\sqrt{2}$-optimal interpolatory algorithm is derived, called "nearly optimal", thus improving over the 2-optimality of "almost optimal" algorithms available in the literature. Using these results, reduced-order model sets with nominal models in $\mathcal{R}H_\infty$ are derived which tightly include the optimal model set, and their optimality levels are evaluated. Since the optimality level of reduced-order model sets can be made near to $\sqrt{2}$ by increasing their order, the model order selection can be performed by suitably trading off between model set complexity and optimality level degradation.

# 3 Set Membership $H_\infty$ identification theory

In this section, the main concepts of Set Membership Identification Theory (SMIT) are introduced, with explicit reference to the specific $H_\infty$ setting investigated in the paper.

**Experimental information**

The noisy measurements are represented as

$$y^N = F_N(S^o) + e^N$$

where $y^N = [y_0 \dots y_{N-1}]^T \in \Re^N$ is a known vector depending on the actual measurements, $F_N$ is a known operator called "information operator" indicating how the measurements depend on $S^o$, and $e^N \in \Re^N$ is an unknown vector representing the measurement noise. The following experimental settings are considered.

- Time domain measurements of $N$ samples of the output $y$ of the system $S^o$, initially at rest, fed by a known one-sided input $u$ ($u_\ell = 0$ for $\ell < 0$, $u_0 \neq 0$):

$$y_\ell = \sum_{k=0}^{\ell} h_k^{S^o} u_{\ell-k} + e_\ell, \quad \ell = 0, \dots, N-1. \tag{3.1}$$

  In this case, the information operator is $F_N(S) = F_N h^S$, where

$$F_N = \begin{bmatrix} U_N & 0_{N\times\infty} \end{bmatrix} \in \Re^{N\times\infty} \tag{3.2}$$

  with $U_N \in \Re^{N\times N}$ the lower triangular Toeplitz matrix formed by the input $u$, and $h^S = \{h_0^S, h_1^S, \dots\}$ the impulse response of $S$.

- Measurements of real and imaginary part of the complex-valued samples of the system frequency response $S^o(\omega_k)$, $k = 1, \dots, N/2$:

$$\begin{aligned} y_{2k-2} &= \Re e\,(S^o(\omega_k)) + e_{2k-2}\,, \\ y_{2k-1} &= \Im m\,(S^o(\omega_k)) + e_{2k-1}\,. \end{aligned} \tag{3.3}$$

  In this case, the information operator is $F_N(S) = F_N h^S$, where

$$F_N = \begin{bmatrix} \Omega^T(\omega_1) & \cdots & \Omega^T(\omega_{N/2}) \end{bmatrix}^T \in \Re^{N\times\infty}\,, \tag{3.4}$$

$$\Omega(\omega) = \begin{bmatrix} \Omega_1(\omega) \\ \Omega_2(\omega) \end{bmatrix} = \begin{bmatrix} \Re e\,(\Psi(\omega)) \\ \Im m\,(\Psi(\omega)) \end{bmatrix} \in \Re^{2\times\infty}\,, \tag{3.5}$$

$$\Psi(\omega) = \begin{bmatrix} 1 & e^{-j\omega} & e^{-j2\omega} & \cdots \end{bmatrix} \in \mathcal{C}^{1\times\infty}\,. \tag{3.6}$$

- Mixed time and frequency domain measurements, consisting of both (3.1) and (3.3), can be considered as well. In such a case, the information operator can be obtained by stacking together the information operators (3.2) and (3.4). □

**SMIT assumptions**

- *Prior information on $S^o$:*

$$S^o \in K(L,\rho) = \left\{S \in H_\infty : \left|h_\ell^S\right| \leq L\rho^\ell, \ \ \forall \ell \geq 0\right\} \qquad (3.7)$$

where $L > 0$ and $0 < \rho < 1$ are known constants, $H_\infty$ is the Banach space of causal, SISO, linear time-invariant, discrete-time, BIBO stable dynamical systems equipped with the norm

$$\|S\|_\infty = \|S(z)\|_\infty = \sup_{0 \leq \omega \leq 2\pi} |S(\omega)| \ . \qquad (3.8)$$

In the paper, any $S \in H_\infty$ is represented as either $h^S$ or $S(z) = \sum_{k=0}^{\infty} h_k^S z^{-k}$ or $S(\omega) = \sum_{k=0}^{\infty} h_k^S e^{-j\omega k}$, according to the context.

In many papers of the literature, it is assumed that $S^o \in K_f(L,\rho) = \{S \in H_\infty : \|S(\rho \cdot z)\|_\infty \leq L\}$. It is well known that $K_f(L,\rho) \subset K(L,\rho)$, so that, for the same values of $L$ and $\rho$, the assumption in the present paper is weaker.

- *Prior information on noise:*

$$e^N \in \mathcal{B}_e = \left\{\tilde{e}^N = [\tilde{e}_0 \ \ldots \ \tilde{e}_{N-1}]^T \in \Re^N : \|A\tilde{e}^N\|_\infty^{W_e} \leq \varepsilon\right\} \qquad (3.9)$$

where $\|A\tilde{e}^N\|_\infty^{W_e} = \|W_e^{-1}A\tilde{e}^N\|_\infty = \max_{0 \leq k \leq l-1} w_{e,k}^{-1} \left|\left(A\tilde{e}^N\right)_k\right|$, $A \in \Re^{l \times N}$ is a given matrix and $W_e = \mathrm{diag}\,(w_{e,0}, \ldots, w_{e,l-1}) \in \Re^{l \times l}$ is a given weighting matrix with $w_{e,k} > 0 \ \forall k$.

By taking $A = W_e = I_{N \times N}$, such an assumption accommodates for constant magnitude bound on noise: $|\tilde{e}_k| \leq \varepsilon \ \forall k$, as typically done in most of SMIT literature. By suitably choosing $W_e$, it is possible to consider noise bounds dependent on $k$: $|\tilde{e}_k| \leq w_{e,k} \cdot \varepsilon \ \forall k$, to account, e.g., for relative measurement errors. By suitably choosing $A$, it is possible to account for information on deterministic uncorrelation properties of the noise, see, e.g., [29, 38]. □

A key role in SMIT is played by the Feasible Systems Set, often called "unfalsified systems set", i.e., the set of all systems consistent with prior information and measured data.

**Definition 3.1.** *Feasible Systems Set:*

$$FSS\left(K, \mathcal{B}_e, F_N, y^N\right) = \left\{S \in K : y^N = F_N(S) + \tilde{e}^N, \ \ \tilde{e}^N \in \mathcal{B}_e\right\} \ .$$

□

If prior assumptions ($S^o \in K$ on system and $e^N \in \mathcal{B}_e$ on noise) are "true", $FSS$ includes $S^o$, an important property in view of subsequent use for robust control design. As required in any identification theory, the problem of checking the "truth" of priors arises. Indeed, the only thing that can be actually done is to check if prior assumptions are invalidated by data, evaluating if no unfalsified system exists, i.e., if $FSS$ is empty. However, it is usual to introduce the concept of prior assumption validation as follows.

**Definition 3.2.** *Validation of prior assumptions:*
prior assumptions are validated by data if $FSS\left(K,\mathcal{B}_e,F_N,y^N\right) \neq \emptyset$. □

The fact that the priors are validated by present data does not exclude that they may be invalidated by future data, as it is always the case in validation problems, [33]. In the paper, $FSS$ will be assumed non-empty. If empty, the prior assumptions on system and noise are invalidated by data and have to be suitably modified to give a non-empty $FSS$.
The $FSS\left(K,\mathcal{B}_e,F_N,y^N\right)$ summarizes the overall information on the system to be identified, i.e., prior assumptions on system and noise and information coming from experimental data, thus describing the uncertainty about the system to be identified. Consistent with the robustness paradigm, control should be designed to be robust versus such an uncertainty model set. However, in most cases $FSS$ is not represented in a suitable form to be used by robust control design techniques, and model sets with such a property have to be looked for, e.g., described by linear fractional transformations. Moreover, to be consistent with robust control design philosophy, *model sets including the set of unfalsified systems* have to be looked for. This is formalized by the following definition.

**Definition 3.3.** *Model set for $S^o$:*
a set of models $\mathcal{M} \subseteq H_\infty$ is called a model set for $S^o$ if $\mathcal{M} \supseteq FSS$. □

Indeed, more restricted classes of model sets could be of interest, belonging, e.g., to $\mathcal{R}H_\infty$: this case will be analyzed later. In the sequel, additive model sets of the form

$$\mathcal{M} = \{M + \Delta : \|\Delta\|_\infty \leq \gamma\} \subset H_\infty \tag{3.10}$$

are considered, where $M$ is called a nominal model and $\gamma$ is an upper bound on the $H_\infty$ identification error $\|S^o - M\|_\infty = \sup_{0\leq\omega\leq 2\pi}|S^o(\omega) - M(\omega)|$.

The nominal model can be obtained by some identification algorithm, i.e., an operator $\phi$ mapping the available information, represented by the quadruple $\left(K,\mathcal{B}_e,F_N,y^N\right)$, into a model $\hat{M} \in K(L,\rho)$:

$$\phi\left(K,\mathcal{B}_e,F_N,y^N\right) = \hat{M}\ .$$

For notational simplicity, the dependence on $y^N$ only will be evidenced if necessary.

Given an identified nominal model $\hat{M}$, under SMIT assumptions it results that

$$\|S^o - \hat{M}\|_\infty \leq \sup_{S\in FSS} \|S - \hat{M}\|_\infty$$

where the R.H.S. represents the tightest bound on the identification error of the model $\hat{M}$. Given an algorithm $\phi$ producing a model $\hat{M} = \phi\left(y^N\right)$, the additive model set for $S^o$ of form (3.10) with minimal $\gamma$ is given by

$$\hat{\mathcal{M}} = \left\{ \hat{M} + \Delta : \|\Delta\|_\infty \leq \sup_{S \in FSS} \|S - \hat{M}\|_\infty \right\} . \tag{3.11}$$

For given information summarized by the $FSS$, the model set $\hat{\mathcal{M}}$ is univocally determined by its nominal model $\hat{M}$, thus motivating the following definition of identification error.

**Definition 3.4.** *Algorithm and model set identification error:*

$$E\left(\phi\right) = E(\hat{\mathcal{M}}) = \sup_{S \in FSS} \left\| S - \phi\left(y^N\right) \right\|_\infty .$$

□

In SMIT literature, other worst-case errors are often defined and investigated, see, e.g., the surveys [27, 28, 17]. In particular, the global identification error $E_g\left(\phi\right)$ is often considered, given by

$$E_g\left(\phi\right) = \sup_{S \in K} \sup_{\tilde{e}^N \in \mathcal{B}_e} \left\| S - \phi(F_N(S) + \tilde{e}^N) \right\|_\infty .$$

It is known that $E_g\left(\phi\right) = \sup_{y^N} E\left(\phi\right)$ [24], so that

$$E\left(\phi\right) \leq E_g\left(\phi\right) . \tag{3.12}$$

The error $E\left(\phi\right)$ considered here, sometimes indicated as "local a posteriori error", is of interest in many respects. First of all, since $FSS$ is the smallest model set guaranteeing, under SMIT assumptions, to include $S^o$, $E\left(\phi\right)$ is the minimal bound on the error $\|S^o - \hat{M}\|_\infty$ that can be guaranteed by using the model $\hat{M}$, identified by algorithm $\phi$ operating on available information $\left(K, \mathcal{B}_e, F_N, y^N\right)$. Moreover, $E\left(\phi\right)$ *is not worst-case* with respect to the noise, thus accounting for possible uncorrelation properties of actual noise. This does not hold for the global error $E_g\left(\phi\right)$, which accounts for worst-case noise. For example, the global error $E_g\left(\phi\right)$ of interpolatory algorithms asymptotically converges to zero only if the noise level $\varepsilon$ goes to zero, while the local a posteriori error $E\left(\phi\right)$ may converge to zero for finite values of $\varepsilon$ (see [19] for details).

As just noticed, the model set $\hat{\mathcal{M}}$ is the minimal radius ball with center $\hat{M}$ containing the $FSS$. In order to minimize the uncertainty size, a minimal radius ball containing the $FSS$ can be looked for by appropriately choosing the nominal model $\hat{M}$, thus leading to the following optimality concepts.

**Definition 3.5.** *Optimal algorithm and model set, radius of information:*

i) a model set $\mathcal{M}^c$ is called optimal if, for given $(K, \mathcal{B}_e, F_N, y^N)$:

$$E(\mathcal{M}^c) = \inf_{\phi} E(\phi) = r\left(K, \mathcal{B}_e, F_N, y^N\right)$$

where

$$r\left(K, \mathcal{B}_e, F_N, y^N\right) = \inf_{\phi} \sup_{S \in FSS(K, \mathcal{B}_e, F_N, y^N)} \left\|S - \phi\left(K, \mathcal{B}_e, F_N, y^N\right)\right\|_{\infty}$$

is called radius of information;

ii) an algorithm $\phi^c$ is called optimal if, for all $(K, \mathcal{B}_e, F_N, y^N)$:

$$\sup_{S \in FSS(K, \mathcal{B}_e, F_N, y^N)} \left\|S - \phi^c\left(K, \mathcal{B}_e, F_N, y^N\right)\right\|_{\infty} = r\left(K, \mathcal{B}_e, F_N, y^N\right) .$$

□

A basic result in Information-Based Complexity relates optimal algorithms and the Chebicheff center $M^c$ of $FSS$, defined as

$$M^c = \arg \inf_{M \in K(L,\rho)} \sup_{S \in FSS} \|S - M\|_{\infty} .$$

Indeed, the algorithm $\phi^c\left(y^N\right) = M^c$, assuming that it exists, is an optimal algorithm [37] and the corresponding model set

$$\mathcal{M}^c = \{M^c + \Delta : \|\Delta\|_{\infty} \leq r\}$$

is an optimal model set. Algorithm $\phi^c$ is called a *central algorithm.* In general, computing the center $M^c$ is difficult and/or not suitable for analysis or design purposes. This motivates interest in deriving algorithms having lower complexity, possibly providing low-order models belonging to $\mathcal{R}H_{\infty}$, at the expense of some degradation in the accuracy of the identified model set. The following definition is introduced to give a measure of such a degradation.

**Definition 3.6.** *Algorithm optimality level:*
an algorithm $\phi$ is called $\alpha$-optimal if there exists $\alpha \in \Re$ such that

$$E\left(\phi\left(K, \mathcal{B}_e, F_N, y^N\right)\right) \leq \alpha \cdot r\left(K, \mathcal{B}_e, F_N, y^N\right), \quad \forall \left(K, \mathcal{B}_e, F_N, y^N\right) .$$

□

The minimal value of $\alpha$ for which $\phi$ is $\alpha$-optimal is called deviation of $\phi$, [37]. It is clearly of interest to find algorithms with deviation as close as possible to 1. In the $H_{\infty}$ case considered in the present paper, this cannot be realized by linear algorithms, e.g., least squares. In fact, linear algorithms with finite deviation exist only if the identification error is measured by a Hilbert norm, see [14], so that linear algorithms may have $H_\infty$ identification error arbitrarily large with respect to the theoretically minimal one. On the contrary, finite deviation can be obtained by quite simple nonlinear algorithms such as interpolatory algorithms, defined as follows.

**Definition 3.7.** *Interpolatory algorithm:*
an algorithm $\phi^I$ is interpolatory if $\phi^I\left(y^N\right) = M^I \in FSS$. □

Interpolatory algorithms are 2-optimal for any norm, and for this reason they are often called "almost optimal", see, e.g., [32, 3].

An $\alpha$-optimal algorithm guarantees that, by processing any possible information $\left(K, \mathcal{B}_e, F_N, y^N\right)$, the identified model set $\hat{\mathcal{M}}$ has an identification error $E(\hat{\mathcal{M}}) \leq \alpha \cdot r$. However, for given information $\left(K, \mathcal{B}_e, F_N, y^N\right)$, the actual ratio $E(\hat{\mathcal{M}})/r$ may be significantly lower than $\alpha$. Then, given an identified model set, it is of interest to evaluate the actual value of this ratio, as formalized in the following definition.

**Definition 3.8.** *Model set optimality level:*
the optimality level $\alpha(\hat{\mathcal{M}})$ of a model set $\hat{\mathcal{M}}$ is

$$\alpha(\hat{\mathcal{M}}) = E(\hat{\mathcal{M}})/r \ .$$

□

Note that $\alpha(\hat{\mathcal{M}}) \geq 1$. Thus, an optimality level close to 1 means a model set close to optimality.

# 4 Validation of prior assumptions

From prior information (3.7)–(3.9) and measurements (3.1) or (3.3), in the present $H_\infty$ setting the $FSS$ is given by

$$FSS = \left\{S \in K : \|A \cdot (y^N - F_N h^S)\|_\infty^{W_e} \leq \varepsilon\right\}$$

where $F_N$ is given in (3.2) or (3.4) for the time or the frequency domain data respectively.

The problem of checking the validity of prior assumptions, i.e., to check if $FSS$ is non-empty according to Definition 3.2, is now considered.

Given a system $S \in H_\infty$ with impulse response $h^S = \left\{h_0^S, h_1^S, \dots\right\}$, let $S^\nu$ be the $FIR_\nu$ system having the same first $\nu$ impulse response samples of $S$, i.e., $h^{S^\nu} = \left\{h_0^S, h_1^S, \dots, h_{\nu-1}^S, 0, 0, \dots\right\}$. Then, we have the following validation result.

**Result 4.1 ([21]).** *For given positive integer $\nu$, let $\varepsilon^*$ be the solution of the linear programming problem*

$$\varepsilon^* = \min_{\tilde{\varepsilon}, S^\nu : \begin{cases} S^\nu \in K \\ \|A \cdot (y^N - F_N h^{S^\nu})\|_\infty^{W_e} \leq \tilde{\varepsilon} \end{cases}} \tilde{\varepsilon} .$$

i) *A sufficient condition for prior assumptions being validated by data is*

$$\varepsilon^* \leq \varepsilon .$$

ii) *A necessary condition for prior assumptions being validated by data is*

$$\varepsilon^* \leq \varepsilon + \delta_\nu \|W_e^{-1} A\|_{\infty,\infty}$$

*where* $\delta_\nu = \dfrac{L\rho^\nu}{1-\rho}$ *and* $\left\|W_e^{-1} A\right\|_{\infty,\infty} = \max\limits_{0 \leq i \leq l-1} \sum\limits_{j=0}^{N-1} \left|\left(W_e^{-1} A\right)_{ij}\right|$.

iii) *In the case of time domain data, choosing $\nu \geq N$, a necessary and sufficient condition for prior assumptions being validated by data is*

$$\varepsilon^* \leq \varepsilon .$$

Note that the gap between the necessary and sufficient conditions i) and ii) can be made as small as desired by increasing $\nu$. Indeed, in case of time domain measurements, no gap exists just for $\nu = N$.

The previous validation result can be used for choosing values of terms representing the prior information on system and noise. Assuming that matrices $A$ and $W_e$ are selected as described in Section 3 in order to accommodate for information on deterministic uncorrelation properties and non-constant magnitude of the noise, the values of $L$, $\rho$ and $\varepsilon$ have to be chosen so that prior assumptions are not invalidated by data. By choosing $\nu$ such that $\delta_\nu \|W_e^{-1} A\|_{\infty,\infty} \ll \varepsilon^*$, the function

$$\varepsilon^*(L, \rho) = \min_{\tilde{\varepsilon}, S^\nu : \begin{cases} S^\nu \in K(L, \rho) \\ \|A \cdot (y^N - F_N h^{S^\nu})\|_\infty^{W_e} \leq \tilde{\varepsilon} \end{cases}} \tilde{\varepsilon}$$

individuates in the space $(L, \rho, \varepsilon)$ a surface separating validated values of $L$, $\rho$, $\varepsilon$ from falsified ones, see Figure 4.1. Clearly, $L$, $\rho$, $\varepsilon$ have to be chosen in the validated region with some "caution" (i.e., not too near the separation surface) and using possible information on the experimental setting. For example, if one is confident that the noise affecting output measurements is due to the errors of the measurement devices and not to the effects of unmeasured inputs, the value of $\varepsilon$ may be obtained by evaluating the instrumentation accuracy. On the other hand, useful information on $L$ and $\rho$ values may be provided by the decay rate of the impulse response of models identified by "untuned" algorithms which do not make use of prior assumptions, such as projection or standard prediction error algorithms.

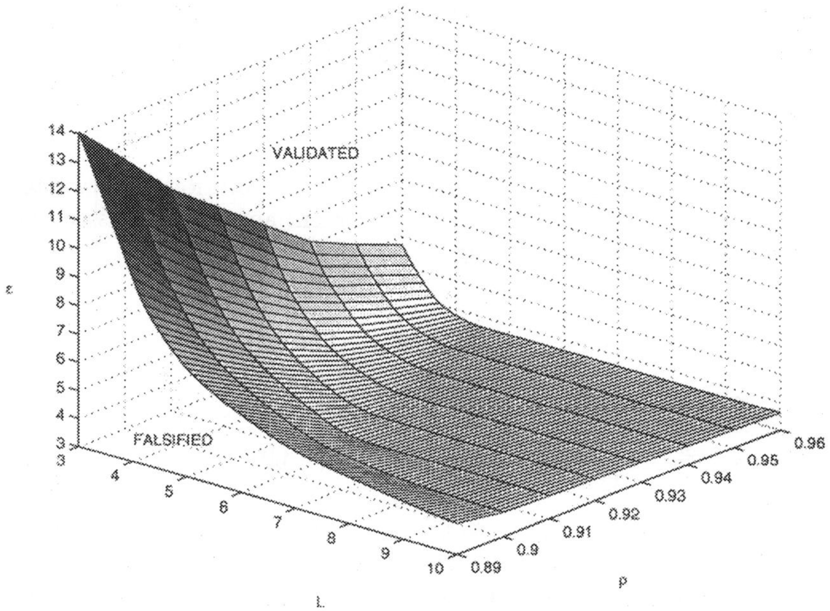


FIGURE 4.1. Surface $\varepsilon^*(L,\rho)$ separating falsified and validated $(L,\rho,\varepsilon)$ values

# 5 Value set computation

The systems in the feasible systems set $FSS$ generate a set of feasible transfer functions. Such a set of transfer functions is now investigated by evaluating the value set for a given frequency $\omega$, i.e., the set of values in the complex plane of the feasible transfer functions at the frequency $\omega$, as formalized in the following definition[1].

**Definition 5.1.** *Value set:*
for given $\omega \in [0, 2\pi]$, the value set $V(\omega)$ is defined by

$$V(\omega) = \Omega(\omega)[FSS] = \left\{s \in \Re^2 : s = \Omega(\omega) \cdot h^S,\ S \in FSS\right\} \subset \Re^2$$

where the matrix $\Omega(\omega)$ is defined in (3.5). □

For given $\omega$, $V(\omega)$ is convex, since $FSS$ is convex and the mapping $\Omega(\omega)$ is linear. Moreover, the value set is bounded $\forall \omega \in [0, 2\pi]$, since $\forall S \in K$ it results that $\sup_{0 \le \omega \le 2\pi} |S(\omega)| \le L/(1-\rho)$.

The Chebicheff center $c_2[V(\omega)]$ in the Euclidean norm of the value set

[1]Throughout the paper, any number $x \in C$ is considered as a complex-valued scalar or a two-dimensional real vector $[\ \Re e(x)\ \ \Im m(x)\ ]^T$. In particular, the notations $|x|$ or $\|x\|_2$ are used for the magnitude of $x$, according to the context.

$V(\omega)$, and the corresponding radius $r_2[V(\omega)]$, are given by

$$c_2[V(\omega)] = \arg \inf_{s\in\Re^2} \sup_{v\in V(\omega)} \|s-v\|_2 ,$$
$$r_2[V(\omega)] = \inf_{s\in\Re^2} \sup_{v\in V(\omega)} \|s-v\|_2 .$$

The exact description of $V(\omega)$ and the exact computation of its center and radius are NP-hard problems, but convergent approximations can be computed as shown in the following.

Let us define

$$\delta_\nu = \frac{L\rho^\nu}{1-\rho},$$
$$\underline{FSS}^\nu = \left\{S^\nu \in K : \|A\cdot(y^N - F_N h^{S^\nu})\|_\infty^{W_e} \le \varepsilon\right\}, \tag{5.1}$$
$$\overline{FSS}^\nu = \left\{S^\nu \in K : \|A\cdot(y^N - F_N h^{S^\nu})\|_\infty^{W_e+\widetilde{W}} \le \varepsilon\right\} \tag{5.2}$$

where $\widetilde{W} = \delta_\nu \cdot \mathrm{diag}(\|a_1\|_1, \ldots, \|a_l\|_1) \in \Re^{l\times l}$ with $a_k$ the $k$-th row of $A$. Let $\underline{V}^\nu(\omega) = \Omega(\omega)[\underline{FSS}^\nu] \subset \Re^2$. Instead of describing this set exactly, e.g., by means of all its vertices, two polytopes $\underline{VI}_m^\nu(\omega)$ and $\underline{VO}_m^\nu(\omega)$ with $m$ vertices are constructed, inner- and outer-bounding $\underline{V}^\nu(\omega)$ respectively. The polytope $\underline{VI}_m^\nu$ is described by its vertices $v_k(\underline{VI}_m^\nu)$, $k = 1, \ldots, m$, computed as tangency points of $\underline{V}^\nu$ with the supporting lines $l_k$ with slope $s_k = 2\pi k/m$. Such points can be obtained by solving the following $m$ linear programming problems

$$v_k(\underline{VI}_m^\nu) = \Omega(\omega) \cdot \arg \min_{S^\nu \in \underline{FSS}^\nu} [-\Omega_1(\omega)\cdot\sin(s_k) + \Omega_2(\omega)\cdot\cos(s_k)] \cdot h^{S^\nu} \tag{5.3}$$

with $\Omega(\omega) = \begin{bmatrix} \Omega_1(\omega) \\ \Omega_2(\omega) \end{bmatrix}$ and $\underline{FSS}^\nu$ defined in (3.5) and (5.1) respectively. The polytope $\underline{VO}_m^\nu$ is described by its vertices $v_k(\underline{VO}_m^\nu)$, $k = 1, \ldots, m$, computed as intersection points of previously defined supporting lines $l_k$ and $l_{\mathrm{mod}(k,m)+1}$ of $\underline{V}^\nu$, with $\mathrm{mod}(k,m) = k$ if $k < m$ and 0 otherwise.

The same procedure applies to $\overline{V}^\nu(\omega) = \Omega(\omega)[\overline{FSS}^\nu] \subset \Re^2$ by substituting $\overline{FSS}^\nu$ for $\underline{FSS}^\nu$ in (5.3). Let $\overline{VO}_m^\nu(\omega)$ denote the outer approximation of $\overline{V}^\nu(\omega)$ and $\overline{\overline{VO}}_m^\nu(\omega) = \left\{s \in \Re^2 : s = \bar{s} + \tilde{s},\ \bar{s} \in \overline{VO}_m^\nu(\omega), \|\tilde{s}\|_2 \le \delta_\nu\right\}$.

**Result 5.2 ([20, 23]).**

i) *For given $\omega \in [0, 2\pi]$ and positive integers $\nu$ and $m$, it results that*

$$\underline{VI}_m^\nu(\omega) \subseteq V(\omega) \subseteq \overline{\overline{VO}}_m^\nu(\omega) ,$$

$$r_2[\underline{VI}_m^\nu(\omega)] \le r_2[V(\omega)] \le r_2[\overline{\overline{VO}}_m^\nu(\omega)] \le r_2[\overline{VO}_m^\nu(\omega)] + \delta_\nu .$$

ii) *For given* $\omega \in [0, 2\pi]$,

$$\lim_{\nu,m\to\infty} r_2\left[\underline{VI}_m^\nu(\omega)\right] = \lim_{\nu,m\to\infty}\left\{r_2[\overline{VO}_m^\nu(\omega)] + \delta_\nu\right\} = r_2\left[V(\omega)\right],$$

$$\lim_{\nu,m\to\infty} c_2\left[\underline{VI}_m^\nu(\omega)\right] = \lim_{\nu,m\to\infty} c_2[\overline{VO}_m^\nu(\omega)] = c_2\left[V(\omega)\right].$$

This result allows one also to evaluate the tightness of obtained approximations. By increasing $m$ and $\nu$, any given accuracy can be achieved at the expense of computational complexity. The choice $m = 16$ guarantees an accuracy in evaluating $r_2\left[\underline{V}^\nu(\omega)\right]$ or $r_2[\overline{V}^\nu(\omega)]$ better than 2% $\forall\nu$ and $\forall\omega$, while the value of $\nu$ has to be chosen such that $\delta_\nu$ is "small" with respect to $r_2\left[V(\omega)\right]$. In any case the overall accuracy, due to given values of $m$ and $\nu$, can be evaluated a posteriori, by computing for example the bound

$$\frac{r_2\left[\overline{VO}_m^\nu(\omega)\right] + \delta_\nu - r_2\left[\underline{VI}_m^\nu(\omega)\right]}{r_2\left[\underline{VI}_m^\nu(\omega)\right]}$$

on the relative error of the radius $r_2\left[V(\omega)\right]$ for a set of frequencies $\omega$ of interest.

# 6 Identification error evaluation and nearly optimal algorithm

Now, the computation of the model set identification error $E(\hat{\mathcal{M}})$ is addressed. Exact evaluation is hard, but convergent upper and lower bounds may be derived. Indeed, the results of [36, 7] could be used to this purpose. However, once the convergent inner and outer approximations of the value set $V(\omega)$ described in Section 5 are available, tighter bounds on $E(\hat{\mathcal{M}})$ are provided by the following result.

**Result 6.1 ([20, 23]).** *For given algorithm* $\phi$ *and nominal model* $\hat{M} = \phi\left(K, \mathcal{B}_e, F_N, y^N\right)$, *the identification error of the corresponding model set* $\hat{\mathcal{M}}$ *is bounded as*

$$\underline{E}_m^\nu(\hat{\mathcal{M}}) \le E(\hat{\mathcal{M}}) \le \overline{E}_m^\nu(\hat{\mathcal{M}})$$

*where* $\nu$ *and* $m$ *are given positive integers and*

$$\begin{aligned} \underline{E}_m^\nu(\hat{\mathcal{M}}) &= \sup_{0\le\omega\le2\pi}\ \max_{k=1,\dots,m}\left\|v_k\left(\underline{VI}_m^\nu(\omega)\right) - \hat{M}(\omega)\right\|_2, \\ \overline{E}_m^\nu(\hat{\mathcal{M}}) &= \sup_{0\le\omega\le2\pi}\ \max_{k=1,\dots,m}\left\|v_k(\overline{VO}_m^\nu(\omega)) - \hat{M}(\omega)\right\|_2 + \delta_\nu. \end{aligned} \tag{6.1}$$

*Moreover,*

$$\lim_{\nu,m\to\infty} \overline{E}_m^\nu(\hat{\mathcal{M}}) = \lim_{\nu,m\to\infty} \underline{E}_m^\nu(\hat{\mathcal{M}}) = E(\hat{\mathcal{M}}).$$

Note that in the literature some lower bound and several upper bounds on the model set identification error $E(\hat{\mathcal{M}})$ are available, whose degree of conservativeness is not known, see, e.g., [12, 31, 10, 11, 5, 4].

It is of interest to find model sets making the identification error as small as possible. On the basis of the above result, a possible approach to approximate the optimal algorithm could consist in finding a model set $\hat{\mathcal{M}}_\nu^{FIR}$ with nominal model $\hat{M}_\nu^{FIR}$ that minimizes the upper bound (6.1) on $E(\hat{\mathcal{M}}_\nu^{FIR})$ within the class of $FIR_\nu$ models. By suitably increasing $\nu$ and $m$, model sets with optimality level as close as desired to 1 can be obtained. However, for $\nu$ larger than a few decades, the computational complexity of the involved nonlinear optimization problem may become prohibitive. The next result gives an algorithm whose optimality level is essentially $\sqrt{2}$, thus improving on the 2-optimality of almost optimal algorithms available in the literature.

Let a set of ordered frequencies $T_\omega = \{\omega_k : \omega_k \in [0, 2\pi]\,,\ k = 1, \cdots, q\}$ be given, such that $\lim_{q\to\infty} \max_{k=1,\dots q-1} (\omega_{k+1} - \omega_k) = 0$. This property is satisfied, for example, by an equispaced frequency set. Within the class of $FIR_\nu$ models, a model $\hat{M}_\nu^{no} \in FSS$ can be derived by solving the following optimization problem

$$h^{\hat{M}_\nu^{no}} = \arg \min_{S^\nu \in FSS} \left\| s^* - \Omega^* \cdot h^{S^\nu} \right\|_\infty \tag{6.2}$$

where

$$s^* = \begin{bmatrix} s^*(\omega_1) \\ \vdots \\ s^*(\omega_q) \end{bmatrix} \in \Re^{4q}, \qquad s^*(\omega) = \begin{bmatrix} \max\limits_{s=\left[\begin{smallmatrix} s_1 \\ s_2 \end{smallmatrix}\right] \in \overline{VO}_m^\nu(\omega)} s_1 \\ \max\limits_{s=\left[\begin{smallmatrix} s_1 \\ s_2 \end{smallmatrix}\right] \in \overline{VO}_m^\nu(\omega)} s_2 \\ \min\limits_{s=\left[\begin{smallmatrix} s_1 \\ s_2 \end{smallmatrix}\right] \in \overline{VO}_m^\nu(\omega)} s_1 \\ \min\limits_{s=\left[\begin{smallmatrix} s_1 \\ s_2 \end{smallmatrix}\right] \in \overline{VO}_m^\nu(\omega)} s_2 \end{bmatrix} \in \Re^4,$$

$$\Omega^* = \begin{bmatrix} \Omega^*(\omega_1) \\ \vdots \\ \Omega^*(\omega_q) \end{bmatrix} \in \Re^{4q\times\nu}, \qquad \Omega^*(\omega) = \begin{bmatrix} \Omega^\nu(\omega) \\ \Omega^\nu(\omega) \end{bmatrix} \in \Re^{4\times\nu}$$

with $m$ a given positive integer and $\Omega^\nu(\omega) \in \Re^{2\times\nu}$ given by the first $\nu$ columns of the matrix $\Omega(\omega)$ defined in (3.5). Note that this problem can be easily solved by linear programming techniques [22]. This way, an interpolatory algorithm is obtained, given by

$$\phi^{no}\left(K, \mathcal{B}_e, F_N, y^N\right) = \hat{M}_\nu^{no}\,.$$

Note that if the interpolation constraint $S^\nu \in FSS$ is not imposed in the optimization problem (6.2), the obtained algorithm may have lower optimality level, but is not guaranteed to be interpolatory. The algorithm $\phi^{no}$, being interpolatory, enjoys the interesting convergence properties of interpolatory algorithms (see, e.g., [12, 10, 5, 19, 3] for details).

The following result shows that, by choosing $\nu, m$ and $q$ sufficiently large, the algorithm $\phi^{no}$ is essentially $\sqrt{2}$-optimal, i.e., its overbound on the identification error of the optimal algorithm can be made as close as desired to $\sqrt{2}$.

**Result 6.2 ([22]).** *The algorithm* $\phi^{no}\left(K, \mathcal{B}_e, F_N, y^N\right) = \hat{M}_\nu^{no}$ *is such that*

$$E(\hat{\mathcal{M}}_\nu^{no}) \leq \sqrt{2}r + g\left(\nu, m, q\right), \text{ where } \lim_{\nu,m,q\to\infty} g\left(\nu, m, q\right) = 0 \, .$$

Note that the actual optimality level of the model set $\hat{\mathcal{M}}_\nu^{no}$, with nominal model $\hat{M}_\nu^{no} = \phi^{no}\left(K, \mathcal{B}_e, F_N, y^N\right)$ identified from given information $\left(K, \mathcal{B}_e, F_N, y^N\right)$, may be lower than $\sqrt{2}$ and can be evaluated using Result 6.3. Indeed, as shown in numerical examples reported in [20, 21], typically $\alpha(\hat{\mathcal{M}}_\nu^{no})$ results quite close to 1, as expected from the fact that $\hat{M}_\nu^{no}$ is constructed as a suitable approximation of the central model. For this reason, $\phi^{no}$ is called "nearly optimal".

The next result provides lower and upper bounds on the optimality level $\alpha(\hat{\mathcal{M}})$ of any given model set $\hat{\mathcal{M}}$.

**Result 6.3 ([20]).** *For given model set* $\hat{\mathcal{M}}$ *and positive integers* $\nu$ *and* $m$, *it results that*

$$\underline{\alpha}_m^\nu(\hat{\mathcal{M}}) \leq \alpha(\hat{\mathcal{M}}) \leq \overline{\alpha}_m^\nu(\hat{\mathcal{M}})$$

*where*

$$\underline{\alpha}_m^\nu(\hat{\mathcal{M}}) = \max\left\{1, \underline{E}_m^\nu(\hat{\mathcal{M}}) \Big/ \left[\sup_{0\leq\omega\leq 2\pi} \max_{k=1,\dots,m} \left\| v_k(\overline{VO}_m^\nu(\omega)) - \hat{M}_\nu^{no}(\omega) \right\|_2 + \delta_\nu\right]\right\},$$

$$\overline{\alpha}_m^\nu(\hat{\mathcal{M}}) = \overline{E}_m^\nu(\hat{\mathcal{M}}) \Big/ \sup_{0\leq\omega\leq 2\pi} r_2\left[\underline{VI}_m^\nu(\omega)\right] .$$

# 7 Frequency-shaped model sets

So far, given an algorithm $\phi$ and the corresponding nominal model $\hat{M} = \phi\left(y^N\right)$, model sets $\hat{\mathcal{M}}$ of the form (3.11) have been considered, where the uncertainty $\Delta$ has a constant bound in frequency

$$\left|\Delta\left(\omega\right)\right| \leq E(\hat{\mathcal{M}}), \quad \forall \omega \in [0, 2\pi] \; .$$

However, model sets with a frequency shaped bound on the uncertainty $\Delta$ of the form

$$\hat{\mathcal{M}}^W = \left\{\hat{M} + \Delta : \left|\Delta\left(\omega\right)\right| \leq W\left(\omega\right), \, \forall \omega \in [0, 2\pi]\right\} \tag{7.1}$$

may be more convenient for $H_\infty$ control design. Indeed, the following inclusion properties hold:

- $\hat{\mathcal{M}}^W$ is a model set for $S^o$ if and only if

$$W(\omega) \geq \sup_{S \in FSS} \left| S(\omega) - \hat{M}(\omega) \right| = W^*(\omega), \quad \forall \omega \in [0, 2\pi] \ .$$

- $\hat{\mathcal{M}}^{W^*}$ is the smallest model set for $S^o$ of the form (7.1); moreover,

$$\hat{\mathcal{M}}^{W^*} \subseteq \hat{\mathcal{M}} \,,$$

$$E(\hat{\mathcal{M}}^{W^*}) = E(\hat{\mathcal{M}}) \,.$$

Thus, though the two model sets $\hat{\mathcal{M}}^{W^*}$ and $\hat{\mathcal{M}}$ have the same identification error, $\hat{\mathcal{M}}^{W^*}$ is smaller than $\hat{\mathcal{M}}$, so that designing $H_\infty$ robust control for $\mathcal{M}^{W^*}$ instead of $\hat{\mathcal{M}}$ may give better guaranteed performances.

Exact computation of $W^*(\omega)$ for a given frequency $\omega$ is not easy, but convergent upper and lower bounds on $W^*(\omega)$ are provided by the following result.

**Result 7.1 ([21]).** *For given* $\omega \in [0, 2\pi]$, $\hat{M}(\omega)$ *and positive integers* $\nu$ *and* $m$, $W^*(\omega)$ *is bounded as*

$$\underline{W}_m^\nu(\omega) \leq W^*(\omega) \leq \overline{W}_m^\nu(\omega)$$

*where*

$$\begin{aligned} \underline{W}_m^\nu(\omega) &= \max_{k=1,\dots,m} \left\| v_k(\underline{VI}_m^\nu(\omega)) - \hat{M}(\omega) \right\|_2 \,, \\ \overline{W}_m^\nu(\omega) &= \max_{k=1,\dots,m} \left\| v_k(\overline{VO}_m^\nu(\omega)) - \hat{M}(\omega) \right\|_2 + \delta_\nu \,. \end{aligned}$$

*Moreover,*

$$\lim_{\nu,m \to \infty} \overline{W}_m^\nu(\omega) = \lim_{\nu,m \to \infty} \underline{W}_m^\nu(\omega) = W^*(\omega) \ .$$

Note that most robust control design techniques require that the bound on the uncertainty $\Delta$ is chosen as the magnitude of a stable rational function $W_\Delta(\omega)$, possibly of low order, since this order affects the complexity of the designed controller. Thus, if the identified model set is intended to be used for robust control design, $W_\Delta(\omega)$ has to be computed, tightly overbounding $W(\omega)$ for a suitable set of frequency values. A systematic approach for such overbounding can be found in [34].

# 8 Conclusions

The overall Set Membership $H_\infty$ identification procedure resulting from the presented results can be summarized as follows. The computations of the $\sup_\omega$ required by Results 6.1 and 6.3 can be carried on by gridding $\omega$ on a sufficiently coarse set of frequencies $T_\omega = \{\omega_k : \omega_k \in [0, 2\pi], \ k = 1, \cdots, q\}$.

**Set Membership $\mathbf{H}_\infty$ identification procedure:**

1. For any $\omega \in T_\omega$, compute inner and outer approximations $\underline{VI}_m^\nu(\omega)$ and $\overline{VO}_m^\nu(\omega)$ of the value set $V(\omega)$ by solving the linear programming problems (5.3).

2. Compute the nearly optimal model $\hat{M}_\nu^{no}$ by solving the linear programming problem (6.2).

3. For given $n$, compute by standard methods a reduced order model $\hat{M}_n^r$ of $\hat{M}_\nu^{no}$. Then, using $\hat{M}_n^r$ as starting point, perform the nonlinear optimization $\hat{p} = \arg\min_p \overline{E}_m^\nu(\mathcal{M}_n(p))$, where $M_n(p)$ is a class of parametric models with transfer function of order $n$ and $\overline{E}_m^\nu(\mathcal{M}_n(p))$ is computed from Result 6.1.

4. Evaluate the optimality level of $\mathcal{M}_n(\hat{p})$ from Result 6.3 and choose the order $n$ by trading off between model set complexity and achievable optimality level.

5. For the chosen model, evaluate $\overline{W}_m^\nu(\omega)$, $\omega \in T_\omega$, from Result 7.1 and compute, using, e.g., the method in [34], a stable low-order rational function $W_\Delta(\omega)$ tightly overbounding $\overline{W}_m^\nu(\omega)$ $\forall \omega \in T_\omega$.

6. The identified model set is then given by

$$\hat{\mathcal{M}}_n^{W_\Delta} = \left\{ \hat{M}_n(\hat{p}) + \Delta : |\Delta(\omega)| \leq W_\Delta(\omega), \ \forall \omega \in [0, 2\pi] \right\} .$$

□

Numerical examples, including comparisons with alternative approaches of the literature, can be found in [20, 21]. The algorithms for performing the above procedure are available under MATLAB environment at the following ftp address: `www.polito.it/~taragna/tools/identification`

# Acknowledgement

This research was supported in part by funds of Ministero dell'Università e della Ricerca Scientifica e Tecnologica under the Project "Robustness techniques for control of uncertain systems".

## References


[1] L. Andersson, A. Rantzer, and C. Beck, "Model comparison and simplification," *International Journal of Robust and Nonlinear Control*, vol. 9, pp. 157–181, 1999.

[2] C. L. Beck, J. Doyle, and K. Glover, "Model reduction of multidimensional and uncertain systems," *IEEE Transactions on Automatic Control*, vol. AC-41, no. 10, pp. 1466–1477, 1996.

[3] J. Chen and G. Gu, *Control-Oriented System Identification: An $H_\infty$ Approach.* New York: John Wiley & Sons, Inc., 2000.

[4] J. Chen and C. N. Nett, "The Carathéodory–Fejér problem and $H_\infty/\ell_1$ identification: A time domain approach," *IEEE Transactions on Automatic Control*, vol. AC-40, no. 4, pp. 729–735, 1995.

[5] J. Chen, C. N. Nett, and M. K. H. Fan, "Worst case system identification in $H_\infty$: Validation of *a priori* information, essentially optimal algorithms, and error bounds," *IEEE Transactions on Automatic Control*, vol. AC-40, no. 7, pp. 1260–1265, 1995.

[6] A. Garulli, A. Tesi, and A. Vicino, Eds., *Robustness in Identification and Control.* Lecture Notes in Control and Information Sciences, vol. 245, Godalming, UK: Springer-Verlag, 1999.

[7] L. Giarré and M. Milanese, "SM identification of approximating models for $H_\infty$ robust control," *International Journal of Robust and Nonlinear Control*, vol. 9, pp. 319–332, 1999.

[8] L. Giarré, M. Milanese, and M. Taragna, "$H_\infty$ identification and model quality evaluation," *IEEE Transactions on Automatic Control*, vol. AC-42, no. 2, pp. 188–199, 1997.

[9] G. Gu and J. Chen, "A nearly interpolatory algorithm for $H_\infty$ identification with mixed time and frequency response data," *IEEE Transactions on Automatic Control*, vol. AC-46, no. 3, pp. 464–469, 2001.

[10] G. Gu and P. P. Khargonekar, "A class of algorithms for identification in $H_\infty$," *Automatica*, vol. 28, no. 2, pp. 299–312, 1992.

[11] G. Gu, D. Xiong, and K. Zhou, "Identification in $H_\infty$ using Pick's interpolation," *Systems & Control Letters*, vol. 20, pp. 263–272, 1993.

[12] A. J. Helmicki, C. A. Jacobson, and C. N. Nett, "Control oriented system identification: A worst-case/deterministic approach in $H_\infty$," *IEEE Transactions on Automatic Control*, vol. AC-36, no. 10, pp. 1163–1176, 1991.

[13] I. Jikuya and H. Kimura, "Representation and reduction of model sets," in *Proc. of the 38th IEEE Conference on Decision and Control*, Phoenix, AZ, 1999, pp. 1482–1487.

[14] M. A. Kon and R. Tempo, "On linearity of spline algorithms," *Journal of Complexity*, vol. 5, no. 2, pp. 251–259, 1989.

[15] L. Ljung, *System Identification: Theory for the User*, 2nd ed. Upper Saddle River, NJ: Prentice-Hall, 1999.

[16] L. Ljung and L. Guo, "The role of model validation for assessing the size of the unmodeled dynamics," *IEEE Transactions on Automatic Control*, vol. AC-42, no. 9, pp. 1230–1239, 1997.

[17] P. M. Mäkilä, J. R. Partington, and T. K. Gustafsson, "Worst-case control-relevant identification," *Automatica*, vol. 31, no. 12, pp. 1799–1819, 1995.

[18] M. Milanese, J. Norton, H. Piet-Lahanier, and É. Walter, Eds., *Bounding Approaches to System Identification.* New York: Plenum Press, 1996.

[19] M. Milanese and M. Taragna, "Inputs for convergent SM identification with approximated models," in *Proc. of the 37th IEEE Conference on Decision and Control*, Tampa, FL, 1998, pp. 4458–4463.

[20] M. Milanese and M. Taragna, "Suboptimality evaluation of approximated models in $H_\infty$ identification," in *Proc. of the 38th IEEE Conference on Decision and Control*, Phoenix, AZ, 1999, pp. 1494–1499.

[21] M. Milanese and M. Taragna, "Set Membership identification for $H_\infty$ robust control design," in *Proc. of 12th IFAC Symposium on System Identification SYSID 2000*, Santa Barbara, CA, 2000.

[22] M. Milanese and M. Taragna, "Nearly optimal model sets in $H_\infty$ identification," in *Proc. of the European Control Conference 2001*, Porto, Portugal, 2001, pp. 1704–1709.

[23] M. Milanese and M. Taragna, "Optimality, approximation, and complexity in Set Membership $H_\infty$ identification," *IEEE Transactions on Automatic Control*, vol. AC-47, no. 10, pp. 1682-1690, 2002.

[24] M. Milanese and R. Tempo, "Optimal algorithms theory for estimation and prediction," *IEEE Transactions on Automatic Control*, vol. AC-30, pp. 730–738, 1985.

[25] M. Milanese, R. Tempo, and A. Vicino, Eds., *Robustness in Identification and Control.* New York: Plenum Press, 1989.

[26] M. Milanese and A. Vicino, "Optimal estimation theory for dynamic systems with set membership uncertainty: An overview," *Automatica*, vol. 27, no. 6, pp. 997–1009, 1991.

[27] M. Milanese and A. Vicino, "Information-based complexity and non-parametric worst-case system identification," *Journal of Complexity*, vol. 9, pp. 427–446, 1993.

[28] B. Ninness and G. C. Goodwin, "Estimation of model quality," *Automatica*, vol. 31, no. 12, pp. 1771–1797, 1995.

[29] F. Paganini, "A set-based approach for white noise modeling," *IEEE Transactions on Automatic Control*, vol. AC-41, no. 10, pp. 1453–1465, 1996.

[30] P. A. Parrilo, M. Sznaier, and R. S. Sánchez Peña, "Mixed time/ frequency-domain based robust identification," *Automatica*, vol. 34, pp. 1375–1389, 1998.

[31] J. R. Partington, "Robust identification and interpolation in $H_\infty$," *International Journal of Control*, vol. 54, no. 5, pp. 1281–1290, 1991.

[32] J. R. Partington, *Interpolation, Identification, and Sampling.* London Mathematical Society Monographs New Series, vol. 17, New York: Clarendon Press - Oxford, 1997.

[33] K. R. Popper, *Conjectures and Refutations: The Growth of Scientific Knowledge.* London, UK: Rontedge and Kegan Paul, 1969.

[34] R. E. Scheid and D. S. Bayard, "A globally optimal minimax solution for spectral overbounding and factorization," *IEEE Transactions on Automatic Control*, vol. AC-40, no. 4, pp. 712–716, 1995.

[35] R. S. Smith and M. Dahleh, Eds., *The Modeling of Uncertainty in Control Systems.* Lecture Notes in Control and Information Sciences, vol. 192, London, UK: Springer-Verlag, 1994.

[36] M. Taragna, "Uncertainty model identification for $H_\infty$ robust control," in *Proc. of the 37th IEEE Conference on Decision and Control*, Tampa, FL, 1998, pp. 3403–3405.

[37] J. F. Traub, G. W. Wasilkowski, and H. Woźniakowski, *Information-Based Complexity.* New York: Academic Press, 1988.

[38] S. R. Venkatesh and M. A. Dahleh, "Identification in the presence of classes of unmodeled dynamics and noise," *IEEE Transactions on Automatic Control*, vol. AC-42, no. 12, pp. 1620–1635, 1997.

[39] M. Vidyasagar, *A Theory of Learning and Generalization with Application to Neural Networks and Control Systems.* Springer-Verlag, 1996.

[40] E. Weyer, "Finite sample properties of system identification of ARX models under mixing conditions," *Automatica*, vol. 36, no. 9, pp. 1291–1299, 2000.

[41] T. Zhou and H. Kimura, "Structure of model uncertainty for a weakly corrupted plant," *IEEE Transactions on Automatic Control*, vol. AC-40, no. 4, pp. 639–655, 1995.

Mario Milanese
Dipartimento di Automatica e Informatica
Politecnico di Torino
Corso Duca degli Abruzzi 24, Torino 10129, Italy
milanese@polito.it

Michele Taragna
Dipartimento di Automatica e Informatica
Politecnico di Torino
Corso Duca degli Abruzzi 24, Torino 10129, Italy
taragna@polito.it

# System Identification: A Learning Theory Approach

M. Vidyasagar and Rajeeva L. Karandikar

*Dedicated with admiration and in friendship to Hidenori Kimura sensei on the occasion of his sixtieth birthday*

ABSTRACT In this paper, the problem of system identification is formulated as a problem in statistical learning theory. It is shown that if a particular type of uniform convergence result holds, then the traditional approach of choosing the current model to minimize the error on the observed data will eventually converge to an optimal model within the specified class. More important, the reformulation of system identification as a learning theory problem leads to *finite time estimates* for the rate at which the estimated model converges to the optimal model. As an illustration of the approach, a result is derived showing that in the case of exponentially stable systems with fading memory, the desired uniform convergence result holds, so that the learning theory approach is applicable, provided the adjustable parameters enter the system model in a linear fashion.

**Keywords:** System identification, Statistical learning, Pseudo-dimension, Uniform convergence, Empirical means

## 1 Introduction

The problem of system identification is so well-studied that one might ask why a new approach is necessary at all. An answer is that, by tradition, system identification theory has been dedicated to the derivation of *asymptotic* results, that is, results that tell us what happens as the number of measurements approaches infinity. On the other hand, suppose one could somehow derive *finite-time* estimates, which tell us just how far the current model is from the 'true' system. Then it would be possible to combine robust control theory with such finite-time estimates to conclude that, after a finite number of measurements, a controller designed so as to stabilize the current model would also stabilize the true but unknown system.

The topic of statistical learning theory is somewhat more recent than system identification theory. But learning theory has its genesis in the study of the convergence properties of empirical processes, which is a subject about as old as identification theory. Moreover, finite-time estimates are the *raison d'être* of learning theory, unlike in identification theory. Thus, *a priori*, learning theory appears to be a more natural approach to the derivation of finite-time estimates of the rate of convergence of identification algorithms.

The idea of formulating system identification as a learning theory problem is of recent vintage; see [11, 12, 13, 15, 3]. In [15, 3] the identification problem is studied when the system to be identified is linear with additive noise. It is shown that, in some cases, the kind of asymptotic results available from traditional identification theory are often misleading in some sense. In [11, 12, 13] the class of systems studied are nonlinear, but the adjustable parameters enter into the model in a linear fashion. Indeed, these papers are preliminary versions of the present paper.

# 2 Problem formulation

The aim of system identification is to fit given data, usually supplied in the form of a time series, with models from within a given model class. Let us denote the time series by $\{(y_t, u_t)\}_{t=-\infty}^{\infty}$, where $u_t$ and $y_t$ denote respectively the input and output of the system at time $t$. Let us denote by $\{h(\theta), \theta \in \Theta\}$ the family of input-output models that are to be used to fit this time series. The notation is now made precise.

For the class of systems under study, the output set is some $Y \subseteq \Re^k$, while the input set is some $U \subseteq \Re^\ell$ for some $k$ and $\ell$. To avoid technicalities, let us suppose that the inputs are restricted to belong to a *bounded* set $U$; this assumption ensures that any random variable assuming values in $U$ has bounded moments of all orders. There is also a 'loss function' $\ell : Y \times Y \to [0, 1]$ which is used to measure how well the predicted output matches the actual output.

To set up the time series that forms the input to identification, let us define $\mathcal{U} := \prod_{-\infty}^{\infty} U$, and define $\mathcal{Y}$ analogously. Equip the doubly infinite Cartesian product $\mathcal{Y} \times \mathcal{U} := \prod_{-\infty}^{\infty}(Y \times U)$ with the product Borel $\sigma$-algebra, and call it $\mathcal{S}^\infty$. Next, introduce a probability measure $\tilde{P}_{\mathbf{y},\mathbf{u}}$ on the measurable space $(\mathcal{Y} \times \mathcal{U}, \mathcal{S}^\infty)$. Following our earlier practice, let us define a 'stochastic process' $\{(y_t, u_t)\}_{t=-\infty}^{\infty}$ as a measurable map from $(\mathcal{Y} \times \mathcal{U}, \mathcal{S}^\infty, \tilde{P}_{\mathbf{y},\mathbf{u}})$ into $\mathcal{Y} \times \mathcal{U}$. Let the coordinate random variables $(y_t, u_t)$ be thought of as the components of the time series at time $t$, and let us assume that the time series is stationary, i.e., that the probability measure $\tilde{P}_{\mathbf{y},\mathbf{u}}$ is shift-invariant. Let $\tilde{P}_{y,\mathbf{u}}$ denote the one-dimensional marginal probability associated with $\tilde{P}_{\mathbf{y},\mathbf{u}}$ on $Y$, and note that $\tilde{P}_{y,\mathbf{u}}$ is a probability measure on the set $Y \times \mathcal{U}$. Let $U_{-\infty}^0$ denote the one-sided infinite Cartesian product $U_{-\infty}^0 := \prod_{-\infty}^{0} U$, and for a given two-sided infinite sequence $\mathbf{u} \in \mathcal{U}$, define

$$\mathbf{u}_t := (u_{t-1}, u_{t-2}, u_{t-3}, \dots) \in U_{-\infty}^0.$$

Thus $\mathbf{u}_t$ denotes the 'infinite past' of the input sequence at time $t$. With this preliminary notation, we can set up the problem under study.

The input to the identification process is a time series $\{(y_t, u_t)\}_{t\geq 1}$ generated through a stochastic process, as described above. Thus, while it is assumed that the stochastic process stretches into the infinite past, the identification algorithm has a definite starting point, which is taken as time $t = 0$. To fit this time series, we use a family of models $\{h(\theta), \theta \in \Theta\}$, where each $h(\theta)$ denotes an input-output mapping from $U_{-\infty}^{0}$ to $Y$, and the parameter $\theta$ captures the variations in the model family. Thus for the system parametrized by $\theta$, the output at time $t$ in response to the input sequence $\mathbf{u} \in \mathcal{U}$ is given by $h(\theta) \cdot \mathbf{u}_t$. Note that this definition automatically guarantees that each system in the family of models is time-invariant.

For each parameter $\theta \in \Theta$, define the objective function

$$J(\theta) := E[\ell(y_t, h(\theta) \cdot \mathbf{u}_t), \tilde{P}_{\mathbf{y},\mathbf{u}}]. \tag{2.1}$$

Thus $J(\theta)$ is the expected value of the loss incurred by using the model output $h(\theta) \cdot \mathbf{u}_t$ to predict the actual output $y_t$. Note that, since the only value of $\mathbf{y}$ that appears within the expected value is $y_t$, we can actually replace the measure $\tilde{P}_{\mathbf{y},\mathbf{u}}$ by $\tilde{P}_{y,\mathbf{u}}$. In other words, we can also write

$$J(\theta) := E[\ell(y_t, h(\theta) \cdot \mathbf{u}_t), \tilde{P}_{y,\mathbf{u}}].$$

Thus the expectation is taken with respect to the 'one-dimensional' marginal measure $\tilde{P}_{y,\mathbf{u}}$ on $Y \times \mathcal{U}$.

One of the most commonly used loss functions is the squared error; thus

$$\ell(y, z) :=\| y - z \|^2,$$

where $\| \cdot \|$ is the usual Euclidean or $\ell_2$-norm on $\Re^k$. In this case $J(\theta)$ is the expected value of the mean squared prediction error when the map $h(\theta)$ is used to predict $y_t$. Note that, by the assumption of stationarity, the quantity on the right side of (2.1) is independent of $t$.

Now we are in a position to state the problem under study.

**The Identification Problem.** Determine a $\theta \in \Theta$ that minimizes the error measure $J(\theta)$. □

Therefore the identification problem is to find the best possible model parameter $\theta^*$, if it exists, in the sense that $J(\theta^*) = \inf_{\theta \in \Theta} J(\theta) =: J^*$. The model parameter $\theta*$ is the 'best' in the sense that the predicted output $h(\theta^*) \cdot \mathbf{u}_t$ is the best possible approximation to the actual output $y_t$, where the goodness of approximation is measured by the expected value of the loss function $\ell(y_t, h(\theta) \cdot \mathbf{u}_t)$.

A weaker form of the identification problem is to determine a sequence of parameters $\{\theta_t\}$ such that $J(\theta_t) \rightarrow \inf_{\theta \in \Theta} J(\theta) =: J^*$ as $t \rightarrow \infty$. In this weaker formulation, the sequence of models $\{\theta_t\}$ asymptotically approaches the best possible performance $J^*$. However, there is no requirement that the sequence $\{\theta_t\}$ should have any convergence properties.

Suppose the measured output $y_t$ corresponds to a noise-corrupted output of a 'true' system $f_{\text{true}}$, and that $\ell$ is the squared error, as above. In such a case, the problem formulation becomes the following: Suppose the input sequence $\{u_t\}_{-\infty}^{\infty}$ is i.i.d. according to some law $P$, and that $\{\eta_t\}_{-\infty}^{\infty}$ is a measurement noise sequence that is zero mean and i.i.d. with law $Q$. Suppose in addition that $u_i, \eta_j$ are independent for each $i, j$. Now suppose that

$$y_t = f_{\text{true}} \cdot \mathbf{u}_t + \eta_t, \ \forall t. \tag{2.2}$$

In such a case, the expected value in (2.1) can be expressed in terms of the probability measure $Q \times P^\infty$, and becomes

$$\begin{aligned} J(\theta) &= E[\| \, (f_{\text{true}} - h(\theta)) \cdot \mathbf{u}_t + \eta_t \, \|^2, Q \times P^\infty] \\ &= E[\| \, \tilde{h}(\theta) \cdot \mathbf{u}_t \, \|^2, P^\infty] + E[\| \, \eta \, \|^2, Q], \end{aligned}$$

where $\tilde{h}(\theta) := h(\theta) - f_{\text{true}}$. Since the second term is independent of $\theta$, we effectively minimize only the first term. In other words, by minimizing $J(\theta)$ with respect to $\theta$, we will find the best approximation to the true system $f_{\text{true}}$ in the model family $\{h(\theta), \theta \in \Theta\}$. Note that it is *not* assumed that the true system $f_{\text{true}}$ belongs to $\{h(\theta), \theta \in \Theta\}$.

# 3 A general result

One can divide the main challenges of system identification into three successively stronger questions, as follows: As more and more data is provided to the identification algorithm:

1. Does the estimation error between the outputs of the identified model and the actual time series approach the minimum possible estimation error achievable by any model within the given model class? In other words, if $\theta_t$ denotes the parameter estimate at time $t$, does $J(\theta_t)$ approach $J^* := \inf_{\theta \in \Theta} J(\theta)$?

2. Assuming that the data is generated by a 'true' system whose output is corrupted by measurement noise, does the identified model converge to the best possible approximation of the 'true' system within the model class? In other words, suppose we define some kind of metric distance $\rho$ between pairs of input-output maps. Does $\rho[f_{\text{true}}, h(\theta_t)]$ approach the quantity $\inf_{\theta \in \Theta} \rho[f_{\text{true}}, h(\theta)]$?

3. Assuming that the true system belongs to the model class, does the estimated model converge to the true system? In other words, suppose $f_{\text{true}} = h(\theta_{\text{true}})$ for some 'true' parameter vector $\theta_{\text{true}}$. Does $\theta_t \rightarrow \theta_{\text{true}}$ as $t \rightarrow \infty$?

From a technical standpoint, Questions 2 and 3 are easier to answer than Question 1. Since identification is carried out recursively, the output of the identification algorithm is a sequence of estimates $\{\theta_t\}_{t\geq 1}$, or equivalently, a sequence of estimated models $\{h(\theta_t)\}_{t\geq 1}$. Traditionally a positive answer to Question 2 is assured by assuming that $\Theta$ is a *compact* set, which in turn ensures that the sequence $\{\theta_t)\}$ contains a convergent subsequence. Alternatively, it is possible to use some kind of 'regularization' whereby the objective function $J(\theta)$ is augmented by an additional term $\| \theta \|^2$. Adding this term to $J(\theta)$ effectively ensures that all iterations $\theta_t$ are contained within a sphere of finite radius. Either way, the aim is to ensure that $\{\theta_t\}$ contains a convergent subsequence. If the answer to Question 1 is 'yes,' and if $\theta^*$ is a limit point of the sequence, then by appealing to some continuity arguments it can be shown that $J(\theta^*) = J^*$. In turn this implies that the expected value of the loss function $\ell(f_{\text{true}} \cdot \mathbf{u}_t, h(\theta^*) \cdot \mathbf{u}_t)$ equals $J^*$. Suppose now that the loss function $\ell$ is chosen in such a way that $\ell(f_{\text{true}} \cdot \mathbf{u}_t, h(\theta^*) \cdot \mathbf{u}_t)$ measures a distance between $f_{\text{true}}$ and $h(\theta^*)$. For instance, in case $\ell(y, z) = \| y - z \|^2$, then $J(\theta_t)$ is the $L_2$-norm of the difference between the 'true' output $f_{\text{true}} \cdot \mathbf{u}_t$ and the 'predicted' output $h(\theta^*) \cdot \mathbf{u}_t$. This defines a pseudometric on the set of systems.[1] In this case, it readily follows that $h(\theta^*)$ is the 'best possible' fit to the true system $f_{\text{true}}$ with respect to this pseudometric distance. Coming now to Question 3, suppose $\theta_{\text{true}}$ is the parameter of the 'true' model, and let $h(\theta_{\text{true}})$ denote the 'true' system. Suppose $\theta^*$ is a limit point of the sequence $\{\theta_t\}$. The traditional way to ensure that $\theta_{\text{true}} = \theta^*$ is to assume that the input to the true system is 'persistingly exciting' or 'sufficiently rich,' so that $f_{\text{true}} \cdot u_t = h(\theta^*) \cdot u_t \; \forall t$ implies that $f_{\text{true}} = h(\theta^*)$.

With this background, the emphasis in this paper is on providing an affirmative answer to Question 1. A sufficient condition for this is given in Theorem 3.1 below. Observe that the probability measure $\tilde{P}_{\mathbf{y},\mathbf{u}}$ in (2.1) is in general *unknown*. This is because $\tilde{P}_{\mathbf{y},\mathbf{u}}$ often corresponds to the 'true but unknown' system, as in (2.2). Thus it is not reasonable to assume that $\tilde{P}_{\mathbf{y},\mathbf{u}}$ is known. In turn, this implies that in general it is *not possible* to compute the objective function $J(\theta)$ in (2.1) directly, for a given choice of the parameter vector $\theta$. An indirect method must therefore be found to minimize $J(\theta)$.

A natural algorithm is the so-called 'minimum empirical cost' algorithm described next. For each $t \geq 1$ and each $\theta \in \Theta$, define the empirical error

$$\hat{J}_t(\theta) := \frac{1}{t} \sum_{i=1}^{t} \ell[y_t, h(\theta) \cdot \mathbf{u}_t].$$

[1] It is a pseudometric in general, and not a metric, because it is possible for the $L_2$-norm of the difference to be zero even if $f_{\text{true}} \neq h(\theta^*)$.

Note that, unlike $J(\theta)$, the quantity $\hat{J}_t(\theta)$ can indeed be computed on the basis of the data available at time $t$. At time $t$, choose $\theta_t^*$ so as to minimize $\hat{J}_t(\theta)$; that is,

$$\theta_t^* = \operatorname{Arg}\min_{\theta \in \Theta} \hat{J}_t(\theta).$$

For this algorithm, we have the following result.

**Theorem 3.1.** *Let*

$$J^* := \inf_{\theta \in \Theta} J(\theta).$$

*Define the quantity*

$$q(t, \epsilon) := \tilde{P}_{\mathbf{y},\mathbf{u}}\{\sup_{\theta \in \Theta} |\hat{J}_t(\theta) - J(\theta)| > \epsilon\}. \tag{3.1}$$

*Suppose it is the case that $q(t, \epsilon) \to 0$ as $t \to \infty$. Then for the minimum empirical cost algorithm we have*

$$\tilde{P}_{\mathbf{y},\mathbf{u}}\{\hat{J}_t(\theta_t^*) > J^* + \epsilon\} \to 0 \text{ as } t \to \infty.$$

*In particular, given any $\epsilon, \delta > 0$, choose an integer $t_0(\epsilon, \delta)$ such that*

$$q(t, \epsilon/3) \le \delta \; \forall t \ge t_0(\epsilon, \delta). \tag{3.2}$$

*Then*

$$\tilde{P}_{\mathbf{y},\mathbf{u}}\{\hat{J}_t(\theta_t^*) > J^* + \epsilon\} \le \delta \; \forall t \ge t_0(\epsilon, \delta).$$

**Proof.** Suppose $q(t, \epsilon) \to 0$ as $t \to \infty$. Given $\epsilon, \delta > 0$, choose $t_0$ large enough that (3.2) holds. Then

$$\tilde{P}_{\mathbf{y},\mathbf{u}}\{\sup_{\theta \in \Theta} |\hat{J}_t(\theta) - J(\theta)| > \epsilon/3\} < \delta \; \forall t \ge t_0(\epsilon, \delta). \tag{3.3}$$

Select a $\theta_\epsilon \in \Theta$ such that $J(\theta_\epsilon) \le J^* + \epsilon/3$. Such a $\theta_\epsilon$ exists in view of the definition of $J^*$. Then, in view of (3.3), we can say with confidence $1 - \delta$ that

$$\hat{J}(\theta_t) \ge J(\theta_t) - \epsilon/3, \text{ and } \hat{J}(\theta_\epsilon) \le J(\theta_\epsilon) + \epsilon/3.$$

By definition,

$$\hat{J}(\theta_t) \le \hat{J}(\theta_\epsilon).$$

Combining these two inequalities shows that

$$J(\theta_t) \le \hat{J}(\theta_t) + \epsilon/3 \le \hat{J}(\theta_\epsilon) + \epsilon/3 \le J(\theta_\epsilon) + 2\epsilon/3 \le J^* + 2\epsilon/3 + \epsilon/3$$
$$= J^* + \epsilon.$$

This statement holds with confidence $1 - \delta$. □

Several remarks are now made regarding the significance of this theorem.

1. If the quantity $q(t, \epsilon)$ defined in (3.1) approaches zero as $t \to \infty$, then the family of functions $J(\theta), \theta \in \Theta$ is said to have the property of **uniform convergence of empirical means (UCEM)**. Moreover, the number $t_0(\epsilon, \delta)$ needed to ensure that $q(t, \epsilon) < \delta \; \forall t \geq t_0(\epsilon, \delta)$ is called the **sample complexity**. The theorem therefore states that if the family of error measures $\{J(\theta), \theta \in \Theta\}$ has the UCEM property, then the natural algorithm of choosing $\theta_t$ so as to minimize the empirical estimate $\hat{J}(\theta)$ at time $t$ is 'asymptotically optimal.' This type of result is similar in spirit to those in identification theory.

2. Moreover, the number of samples needed to ensure $\epsilon$-optimality to a specified confidence $\delta$ is bounded by the number of samples needed to ensure that the UCEM property holds with accuracy $\epsilon/3$ and confidence $\delta$. In contrast to the asymptotic statement in Item 1 above, this statement is *finite-time* in nature. The finite-time statements can be made quantitatively precise because one of the aims of statistical learning theory is to obtain sample complexity estimates for families that have the UCEM property.

3. Note that the result given in Theorem 3.1 is not by any means the most general possible. In particular, it is possible to show that if $\theta_t$ is chosen so as to 'nearly' minimize the empirical error 'most of the time,' then the minimum empirical cost algorithm is still asymptotically optimal.

Thus the sample complexity of ensuring that $J(\theta_t) \leq J^* + \epsilon$ is at most equal to the sample complexity of $q(m, \epsilon/3)$. This naturally brings up the question as to what kinds of families $\{h(\theta), \theta \in \Theta\}$ have this particular UCEM property, and what their sample complexities are like. These questions are addressed in the next subsection.

# 4 Sufficient conditions for the UCEM property

In a seminal paper [5], Ljung has shown that indeed Question 1 can be answered in the affirmative provided empirical estimates of the performance of each model $h(\theta)$ converge *uniformly* to the corresponding true performance, where the uniformity is with respect to $\theta \in \Theta$. Thus he was among the first to realize the importance of the UCEM property in establishing the convergence properties of identification algorithms. He also showed that this particular uniform convergence property does hold, provided two assumptions are satisfied, namely:

- The model class consists of uniformly exponentially stable systems, and

- The parameter $\theta$ enters the description of the model $h(\theta)$ in a 'differentiable' manner. Coupled with the assumption that $\Theta$ is a compact set, this assumption implies that various quantities have bounded gradients with respect to $\theta$.

It is interesting to note that Ljung's UCEM result predates that of Vapnik and Chervonenkis for real-valued functions as found in [9]. Some related work can be found in [1, 2].

Now it turns out that the assumptions employed in [5] in fact guarantee the finiteness of the VC-dimension of the family of cost functions associated with the model family $\{h(\theta), \theta \in \Theta\}$. See for example [4]. Thus in principle it is possible to derive UCEM results very similar to (if not exactly identical to) those in [5] using the results of [10], Chapter 10. However, it is well-known that the estimates of the (finite) VC-dimension based on differentiability assumptions are extremely conservative. In the present subsection, we shall derive an alternate UCEM result based on more direct arguments.

We begin by listing below the assumptions regarding the family of models employed in identification, and on the time series. Recall the symbol $\tilde{h}(\theta)$ to denote the function $\mathbf{u}_t \mapsto (f_{\text{true}} - h(\theta)) \cdot \mathbf{u}_t$, which maps $U_{-\infty}^{0}$ into $\Re^k$. Define the collection of functions $\mathcal{G}$ mapping $\mathcal{U}$ into $\Re$ as follows:

$$g(\theta) := \mathbf{u}_0 \mapsto \| \ (f - h(\theta)) \cdot \mathbf{u}_0 \ \|^2 : \mathcal{U} \to \Re,$$

$$\mathcal{G} := \{g(\theta) : \theta \in \Theta\}.$$

Now the various assumptions are listed.

A1. There exists a constant $M$ such that

$$|g(\theta) \cdot \mathbf{u}_0| \leq M, \ \forall \theta \in \Theta, \mathbf{u} \in \mathcal{U}.$$

A2. For each integer $k \geq 1$, define

$$g_k(\theta) \cdot \mathbf{u}_t := g(\theta) \cdot (u_{t-1}, u_{t-2}, \dots, u_{t-k}, 0, 0, \dots).$$

With this notation, define

$$\mu_k := \sup_{\mathbf{u} \in \mathcal{U}} \sup_{\theta \in \Theta} |(g(\theta) - g_k(\theta)) \cdot u_0|.$$

Then the assumption is that $\mu_k$ is finite for each $k$ and approaches zero as $k \to \infty$.

A3. Consider the collection of maps $\mathcal{G}_k = \{g_k(\theta) : \theta \in \Theta\}$, viewed as maps from $U^k$ into $\Re$. For each $k$, the family $\mathcal{G}_k$ has finite P-dimension, bounded by $d(k)$. (See [10], Chapter 4 for a definition of the P-dimension.)

Now the significance of each assumption is discussed in turn. Assumption A1 can be satisfied, for example, by assuming that the true system and each system in the family $\{h(\theta), \theta \in \Theta\}$ is BIBO stable (with an upper bound on the gain, independent of $\theta$), and that the set $U$ is bounded (so that $\{u_t\}$ is a bounded stochastic process). Assumption A2 essentially means that each of the systems in the model family has decaying memory (in the sense that the effect of the values of the input in the distant past on the current output becomes negligibly small). In the SISO case, this assumption is satisfied, for example, if

- Each of the models $h(\theta)$ is a linear ARMA model of the form

$$y_t = \sum_{i=1}^{l} a_i(\theta)u_{t-i} + b_i(\theta)y_{t-i}.$$

- The characteristic polynomials

$$\phi(\theta, z) := z^{l+1} - \sum_{i=1}^{l} b_i(\theta)z^{l-i}$$

  all have their zeros inside a circle of radius $\rho < 1$, where $\rho$ is independent of $\theta$.

- The numbers $a_i(\theta)$ are uniformly bounded with respect to $\theta$.

The extension of the above condition to MIMO systems is straightforward and is left to the reader.

Now we can state the main theorem.

**Theorem 4.1.** *Define the quantity $q(t, \epsilon)$ as in (3.1) and suppose Assumptions A1 through A3 are satisfied. Given an $\epsilon > 0$, choose $k(\epsilon)$ large enough that $\mu_k \leq \epsilon/4$ for all $k \geq k(\epsilon)$. Then for all $t \geq k(\epsilon)$ we have*

$$q(t, \epsilon) \leq 8k(\epsilon) \left(\frac{32e}{\epsilon} \ln \frac{32e}{\epsilon}\right)^{d(k(\epsilon))} \exp(-\lfloor t/k(\epsilon) \rfloor \epsilon^2 / 512M^2), \qquad (4.1)$$

*where $\lfloor t/k(\epsilon) \rfloor$ denotes the largest integer part of $t/k(\epsilon)$.*

**Remark.** From Theorem 3.1, it is also possible to quantify the rate of convergence of the performance of the estimated model to the optimal performance. □

**Proof.** Write $g(\theta) = g_k(\theta) + (g(\theta) - g_k(\theta))$, and define

$$q_1^k(t, \epsilon) := \Pr\{\sup_{\theta \in \Theta} \left| \frac{1}{t} \sum_{i=1}^{t} g_k(\theta) \cdot \mathbf{u}_i - E[g_k(\theta) \cdot \mathbf{u}_i, \tilde{P}] \right| > \epsilon\},$$

$$q_2^k(t,\epsilon) := \Pr\{\sup_{\theta\in\Theta} \left| \frac{1}{t}\sum_{i=1}^{t}(g(\theta)-g_k(\theta))\cdot \mathbf{u}_i - E[(g(\theta)-g_k(\theta))\cdot \mathbf{u}_i, \tilde{P}] \right| > \epsilon\},$$

where $\tilde{P}$ is used as a shorthand for $\tilde{P}_{\mathbf{y},\mathbf{u}}$. Then it is easy to see that

$$q(t,\epsilon) \le q_1^k(t,\epsilon/2) + q_2^k(t,\epsilon/2). \tag{4.2}$$

Now observe that if $k$ is sufficiently large that $\mu_k \le \epsilon/4$, then $q_2^k(t,\epsilon) = 0$. This is because, if $|(g(\theta)-g_k(\theta))\cdot \mathbf{u}_i|$ is always smaller than $\epsilon/4$, then its expected value is also smaller than $\epsilon/4$, so that their difference can be at most equal to $\epsilon/2$. Since this is true for all $\mathbf{u}$ and all $\theta$, the above observation follows. Thus if $k(\epsilon)$ is chosen large enough that $\mu_k \le \epsilon/4$ for all $k \ge k(\epsilon)$, then

$$q(t,\epsilon) \le q_1^{k(\epsilon)}(t,\epsilon/2)\ \forall t \ge k(\epsilon),\ \forall \epsilon. \tag{4.3}$$

Hence the rest of the proof consists of estimating $q_1^{k(\epsilon)}(t,\epsilon)$ when $t \ge k(\epsilon)$.

From here onwards, let us replace $k(\epsilon)$ by $k$ in the interests of notational clarity. When $t \ge k$, define $l := \lfloor t/k \rfloor$, and $r = t - kl$. Partition $\{1,\ldots,t\}$ into $k$ intervals, as follows.

$$I_j := \{i, i+k, \ldots, i+lk\} \text{ for } 1 \le j \le r, \text{ and}$$

$$I_j := \{i, i+k, \ldots, i+(l-1)k\} \text{ for } r+1 \le j \le k.$$

Then we can write

$$\frac{1}{t}\sum_{i=1}^{t} g_k(\theta)\cdot \mathbf{u}_i = \frac{1}{t}\sum_{j=1}^{k}\sum_{i\in I_j} g_k(\theta)\cdot \mathbf{u}_i.$$

Now define

$$\alpha_j := \frac{1}{l+1}\left|\sum_{i\in I_j}\left(g_k(\theta)\cdot \mathbf{u}_i - E[g_k(\theta)\cdot \mathbf{u}_i, \tilde{P}]\right)\right|,\ 1 \le j \le r, \text{ and}$$

$$\alpha_j := \frac{1}{l}\left|\sum_{i\in I_j}\left(g_k(\theta)\cdot \mathbf{u}_i - E[g_k(\theta)\cdot \mathbf{u}_i, \tilde{P}]\right)\right|,\ r+1 \le j \le k.$$

Then, noting that $E[g_k(\theta)\cdot \mathbf{u}_i, \tilde{P}]$ is independent of $i$ due to the stationarity assumption, we get

$$\left|\frac{1}{t}\sum_{i=1}^{t} g_k(\theta)\cdot \mathbf{u}_i - E[g_k(\theta)\cdot \mathbf{u}_i, \tilde{P}]\right| \le \left|\sum_{j=1}^{r}\frac{l+1}{t}\alpha_j + \sum_{j=r+1}^{k}\frac{l}{t}\alpha_j\right|.$$

It follows that if $\alpha_j \leq \epsilon$ for each $j$, then the left side of the inequality is also less than $\epsilon$. So the following containment of events holds:

$$\left\{ \sup_{\theta \in \Theta} \left| \frac{1}{t}(g_k \cdot \mathbf{u}_i - E[g_k \cdot \mathbf{u}_i, \tilde{P}]) \right| > \epsilon \right\} \subseteq \bigcup_{j=1}^{k} \{\alpha_j > \epsilon\}.$$

Hence

$$q_1^k(t, \epsilon) \leq \sum_{j=1}^{k} \tilde{P}\{\alpha_j > \epsilon\}. \tag{4.4}$$

Now note that each $g_k \cdot \mathbf{u}_i$ depends on only $u_{i-1}$ through $u_{i-k}$. Hence, in the summation defining each of the $\alpha_j$, the various quantities being summed are independent. Since it is assumed that the family $\{g_k(\theta), \theta \in \Theta\}$ has finite P-dimension $d(k)$, the quantity $q_1^k(t, \epsilon)$ can be estimated using the results of [10], Theorem 7.1, in particular Equation (7.1). A small adjustment is necessary, however. The hypotheses of Theorem 7.1 of [10] assume that all the functions under study assume values in the interval $[0, 1]$, whereas in the present instance the functions $h(\theta) \cdot \mathbf{u}_i$ all assume values in the interval $[-M, M]$. Thus the range of values now has width $2M$ instead of one. With this adjustment, [10], (7.1) implies that

$$\Pr\{\alpha_j > \epsilon\} \leq 8 \left( \frac{16e}{\epsilon} \ln \frac{16e}{\epsilon} \right)^{d(k)} \exp(-(l+1)^2 \epsilon^2 / 128M^2), \text{ for } 1 \leq j \leq r,$$

and

$$\Pr\{\alpha_j > \epsilon\} \leq 8 \left( \frac{16e}{\epsilon} \ln \frac{16e}{\epsilon} \right)^{d(k)} \exp(-l^2 \epsilon^2 / 128M^2), \text{ for } r+1 \leq j \leq k.$$

Since $\exp(-(l+1)^2) < \exp(-l^2)$, the $l+1$ term can be replaced by $l$ in the first inequality as well. Substituting these estimates into (4.4) yields the desired estimate

$$q_1^k(t, \epsilon) \leq 8k \left( \frac{16e}{\epsilon} \ln \frac{16e}{\epsilon} \right)^{d(k)} \exp(-l^2 \epsilon^2 / 128M^2). \tag{4.5}$$

Finally, the conclusion (4.1) is obtained by replacing $\epsilon$ by $\epsilon/2$ in the above expression, and then applying (4.3). □

# 5 Bounds on the P-dimension

In order for the estimate in Theorem 4.1 to be useful, it is necessary for us to derive an estimate for the P-dimension of the family of functions defined by

$$\mathcal{G}_k := \{g_k(\theta) : \theta \in \Theta\}, \tag{5.1}$$

where $g_k(\theta) : U^k \rightarrow \Re$ is defined by

$$g_k(\theta)(\mathbf{u}) := \| (f - h(\theta)) \cdot \mathbf{u}_k \|^2,$$

and

$$\mathbf{u}_k := (\dots, 0, u_k, u_{k-1}, \dots, u_1, 0, 0, \dots).$$

Note that, in the interests of convenience, we have denoted the infinite sequence with only $k$ nonzero elements as $u_k, \dots, u_1$ rather than $u_0, \dots, u_{1-k}$ as done earlier. Clearly this makes no difference. In this subsection, we state and prove such an estimate for the commonly occurring case where each system model $h(\theta)$ is an ARMA model where the parameter $\theta$ enters linearly. Specifically, it is supposed that the model $h(\theta)$ is described by

$$x_{t+1} = \sum_{i=1}^{l} \theta_i \, \phi_i(x_t, u_t), \; y_t = x_t, \tag{5.2}$$

where $\theta = (\theta_1, \dots, \theta_l) \in \Theta \subseteq \Re^l$, and each $\phi_i(\cdot, \cdot)$ is a polynomial of degree no larger than $r$ in the components of $x_t, u_t$.

**Theorem 5.1.** *With the above assumptions, we have that*

$$\begin{aligned} \text{P-dim}(\mathcal{G}_k) &\leq 9l + 2l \lg[2(r^{k+1} - 1)/(r - 1)] \\ &\approx 9l + 2lk \lg(2r) \;\; \textit{if } r > 1. \end{aligned} \tag{5.3}$$

*In case $r = 1$ so that each system is linear, the above bound can be simplified to*

$$\text{P-dim}(\mathcal{G}_k) \leq 9l + 2l \lg(2k). \tag{5.4}$$

**Remark.** It is interesting to note that the above estimate is *linear* in both the number of parameters $l$ and the duration $k$ of the input sequence $\mathbf{u}$, but is only logarithmic in the degree of the polynomials $\phi_i$. In the practically important case of linear ARMA models, even $k$ appears inside the logarithm. □

**Proof.** For each function $g_k(\theta) : U^k \rightarrow \Re$ defined as in (5.1), define an associated function $g'_k : U^k \times [0,1] \rightarrow \{0,1\}$ as follows:

$$g'_k(\theta)(\mathbf{u}, c) := \eta[g_k(\theta)(\mathbf{u}) - c],$$

where $\eta(\cdot)$ is the Heaviside or 'step' function. Then it follows from [10], Lemma 10.1 that

$$\text{P-dim}(\mathcal{G}_k) = \text{VC-dim}(\mathcal{G}'_k).$$

Next, to estimate VC-dim($\mathcal{G}'_k$), we use [10], Corollary 10.2, which states that, if the condition $\eta[g_k(\theta)\mathbf{u} - c] = 1$ can be stated as a Boolean formula involving $s$ polynomial inequalities, each of degree no larger than $d$, then

$$\text{VC-dim}(\mathcal{G}'_k) \leq 2l \lg(4eds). \tag{5.5}$$

Thus the proof consists of showing that the conditions needed to apply this bound hold, and of estimating the constants $d$ and $s$.

Towards this end, let us back-substitute repeatedly into the ARMA model (5.2) to express the inequality

$$\| (f - h(\theta))\mathbf{u}_k \|^2 - c < 0$$

as a polynomial inequality in $\mathbf{u}$ and the $\theta$-parameters. To begin with, we have

$$\begin{aligned} x_{k+1} &= \sum_{i=1}^{l} \theta_i \, \phi_i(x_k, u_k) \\ &= \sum_{i=1}^{l} \theta_i \phi_i \left( \sum_{j=1}^{l} \theta_j \, \phi_j(x_{k-1}, u_{k-1}) \right) \\ &= \ldots . \end{aligned} \tag{5.6}$$

Thus each time one of the functions $\phi_i$ is applied to its argument, the degree with respect to any of the $\theta_j$ goes up by a factor of $r$. In other words, the total degree of $x_{k+1}$ with respect to each of the $\theta_j$ is no larger than $1 + r + r^2 + \cdots + r^k = (r^{k+1} - 1)/(r - 1)$. If $r = 1$, then the degree is simply $k$. Next, we can write

$$\| x_{k+1} \|^2 - c < 0 \Leftrightarrow x'_{k+1} x_{k+1} - c < 0.$$

This is a single polynomial inequality. Moreover, the degree of this polynomial in the components of $\theta$ is at most $2(r^{k+1} - 1)/(r - 1)$ if $r > 1$, and $2k$ if $r = 1$. Thus we can apply the bound (5.5) with and $s = 1$, and

$$d = \begin{cases} \frac{2(r^{k+1}-1)}{r-1} & \text{if } r > 1, \\ 2k & \text{if } r = 1. \end{cases}$$

The desired estimate now follows on noting that $\lg e < 1.5$, so that $\lg(8e) < 4.5$. □

# 6 Conclusions

In this paper, we have presented some *finite-time* results that quantify the rate at which the estimates generated by a minimum empirical error identification algorithm converge to the minimum achievable identification error. These results are applicable to a broad class of problems, since the assumptions are fairly weak. However, the actual estimates themselves are quite conservative, and there is much room for improvement.

Ideally, the objective function $J(\theta)$ should be meaningful in the context of robust control, in the sense that when $J(\theta)$ is small enough, a controller that stabilizes the estimated model $h(\theta)$ should also stabilize the true system $f_{\text{true}}$. The present quadratic objective function does not satisfy this requirement, unless it is assumed ahead of time that the dynamic orders of both the true system $f_{\text{true}}$ and the model family $\{h(\theta), \theta \in \Theta\}$ are bounded, and in addition, the input $\{u_t\}$ is persistently exciting in some sense. The study of adapting the present approach to provide more efficient and natural results on the design of robust controllers is a suitable topic for future research.

# A personal reminiscence

As this volume is a festschrift, a personal note from the first author (MV) may perhaps not be inappropriate. Kimura sensei's contributions to control theory are so well known that additional comments from MV are unnecessary. But it is perhaps less well-known that during a visit to Waterloo in 1986, Kimura sensei *pushed MV into Lake Victoria*, an incident that was at one time very much in circulation within the Japanese control theory community. The passage of time seems to have lessened everyone's memory of this in(ac-)cident.

Seriously, during the Kimurafest it was pointed out that three of Kimura sensei's publications have been cited more than one hundred times. Two of them were singly authored, but MV is proud that the third widely-cited paper is joint [14]. He remembers well the discussions with Kimura sensei that took place in LIDS in MIT, that eventually led to that paper. Incidentally, the collaboration also reflects the greatness of MIT, in that two visitors could collaborate on such a widely-cited paper 'on neutral territory' so to speak.

MV wishes Kimura sensei many more years of happiness and productive scientific research.

## References

[1] P. E. Caines, "Prediction error identification methods for stationary stochastic processes," *IEEE Trans. Autom. Control*, vol. AC-21, no. 4, pp. 500–505, Aug. 1976.

[2] P. E. Caines, "Stationary linear and nonlinear system identification and predictor set completeness," *IEEE Trans. Autom. Control*, vol. AC-23, no. 4, pp. 583–594, Aug. 1978.

[3] M. C. Campi and E. Weyer, "Finite sample properties of system identification methods," *IEEE Trans. Autom. Control*, vol. AC-47, no. 8, pp. 1329–1334, Aug. 2002.

[4] D. Haussler, "Decision theoretic generalizations of the PAC model for neural net and other learning applications," *Information and Computation*, vol. 100, pp. 78–150, 1992.

[5] L. Ljung, "Convergence analysis of parametric identification methods," *IEEE Trans. Autom. Control*, vol. AC-23, no. 5, pp. 770–783, Oct. 1978.

[6] L. Ljung, *System Identification: Theory for the User*. Englewood Cliffs, NJ: Prentice-Hall, 1987.

[7] L. Ljung, *System Identification: Theory for the User*, 2nd ed. Englewood Cliffs, NJ: Prentice-Hall, 1999.

[8] V. N. Vapnik and A. Ya. Chervonenkis, "On the uniform convergence of relative frequencies to their probabilities," *Theory of Probab. Appl.*, vol. 16, no. 2, pp. 264–280, 1971.

[9] V. N. Vapnik and A. Ya. Chervonenkis, "Necessary and sufficient conditions for the uniform convergence of means to their expectations," *Theory of Probab. Appl.*, vol. 26, no. 3, pp. 532–553, 1981.

[10] M. Vidyasagar, *A Theory of Learning and Generalization*. London, UK: Springer-Verlag, 1997.

[11] M. Vidyasagar and R. L. Karandikar, "A learning theory approach to system identification and stochastic adaptive control," presented at *the IFAC Workshop on Adaptation and Learning in Control and Signal Processing*, Como, Italy, Aug. 2001.

[12] M. Vidyasagar and R. L. Karandikar, "System identification: A learning theory approach," in *Proceedings of the 40th IEEE Conference on Decision and Control*, Orlando, FL, Dec. 2001, pp. 2001–2006.

[13] M. Vidyasagar and R. L. Karandikar, "A learning theory approach to system identification and stochastic adaptive control," in *Proceedings of the 15th IFAC World Congress*, Barcelona, Spain, July 2002.

[14] M. Vidyasagar and H. Kimura, "Robust controllers for uncertain multivariable linear systems," *Automatica*, vol. 22, no. 1, pp. 85–94, January 1986.

[15] E. Weyer, "Finite sample properties of system identification of ARX models under mixing conditions," *Automatica*, vol. 36, no. 9, pp. 1291–1299, Sept. 2000.

M. Vidyasagar
Advanced Technology Centre, Tata Consultancy Services
6th Floor, Khan Lateefkhan Estate
Fateh Maidan Road, Hyderabad 500 001, India
sagar@atc.tcs.co.in

Rajeeva L. Karandikar
Indian Statistical Institute
No. 7, S. J. S. Sansawal Marg
Hauz Khas, New Delhi 110 016, India
rlk@isid.ac.in

# Distribution-Free Approach to Probabilistic Model-Set Identification

## Yasuaki Oishi

ABSTRACT A model-set identification algorithm is proposed in a probabilistic framework based on a leave-one-out technique. It provides a nominal model and a bound of its uncertainty for a provided plant assuming that the effect of past inputs decays with a known bound. Because it does not require further assumptions on true plant dynamics or on noise, the risk of making inappropriate assumptions is small. The number of assumptions is shown to be minimum in the sense that identification is impossible after removing the assumption made here. Generalization of the proposed algorithm is considered in several aspects. A simple plant is identified for illustration.

**Keywords:** Model-set identification, Leave-one-out estimation, Linear programming, Statistical learning theory, Robust control

## 1 Dedication

I worked with Professor Hidenori Kimura as his research associate from 1995 to 2001 at the University of Tokyo. It was very lucky for me to have such an experience at the beginning of my research life. He gave me a number of valuable advice and encouragement. Although his words were bitter on some occasions, behind them, there always was his warm consideration. I really think that his help made me what I am.

The subject of model-set identification is special for me because my motivation and perspective on this subject were formed under the influence of Professor Kimura. From 1995 to 1997, I had a chance to help the thesis work of his undergraduate students. The subject they worked on was model-set identification by Zhou and Kimura [22]. I was impressed to see that the set of all models unfalsified by the input-output data is beautifully characterized. Since then, I have been interested in model-set identification, especially on how we should make assumptions on a plant to be identified. In the collaborated work with him [11], we proposed to use statistical learning theory to enable identification under weak assumptions. He was critical, however, of the fact that the proposed method requires an unrealistically large number of input-output data. This criticism was a starting point of the present study. Although I present only my name as the author of this paper, the work itself should be considered as a collaboration with Professor Kimura.

I would like to dedicate this paper to Professor Hidenori Kimura on the occasion of his 60th birthday. I wish him happiness and health in his future.

# 2 Introduction

In order to use the robust control scheme, we need not only a nominal model of the plant to be controlled but also an upper bound of its uncertainty. Model-set identification (or worst-case identification, set-membership identification) is an identification method to obtain these two and has actively been investigated in the last decade by many authors [23, 5, 9, 4, 13, 3, 22] to name a few.

In model-set identification, one makes prior assumptions on the true plant dynamics and on the noise; considers all the models that satisfy these assumptions and are consistent with the observed input-output data; gives a nominal model and a bound of its uncertainty so as to cover the worst case. In the existing algorithms for model-set identification, it is assumed for example that the true plant dynamics can be described by an ARX model of some order, that the noise has a known hard bound, or that the noise is subject to a Gaussian distribution. However, it does not seem to be often the case that the plant dynamics and the noise actually satisfy these strong assumptions. Indeed, many actual plants are infinite-dimensional and have nonlinearity; The noise usually contains a modeling error and is not Gaussian. Even if satisfaction of these assumptions is possible in principle, it is difficult to make appropriate assumptions using an appropriate bound or an appropriate distribution because neither the true plant dynamics nor the noise is directly observable. This problem is serious in model-set identification, which is for capturing the worst case. Suppose that the adopted assumptions are inappropriate and are not satisfied by the actual plant. Then we may miss a possible worst case that does not satisfy those assumptions.

In order to partially resolve this problem, we formulate model-set identification in a probabilistic framework and apply a technique of leave-one-out estimation, which was used by Vapnik to prove statistical properties of a support vector machine [14]. It enables us to make a worst-case estimation with little information on the underlying probability structure. In our approach, this technique is used so that one can perform model-set identification with only one assumption, which is on the decay rate of the past memory in the output. Consequently, the risk of making inappropriate assumptions is suppressed. It is possible to prove that identification is impossible after removing this remaining assumption. In this sense, the number of assumptions is set to the minimum. We also consider in this paper generalization of this approach in several directions.

There are other techniques to perform worst-case estimation with little information on the object especially in the field of statistical learning theory. They have been applied to model-set identification [11] and to conventional identification [1, 2, 21, 19, 20, 18, 16]. See also [7, 12, 8, 15] for their application to analysis and synthesis of control systems. However, it has been often the case that the estimation given by learning theory is conservative and one has to collect millions of data to obtain meaningful information. In this regard, our approach seems to be advantageous because it gives a result from a reasonable number of data as is seen in Section 5.

# 3 Identification with a small number of assumptions

## 3.1 Identification algorithm

In this paper, a plant to be identified is a single-input-single-output discrete-time time-invariant causal system with an input $u_k$ and an output $y_k$, where $k$ is an integer. It is supposed that the input-output relation of the plant is described by an unknown but fixed conditional probability distribution $P_{y_k|u_{k-1},u_{k-2},\ldots}$, which is invariant with $k$. This means that the output $y_k$ is affected by a stochastic noise and its probability distribution varies depending on what inputs $u_{k-1}, u_{k-2}, \ldots$ are provided. Note that the plant is allowed to have quite a large class of dynamics including nonlinear ones. We let the inputs $\{u_k\}$ be a sequence of random numbers, which distribute independently and identically in $[-\mu, \mu]$ for some positive $\mu$. Here we make an assumption crucial for our purpose. Assume the existence of an auxiliary signal $\{\overline{y}_k\}$ such that there holds $|y_k - \overline{y}_k| \leq c$ with probability one for any $k$ and that $\overline{y}_k$ is statistically independent of $\overline{y}_{k'}$ and $u_{k'}$ whenever $|k - k'| \geq K$. The value of $\overline{y}_k$ does not need to be known though those of $c$ and $K$ should be known. Let us call this property *approximate $K$-independence* henceforth. The assumption of this property, which was used in [21], is acceptable because the effect of the past inputs decay exponentially fast in many practical plants and the amount of this effect may be estimated by preliminary experiments. Since the output $y_k$ is directly observable, an assumption on $y_k$ is considered to be easier to make than those on the true plant dynamics or the noise.

We use an FIR model to obtain a nominal model. Even if the plant itself is nonlinear, it is often reasonable to choose a linear nominal model for the sake of convenience in a controller design. For some integer $d$ smaller than $K$, write $\boldsymbol{x}_k = [u_{k-1} \ \ \ldots \ \ u_{k-d}]^{\mathrm{T}}$ and $\boldsymbol{h} = [h_1 \ \ \ldots \ \ h_d]^{\mathrm{T}}$. We approximate the plant output $y_k$ by $\boldsymbol{x}_k^{\mathrm{T}}\boldsymbol{h}$, which is the one-step-ahead predictor associated with the $d$-th order FIR model. Suppose that the input-output data $u_1, u_2, \ldots, u_N$; $y_1, y_2, \ldots, y_N$ are provided. From them we extract $n := \lfloor N/K \rfloor$ pairs $\{(y_{(j)}, \boldsymbol{x}_{(j)})\}_{j=1}^n$, where $y_{(j)} := y_{jK}$ and $\boldsymbol{x}_{(j)} := \boldsymbol{x}_{jK} = [u_{jK-1} \ \cdots \ u_{jK-d}]^{\mathrm{T}}$. Our identification algorithm is the following.

**Algorithm 3.1.**

1. Obtain the pair $(\boldsymbol{h}, b)$ that minimizes $b$ under the conditions $|y_{(j)} - \boldsymbol{x}_{(j)}^{\mathrm{T}}\boldsymbol{h}| \leq b$, $j = 1, \ldots, n$, and write it as $(\boldsymbol{h}^*, b^*)$.

2. For each $i = 1, \ldots, d$, find $\boldsymbol{h}$ that maximizes $|h_i - h_i^*|$ under the conditions $|y_{(j)} - \boldsymbol{x}_{(j)}^{\mathrm{T}}\boldsymbol{h}| \leq b^* + 2c$, $j = 1, \ldots, n$, and write the maximum value as $e_i$.

3. Give $y_k = \boldsymbol{x}_k^{\mathrm{T}}\boldsymbol{h}^*$ as a nominal model and $b^* + 2c + \mu \sum_{i=1}^d e_i$ as an uncertainty bound. □

Optimization in Steps 1 and 2 can be carried out efficiently by means of linear programming. Suppose that we sample a new input-output pair $(y_{k_0}, \boldsymbol{x}_{k_0})$ at $k_0 \geq N + K$ in order to see how accurately $\boldsymbol{x}_{k_0}^{\mathrm{T}}\boldsymbol{h}^*$ predicts the actual output $y_{k_0}$.

**Theorem 3.2.** *There holds* $|y_{k_0} - \boldsymbol{x}_{k_0}^{\mathrm{T}}\boldsymbol{h}^*| \leq b^* + 2c + \mu \sum_{i=1}^d e_i$ *with probability greater than or equal to* $1 - (d+1)/(n+1)$. *Here the probability is measured with respect to the input-output data* $\{(y_k, \boldsymbol{x}_k)\}_{k=1}^N$ *and the newly sampled input-output pair* $(y_{k_0}, \boldsymbol{x}_{k_0})$.

The performance guarantee given by this theorem is not asymptotic, that is, it is meaningful with a finite $n$. Note also that our algorithm captures the worst case, since Theorem 3.2 holds irrespective of the probabilistic properties of the plant and the noise. One may notice that only a small part of input-output data is used in the proposed algorithm. This is because we suppose that two outputs closer in time than $K$ may be dependent on each other to any level; We can expect fresh information to be obtained only at every $K$ time instances.

## 3.2 Proof of Theorem 3.2

**Lemma 3.3.** *Let* $(\boldsymbol{h}^*, b^*)$ *minimize* $b$ *subject to* $|y_{(j)} - \boldsymbol{x}_{(j)}^{\mathrm{T}}\boldsymbol{h}| \leq b$, $j = 1, \ldots, n$, *and let* $(\overline{\boldsymbol{h}}^*, \overline{b}^*)$ *minimize* $b$ *subject to* $|\overline{y}_{(j)} - \boldsymbol{x}_{(j)}^{\mathrm{T}}\boldsymbol{h}| \leq b$, $j = 1, \ldots, n$. *Then there hold* $|b^* - \overline{b}^*| \leq c$ *and* $|h_i^* - \overline{h}_i^*| \leq e_i$ *for each of* $i = 1, \ldots, d$.

**Proof.** Note that $b^*$ is equal to $\inf_{\boldsymbol{h}} \max_{j=1,\ldots,n} |y_{(j)} - \boldsymbol{x}_{(j)}^{\mathrm{T}}\boldsymbol{h}|$ and this infimum is attained at $\boldsymbol{h} = \boldsymbol{h}^*$. The corresponding relation holds on $\overline{b}^*$ and $\overline{\boldsymbol{h}}^*$, too. Since the discrepancy between $\max_j |y_{(j)} - \boldsymbol{x}_{(j)}^{\mathrm{T}}\boldsymbol{h}|$ and $\max_j |\overline{y}_{(j)} - \boldsymbol{x}_{(j)}^{\mathrm{T}}\boldsymbol{h}|$ is less than or equal to $c$ for any $\boldsymbol{h}$, the inequality $|b^* - \overline{b}^*| \leq c$ has to hold. It is also seen that $\max_j |y_{(j)} - \boldsymbol{x}_{(j)}^{\mathrm{T}}\overline{\boldsymbol{h}}^*| - c \leq \max_j |\overline{y}_{(j)} - \boldsymbol{x}_{(j)}^{\mathrm{T}}\overline{\boldsymbol{h}}^*| = \overline{b}^*$, which is combined with $\overline{b}^* \leq b^* + c$ and implies $\max_j |y_{(j)} - \boldsymbol{x}_{(j)}^{\mathrm{T}}\overline{\boldsymbol{h}}^*| \leq b^* + 2c$. By the definition of $e_i$, there has to hold $|h_i^* - \overline{h}_i^*| \leq e_i$. □

To the newly sampled input-output pair $(y_{k_0}, \boldsymbol{x}_{k_0})$, we can associate $\overline{y}_{k_0}$, which approximates $y_{k_0}$. Let us write $(\overline{y}_{k_0}, \boldsymbol{x}_{k_0})$ as $(\overline{y}_{(n+1)}, \boldsymbol{x}_{(n+1)})$ to see that $\{(\overline{y}_{(j)}, \boldsymbol{x}_{(j)})\}_{j=1}^{n+1}$ is a set of $n+1$ independently identically distributed input-output pairs. Now we can show the next lemma, which is based on a technique of *leave-one-out estimation*, which was used by Vapnik [14].

**Lemma 3.4.** *Define* $(\overline{\boldsymbol{h}}^*, \overline{b}^*)$ *as in the previous lemma. The probability of having* $|\overline{y}_{(n+1)} - \boldsymbol{x}_{(n+1)}^{\mathrm{T}}\overline{\boldsymbol{h}}^*| > \overline{b}^*$ *is less than or equal to* $(d+1)/(n+1)$.

**Proof.** Recall that $(\overline{\boldsymbol{h}}^*, \overline{b}^*)$ minimizes $b$ under the conditions $|\overline{y}_{(j)} - \boldsymbol{x}_{(j)}^{\mathrm{T}}\boldsymbol{h}| \leq b$, $j = 1, \ldots, n$. Instead, we consider $(\boldsymbol{h}, b)$ that minimizes $b$ under the same conditions but for $j = 1, \ldots, n+1$. Let us write the minimizing pair as $(\overline{\boldsymbol{h}}^+, \overline{b}^+)$. We will evaluate the probability with which the minimum value of $b$ changes from $\overline{b}^+$ when arbitrarily one of the $n+1$ conditions is removed. If this probability is shown to be less than or equal to $(d+1)/(n+1)$, the claim of the lemma is ensured.

Each of the inequalities $|\overline{y}_{(j)} - \boldsymbol{x}_{(j)}^{\mathrm{T}}\boldsymbol{h}| \leq b$, $j = 1, \ldots, n+1$, can be decomposed into $[\boldsymbol{x}_{(j)}^{\mathrm{T}} \;\; 1] \begin{bmatrix} \boldsymbol{h} \\ b \end{bmatrix} \geq \overline{y}_{(j)}$ and $[-\boldsymbol{x}_{(j)}^{\mathrm{T}} \;\; 1] \begin{bmatrix} \boldsymbol{h} \\ b \end{bmatrix} \geq -\overline{y}_{(j)}$. Among these $2(n+1)$ inequalities, we pick up those satisfied with the equality at the minimizing point $(\overline{\boldsymbol{h}}^+, \overline{b}^+)$. By the Karush–Kuhn–Tucker necessary condition for optimality, the vector $[0 \; \ldots \; 0 \;\; 1]$ can be described as a linear combination with positive coefficients of the vectors $[\boldsymbol{x}_{(j)}^{\mathrm{T}} \;\; 1]$ and $[-\boldsymbol{x}_{(j)}^{\mathrm{T}} \;\; 1]$ that originated from the inequalities picked up above. Since the description may not be unique, we notice the vectors included in common by all of such descriptions. The number of such vectors should be less than or equal to $d+1$. This means that the number of the conditions $|\overline{y}_{(j)} - \boldsymbol{x}_{(j)}^{\mathrm{T}}\boldsymbol{h}| \leq b$ producing those vectors is also less than or equal to $d+1$.

The minimum value of $b$ changes only when the removed condition is one of those considered above. Hence its probability is no greater than $(d+1)/(n+1)$. □

Theorem 3.2 is now proven. Assume that $|\overline{y}_{k_0} - \boldsymbol{x}_{k_0}^{\mathrm{T}}\overline{\boldsymbol{h}}^*| \leq \overline{b}^*$, which holds with probability greater than or equal to $1 - (d+1)/(n+1)$ by Lemma 3.4. Then we can derive from Lemma 3.3

$$\begin{aligned} |y_{k_0} - \boldsymbol{x}_{k_0}^{\mathrm{T}}\boldsymbol{h}^*| \leq & |y_{k_0} - \overline{y}_{k_0}| + |\overline{y}_{k_0} - \boldsymbol{x}_{k_0}^{\mathrm{T}}\overline{\boldsymbol{h}}^*| + |\boldsymbol{x}_{k_0}^{\mathrm{T}}(\overline{\boldsymbol{h}}^* - \boldsymbol{h}^*)| \\ \leq & c + \overline{b}^* + \mu \sum_{i=1}^{d} e_i \leq b^* + 2c + \mu \sum_{i=1}^{d} e_i. \end{aligned}$$

# 4 Discussion

The proposed identification algorithm is investigated in several aspects.

## 4.1 Minimality of the number of assumptions

The proposed identification algorithm relies on the assumption of approximate $K$-independence. It is shown here that identification is impossible after removing this assumption. In this sense, the number of assumptions is set minimum possible in the proposed identification algorithm.

The plant and the inputs are considered as before except that approximate $K$-independence is not assumed. This means that the class of considered plants is taken larger than before. Supposing that the input-output data $\{y_k, u_k\}_{k=1}^N$ are provided, we consider two functions of them, $\widehat{\boldsymbol{h}}$ and $\widehat{b}$. It is possible to regard each pair of these functions as an identification algorithm with the interpretation that $\widehat{\boldsymbol{h}}$ gives a nominal model and $\widehat{b}$ its uncertainty bound. What we want especially is an algorithm that guarantees $|y_{k_0} - \boldsymbol{x}_{k_0}^{\mathrm{T}} \widehat{\boldsymbol{h}}| \le \widehat{b}$ with high probability for a newly sampled input-output pair $(y_{k_0}, \boldsymbol{x}_{k_0})$. However, it is not possible to guarantee this for all the considered plants.

**Theorem 4.1.** *For any algorithm $(\widehat{\boldsymbol{h}}, \widehat{b})$ and any time instant $k_0 > N$, there exists a plant that establishes*

$$\left| y_{k_0} - \boldsymbol{x}_{k_0}^{\mathrm{T}} \widehat{\boldsymbol{h}}(\{y_k, u_k\}_{k=1}^N) \right| \ge \widehat{b}(\{y_k, u_k\}_{k=1}^N) + 1$$

*with probability one.*

To prove this theorem, just consider the plant having the deterministic dynamics

$$y_k = \boldsymbol{x}_k^{\mathrm{T}} \widehat{\boldsymbol{h}}(\{y_\ell, u_\ell\}_{\ell=1+k-k_0}^{N+k-k_0}) + \widehat{b}(\{y_\ell, u_\ell\}_{\ell=1+k-k_0}^{N+k-k_0}) + 1,$$

which is nonlinear but is time-invariant. The inequality in the theorem obviously holds for this plant.

## 4.2 Relaxation of the assumption

Theorem 4.1 states that the assumption of approximate $K$-independence cannot be removed. However, it does not mean that this assumption cannot be relaxed or replaced by another one.

Let us relax this assumption by supposing that the inequality $|y_k - \overline{y}_k| \le c$ holds only in a probabilistic sense, that is, it holds with probability greater than or equal to $1 - p$. This relaxation is practically reasonable because it is sometimes difficult to guarantee $|y_k - \overline{y}_k| \le c$ with 100% confidence. In order that the statement of Theorem 3.2 continues valid with this relaxed assumption, it suffices that there hold the $n+1$ inequalities, $|y_{(j)} - \overline{y}_{(j)}| \leq c$, $j = 1, \ldots, n$, and $|y_{k_0} - \overline{y}_{k_0}| \leq c$. The probability with which at least one of these $n+1$ inequalities fails to hold is $(n+1)p$ at most. Hence the next theorem can be derived from Theorem 3.2.

**Theorem 4.2.** *There holds* $|y_{k_0} - \boldsymbol{x}_{k_0}^{\mathrm{T}} \boldsymbol{h}^*| \leq b^* + 2c + \mu \sum_{i=1}^{d} e_i$ *with probability greater than or equal to* $1 - (d+1)/(n+1) - (n+1)p$.

A major difference from Theorem 3.2 is that the confidence achieves the maximum when $n$ is equal to $\lfloor \sqrt{(d+1)/p} \rfloor$ or $\lfloor \sqrt{(d+1)/p} \rfloor - 1$. When $n$ is larger than this value, increase of the input-output data does not imply increase of the confidence.

### 4.3 Analysis of a given model

Suppose that a linear model $y_k = \boldsymbol{x}_k^{\mathrm{T}} \boldsymbol{h}^0$ has already been obtained for a plant at hand and we want to analyze the quality of this model. The leave-one-out technique used in Section 3 is applicable to this problem, too. In fact, one can obtain a better result for the present problem than for the identification problem because the existence of the model makes the problem easier.

With the notation of Section 3, compute $\max_{j=1,\ldots,n} |y_{(j)} - \boldsymbol{x}_{(j)}^{\mathrm{T}} \boldsymbol{h}^0|$ and regard it as a quality index of the model $y_k = \boldsymbol{x}_k^{\mathrm{T}} \boldsymbol{h}^0$. Let $(y_{k_0}, \boldsymbol{x}_{k_0})$ be a newly sampled input-output pair at $k_0 \geq N + K$. Then the next property holds.

**Theorem 4.3.** *There holds the inequality*

$$|y_{k_0} - \boldsymbol{x}_{k_0}^{\mathrm{T}} \boldsymbol{h}^0| \leq \max_{j=1,\ldots,n} |y_{(j)} - \boldsymbol{x}_{(j)}^{\mathrm{T}} \boldsymbol{h}^0| + 2c$$

*with probability greater than or equal to* $1 - 1/(n+1)$. *Here the probability is measured with respect to the input-output data* $\{(y_k, \boldsymbol{x}_k)\}_{k=1}^{N}$ *and the newly sampled pair* $(y_{k_0}, \boldsymbol{x}_{k_0})$.

**Proof.** Computing $\max_{j=1,\ldots,n} |y_{(j)} - \boldsymbol{x}_{(j)}^{\mathrm{T}} \boldsymbol{h}^0|$ is equivalent to finding the minimum $b$ under the conditions $|y_{(j)} - \boldsymbol{x}_{(j)}^{\mathrm{T}} \boldsymbol{h}^0| \leq b$, $j = 1, \ldots, n$. The difference between this minimization problem and the one considered in Section 3 is that the minimization variable is $b$ here but $(\boldsymbol{h}, b)$ there. Hence the same leave-one-out technique can be used here but the variable dimension is set to unity in place of $d+1$. This proves the above theorem. □

Comparing this theorem to Theorem 3.2, one can see that the result obtained here is better than the previous one. Namely, the upper bound of $|y_{k_0} - \boldsymbol{x}_{k_0}^{\mathrm{T}} \boldsymbol{h}^0|$ does not include the term $\mu \sum_{i=1}^{d} e_i$; The result holds with more confidence $1 - 1/(n+1)$; The computational load to obtain $e_i$ is eliminated. Hence the application here is considered to be more useful than the previous one in many cases. Note also that the relaxation of Section 4.2 is possible in this application, too.

### 4.4 Assumption of a mixing property

We assumed approximate $K$-independence in order to apply the leave-one-out technique to a dependent stochastic process. In the literature, a mixing property is often assumed for the same purpose. See [2, 19, 18, 16] for its use in system identification; See [6, 17] for its application in learning theory. In this subsection, we derive a result similar to the one in Section 3 by assuming a mixing property instead of approximate $K$-independence.

Let $Q$ be the probability measure on the stochastic process $\{y_k, u_k\}_{k=-\infty}^{\infty}$ and $Q_{-\infty}^{0}$ and $Q_1^{\infty}$ be its semi-infinite marginals. With $\sigma_K$ being the $\sigma$-algebra generated by the random variables $(y_k, u_k)$, $k \leq 0$ or $k \geq K$, define the function

$$\beta(K) := \sup_{A \in \sigma_K} \left| Q(A) - (Q_{-\infty}^{0} \times Q_1^{\infty})(A) \right|.$$

If $\beta(K)$ converges to zero as $K \to \infty$, the process $\{y_k, u_k\}_{k=-\infty}^{\infty}$ is said to be *$\beta$-mixing* or *absolutely regular* [10, 6, 17]. Let us assume the $\beta$-mixing property instead of approximate $K$-independence. The rest of the setup and the identification procedure are the same as in Section 3. Here we have the following.

**Theorem 4.4.** *There holds $|y_{k_0} - \boldsymbol{x}_{k_0}^{\mathrm{T}} \boldsymbol{h}^*| \leq b^*$ with probability greater than or equal to $1 - (d+1)/(n+1) - (n+1)\beta(K-d)$.*

**Proof.** Due to Lemma 2 of [10] (also see [6, 17]), there holds the inequality

$$\left| Q\left(|y_{k_0} - \boldsymbol{x}_{k_0}^{\mathrm{T}} \boldsymbol{h}^*| \leq b^*\right) - (Q_0)^{n+1}\left(|y_{k_0} - \boldsymbol{x}_{k_0}^{\mathrm{T}} \boldsymbol{h}^*| \leq b^*\right) \right| \leq (n+1)\beta(K-d),$$

where $Q_0$ is the marginal distribution of one input-output pair $(y_k, \boldsymbol{x}_k)$ and $(Q_0)^{n+1}$ is its $(n+1)$-fold product. With the measure $(Q_0)^{n+1}$, the input-output pairs $\{(y_{(j)}, \boldsymbol{x}_{(j)})\}_{j=1}^{n}$ and $(y_{k_0}, \boldsymbol{x}_{k_0})$ are statistically independent. Hence one can use the technique of Lemma 3.4 and obtains

$$(Q_0)^{n+1}\left(|y_{k_0} - \boldsymbol{x}_{k_0}^{\mathrm{T}} \boldsymbol{h}^*| \leq b^*\right) \geq 1 - (d+1)/(n+1).$$

This ensures the theorem. □

The present result is advantageous to the one in Section 3 because the present bound of modeling error does not include the terms $2c + \mu \sum_{i=1}^{d} e_i$. As the number of input-output data, $N$, grows, one can make the confidence arbitrarily close to unity because, in that case, one can choose a large $K$ to have a $\beta(K-d)$ close to zero. A disadvantage of the present approach is that the definition of $\beta(K-d)$ is abstract and it is not obvious how one can estimate its value for a provided actual plant. This is the reason why we mainly propose the approach of Section 3.

## 4.5 An uncertainty bound relative to the input size

So far we have measured the amount of uncertainty by the discrepancy between the actual output $y_k$ and the predicted one $\boldsymbol{x}_k^{\mathrm{T}}\boldsymbol{h}^*$. However, in robust control, the uncertainty is not an error at a signal level but that at an operator level. In order to improve consistency with robust control, it is preferable to measure the error relative to the input size so that one can interpret the error in terms of the operator norm. We propose in this section an algorithm for this purpose.

The plant and the inputs are as considered in Section 3 except that approximate $K$-independence is assumed as $|y_k - \overline{y}_k| \le c\mu$ in place of $|y_k - \overline{y}_k| \le c$. Write $\mu_k := \sup\{|u_{k-1}|, \ldots, |u_{k-K+1}|\}$ and, especially, $\mu_{(j)} := \mu_{jK}$. The algorithm to be proposed is applicable only when $\mu_{(j)} > 0$ for each $j = 1, \ldots, n$. One may choose the distribution of the inputs so as to ensure this property.

**Algorithm 4.5.**

1. Find $(\boldsymbol{h}, b)$ that minimizes $b$ under the conditions $|y_{(j)} - \boldsymbol{x}_{(j)}^{\mathrm{T}}\boldsymbol{h}| \le b\mu_{(j)}$, $j = 1, \ldots, n$. Write the minimizing pair as $(\boldsymbol{h}^*, b^*)$.
2. For each $i = 1, \ldots, d$, find $\boldsymbol{h}$ that maximizes $|h_i - h_i^*|$ under the conditions $|y_{(j)} - \boldsymbol{x}_{(j)}^{\mathrm{T}}\boldsymbol{h}| \le (b^* + c\mu/\min_{j=1,\ldots,n}\mu_{(j)})\mu_{(j)} + c\mu$, $j = 1, \ldots, n$. Write the attained maximum value as $e_i$.
3. Give $y_k = \boldsymbol{x}_k^{\mathrm{T}}\boldsymbol{h}^*$ as a nominal model and
$$\left(b^* + \frac{c\mu}{\min_{j=1,\ldots,n}\mu_{(j)}} + \sum_{i=1}^{d} e_i\right)\mu_k + c\mu$$
as an uncertainty bound. □

The following theorem gives the performance guarantee to this algorithm.

**Theorem 4.6.** *On the quantities obtained by the algorithm above, we have*

$$|y_{k_0} - \boldsymbol{x}_{k_0}^{\mathrm{T}}\boldsymbol{h}^*| \le \left(b^* + \frac{c\mu}{\min_{j=1,\ldots,n}\mu_{(j)}} + \sum_{i=1}^{d} e_i\right)\mu_{k_0} + c\mu$$

*with probability greater than or equal to* $1 - (d+1)/(n+1)$*. Here the probability is measured with respect to the input-output data* $\{(y_k, \boldsymbol{x}_k)\}_{k=1}^{N}$ *and the newly sampled input-output pair* $(y_{k_0}, \boldsymbol{x}_{k_0})$.

**Proof.** The proof proceeds basically along the same line as Theorem 3.2.

By definition, $|y_{(j)} - \boldsymbol{x}_{(j)}^{\mathrm{T}}\boldsymbol{h}^*|/\mu_{(j)} \le b^*$ for $j = 1, \ldots, n$. We hence have $|\overline{y}_{(j)} - \boldsymbol{x}_{(j)}^{\mathrm{T}}\boldsymbol{h}^*|/\mu_{(j)} \le b^* + c\mu/\mu_{(j)} \le b^* + c\mu/\min_{j=1,\ldots,n}\mu_{(j)}$. Let $(\overline{\boldsymbol{h}}^*, \overline{b}^*)$ be the pair that minimizes $b$ subject to $|\overline{y}_{(j)} - \boldsymbol{x}_{(j)}^{\mathrm{T}}\boldsymbol{h}| \le b\mu_{(j)}$, $j = 1, \ldots, n$.

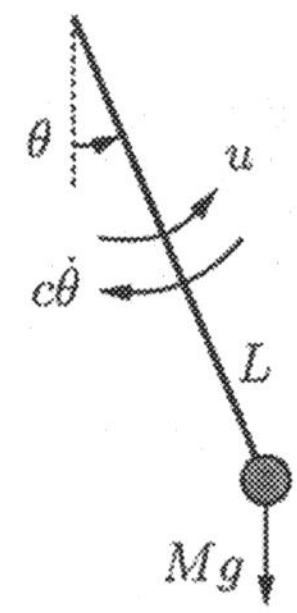


FIGURE 5.1. Identified system

Then there has to hold $\overline{b}^* \leq b^* + c\mu/\min_j \mu_{(j)}$. By repeating a similar inference with exchanging the roles of $y_{(j)}$ and $\overline{y}_{(j)}$, we have $|b^* - \overline{b}^*| \leq c\mu/\min_j \mu_{(j)}$. It is also seen that $|y_{(j)} - \boldsymbol{x}_{(j)}^{\mathrm{T}}\overline{\boldsymbol{h}}^*|/\mu_{(j)} - c\mu/\mu_{(j)} \leq |\overline{y}_{(j)} - \boldsymbol{x}_{(j)}^{\mathrm{T}}\overline{\boldsymbol{h}}^*|/\mu_{(j)} \leq \overline{b}^*$. Since $\overline{b}^* \leq b^* + c\mu/\min_j \mu_{(j)}$, we have $|y_{(j)} - \boldsymbol{x}_{(j)}^{\mathrm{T}}\overline{\boldsymbol{h}}^*| \leq (b^* + c\mu/\min_j \mu_{(j)})\mu_{(j)} + c\mu$ for $j = 1, \ldots, n$. By the definition of $e_i$, we have $|h_i^* - \overline{h}_i^*| \leq e_i$. Furthermore, in a similar way to Lemma 3.4, it is possible to prove that $|\overline{y}_{k_0} - \boldsymbol{x}_{k_0}^{\mathrm{T}}\overline{\boldsymbol{h}}^*| \leq \overline{b}^*\mu_{k_0}$ holds with probability greater than or equal to $1 - (d+1)/(n+1)$.

Evaluate now $|y_{k_0} - \boldsymbol{x}_{k_0}^{\mathrm{T}}\boldsymbol{h}^*|$ as in the proof of Theorem 3.2 but with the inequalities obtained above. Then we have the desired result. □

# 5 Example

A proposed identification algorithm is applied to a simple example plant.

Consider a pendulum shown in Figure 5.1, whose dynamics is described by $ML^2\ddot{\theta} = -MgL\sin\theta - c\dot{\theta} + u$. Here, the mass is taken as $M = 1\,\mathrm{kg}$, the length of the arm as $L = 1\,\mathrm{m}$, the coefficient of the friction as $c = 2\,\mathrm{Nms/rad}$, and the acceleration of the gravity as $g = 9.8\,\mathrm{m/s^2}$. The torque $u(t)$[Nm] is provided as $u(t) = u_k$ for $k \leq t < k+1$, where the input sequence $\{u_k\}$ is a sequence of independent random numbers that distribute uniformly in $[-1, 1]$. The output is the angle $\theta(t)$[rad] sampled at a discrete time $t = k$ and is denoted by $y_k$. Based on a preliminary experiment we assume approximate $K$-independence with $K = 6$ and $c = 0.001$. In fact, one can theoretically confirm that the true plant dynamics satisfies this assumption.

We apply Algorithm 3.1 to this system with the one-step-ahead predictor

$$\boldsymbol{x}_k^{\mathrm{T}}\boldsymbol{h} := \begin{bmatrix} u_{k-1} & u_{k-2} & u_{k-3} \end{bmatrix} \begin{bmatrix} h_1 \\ h_2 \\ h_3 \end{bmatrix}$$

obtained from the third-order FIR model. The number of input-output data, $N$, is chosen as 2400, which implies $n = 400$. The computational time to solve the associated linear programming problems is 1.17 s with Pentium II 450 MHz and 128 MByte memory. As a result, we have $\boldsymbol{h}^* = [0.137 \quad -0.0462 \quad 0.0145]^{\mathrm{T}}$, $b^* = 0.00556$, $e_1 = 0.00349$, $e_2 = 0.00375$, and $e_3 = 0.00387$. By Theorem 3.2, the inequality $|y_{k_0} - \boldsymbol{x}_{k_0}^{\mathrm{T}} \boldsymbol{h}^*| \leq 0.0187$ holds with probability greater than 99.0%.

# 6 Conclusion

In this paper, we saw how one can identify a provided plant in the worst-case sense using the leave-one-out technique. The proposed identification algorithm gives a nominal model and its uncertainty bound with a non-asymptotic probabilistic guarantee. The number of assumptions is kept small in order to reduce the risk of making too optimistic assumptions. Generalization of the algorithm is possible in several directions. The example in Section 5 suggests that the proposed algorithm gives a meaningful result with a reasonable number of input-output data. This is important because many existing learning-theoretic identification algorithms often require a large number of data beyond the practical level.

# Acknowledgement

The author is thankful to Dr. M. Vidyasagar, who acquainted him with the $\beta$-mixing property.

### References

[1] P. L. Bartlett and S. R. Kulkarni, "The complexity of model classes, and smoothing noisy data," *Systems & Control Letters*, vol. 34, no. 3, pp. 133–140, 1998.

[2] M. C. Campi and P. R. Kumar, "Learning dynamical systems in a stationary environment," *Systems & Control Letters*, vol. 34, no. 3, pp. 125–132, 1998.

[3] D. K. De Vries and P. M. J. Van den Hof, "Quantification of uncertainty in transfer function estimation: a mixed probabilistic-worst-case approach," *Automatica*, vol. 31, no. 4, pp. 543–557, 1995.

[4] G. C. Goodwin, M. Gevers, and B. Ninness, "Quantifying the error in estimated transfer functions with application to model order selection," *IEEE Transactions on Automatic Control*, vol. 37, no. 7, pp. 913–928, 1992.

[5] A. J. Helmicki, C. A. Jacobson, and C. N. Nett, "Control oriented system identification: a worst-case/deterministic approach in $H^\infty$," *IEEE Transactions on Automatic Control*, vol. 36, no. 10, pp. 1163–1176, 1991.

[6] R. L. Karandikar and M. Vidyasagar, "Rates of uniform convergence of empirical means with mixing processes," *Statistics & Probability Letters*, vol. 58, no. 3, pp. 297–307, 2002.

[7] P. Khargonekar and A. Tikku, "Randomized algorithms for robust control analysis and synthesis have polynomial complexity," in *Proceedings of the 35th IEEE Conference on Decision and Control*, Kobe, Japan, December 1996, pp. 3470–3475.

[8] V. Koltchinskii, C. T. Abdallah, M. Ariola, P. Dorato, and D. Panchenko, "Improved sample complexity estimates for statistical learning control of uncertain systems," *IEEE Transactions on Automatic Control*, vol. 45, no. 12, pp. 2383–2388, 2000.

[9] M. Milanese and A. Vicino, "Optimal estimation theory for dynamic systems with set membership uncertainty: an overview," *Automatica*, vol. 27, no. 6, pp. 997–1009, 1991.

[10] A. Nobel and A. Dembo, "A note on uniform laws of averages for dependent processes," *Statistics & Probability Letters*, vol. 17, no. 3, pp. 169–172, 1993.

[11] Y. Oishi and H. Kimura, "A learning-theoretic approach to model-set identification," *Asian Journal of Control*, vol. 3, no. 1, pp. 10–20, 2001.

[12] R. Tempo, E. W. Bai, and F. Dabbene, "Probabilistic robustness analysis: explicit bounds for the minimum number of samples," *Systems & Control Letters*, vol. 30, no. 5, pp. 237–242, 1997.

[13] D. N. C. Tse, M. A. Dahleh, and J. N. Tsitsiklis, "Optimal asymptotic identification under bounded disturbances," *IEEE Transactions on Automatic Control*, vol. 38, no. 8, pp. 1176–1190, 1993.

[14] V. N. Vapnik, *Statistical Learning Theory*. New York, NY: John Wiley & Sons, 1998.

[15] M. Vidyasagar, "Randomized algorithms for robust controller synthesis using statistical learning theory," *Automatica*, vol. 37, no. 10, pp. 1515–1528, 2001.

[16] M. Vidyasagar and R. L. Karandikar, "System identification: A learning theory approach," in *Proceedings of the 40th IEEE Conference on Decision and Control*, Orlando, FL, December 2001, pp. 2001–2006; also in this volume.

[17] M. Vidyasagar and R. L. Karandikar, "On the behaviour of learning algorithms with $\beta$-mixing process inputs," submitted for publication.

[18] E. Weyer, "Finite sample properties of system identification of ARX models under mixing conditions," *Automatica*, vol. 36, no. 9, pp. 1291–1299, 2000.

[19] E. Weyer and M. C. Campi, "Finite sample properties of system identification methods," in *Proceedings of the 38th IEEE Conference on Decision and Control*, Phoenix, AZ, December 1999, pp. 510–515.

[20] E. Weyer and M. C. Campi, "Non-asymptotic confidence ellipsoids for the least squares estimate," in *Proceedings of the 39th IEEE Conference on Decision and Control*, Sydney, Australia, December 2000, pp. 2688–2693.

[21] E. Weyer, R. C. Williamson, and I. M. Y. Mareels, "Finite sample properties of linear model identification," *IEEE Transactions on Automatic Control*, vol. 44, no. 7, pp. 1370–1383, 1999.

[22] T. Zhou and H. Kimura, "Structure of model uncertainty for a weakly corrupted plant," *IEEE Transactions on Automatic Control*, vol. 40, no. 4, pp. 639–655, 1995.

[23] Y. C. Zhu, "Estimation of transfer functions: asymptotic theory and a bound of model uncertainty," *International Journal of Control*, vol. 49, no. 6, pp. 2241–2258, 1989.

Yasuaki Oishi
Department of Mathematical Informatics
Graduate School of Information Science and Technology
The University of Tokyo
Hongo 7–3–1, Bunkyo-ku, Tokyo 113–8656, Japan
oishi@mist.i.u-tokyo.ac.jp

# Part III

# Robust Control

# Analysis of Weight Change in $H_\infty$ Control Design


## Brian D. O. Anderson and Xavier Bombois



ABSTRACT $H_\infty$ control design is generally performed iteratively. At each iteration, the weights constraining the desired closed loop transfer functions are adapted. The way in which the weights are adapted is generally purely heuristic. It is consequently very interesting to build some insights about the influence of a weight modification on the obtained (central) controller and, more importantly, on the obtained closed-loop transfer functions. In this paper, we analyze this influence in the case of a classical two-block problem under the assumption of "small" modifications in the weights. The concept *small modification* must be understood in the sense of small enough to allow first order approximation.



**Keywords:** $H_\infty$ control design, $J$-spectral factorization, Weight selection, Robust control design


## 1 Dedication

Hidenori Kimura has been associated with $H_\infty$ control since the 1980s, and he is one of a comparatively small number of workers in the area who can be said to have command of all the different approaches to considering $H_\infty$ control, as is amply evidenced by his fine book [7]. His early work focused on use of interpolation theory to solve control problems, see e.g., [5, 6]. In [7], he married many ideas of a classical kind and a modern system theory kind. The classical ideas included the use of scattering representations for networks or more generally interconnected systems, and frequency domain factorization; here, the notion of Wiener factorization which underpins linear-quadratic control and filtering is replaced by a type of generalization, $J$-spectral factorization. An allied concept especially associated with Kimura Sensei is $J$-lossless factorization, which relates to $J$-spectral factorization is a similar manner to the way Darlington synthesis of networks relates to Wiener or spectral factorization.

Hidenori Kimura is a consistent supporter of the idea that each new view of a problem adds to one's understanding of it. A corollary is that if there are certain aspects of a problem one wishes to understand, there may be preferred views of the problem when one is seeking that understanding. That precept motivated our use in this paper of a tool highly familiar to Kimura Sensei, and yet described rather less frequently in the most modern

$H_\infty$ textbooks, that of $J$-spectral factorization. For we are seeking to isolate some frequency domain properties of $H_\infty$ design, with little concern for tracking state-variable dimensions, let alone state-variable matrices.

It is with admiration for his work and in gratitude for the inspiration it has provided us that we dedicate this paper to Kimura Sensei.

# 2 Introduction

In recent years, $H_\infty$ control design has become a well-known method to design a model-based controller satisfying a number of constraints expressed by amplitude bounds (weights) on the "to-be-designed" closed-loop transfer functions. This method whose theoretical basis can be found in the works [2, 8] has known numerous applications for control design on real-life systems (see e.g., [1]).

The design of a controller using $H_\infty$ control design generally follows an iterative procedure. In a first step, only the sensitivity function is effectively constrained (i.e., the constraints on the other transfer functions are chosen in such a way that they remain ineffective). A first controller is obtained in this way. However, this controller has generally an unsatisfactory performance with respect to the closed-loop transfer functions for which the constraints were (in this first step) too loose to be effective. Consequently, in a second step, the weights on these closed-loop transfer functions are adapted in order to improve the closed-loop behaviour of the controller and a second controller is computed using these adapted weights. This procedure is pursued until the obtained controller is judged satisfactory enough.

The way with which the weights are adapted at each "iteration" is generally purely heuristic. It is consequently very interesting to build some insights about the influence of a weight modification on the obtained (central) controller and, more importantly, on the obtained closed-loop transfer functions. In this work, we analyze this influence in the case of a classical two-block problem (for a scalar model) under the assumption of "small" modifications in the weights. The concept *small modification* must be understood in the sense of small enough to allow first order approximation.

In order to solve an $H_\infty$ control design problem, two different methods are available. The most attractive one with respect to computational efficiency is the one developed in [2], and relies on a state-space formulation. The second one found for example in [3, 4, 7] is based on the $J$-spectral factorization of the augmented plant. This second method is less attractive computationally speaking. However, it can of course be used for analysis purposes. In order to develop our analysis, we will use the second method since this is based on frequency domain expressions which are important for our purpose. In particular, we contemplate variation of weight functions that may change their McMillan degree, and yet in frequency domain terms are small. Such variations almost certainly could not easily be cast as small for state-space descriptions[1], involving as they do degree change.

In this work, the first contribution is to give a frequency domain approximation of the change in the central controller due to a (small) weight modification after a first $H_\infty$ control design step. This approximation is a function of the weight modification frequency response and of the variables involved in the initial control design problem, and is therefore computable before performing the new control design step (i.e., the one with the modified weights). We show also that the modification in the central controller persists outside the frequency band where the weight modification is mainly located. A last contribution is to analyze the influence of this modification of the central controller on the modified closed-loop transfer functions. Note that we restrict attention to scalar systems and to the two-block $H_\infty$ problem. However it is our opinion that the results can be extended to multivariable systems and to the four-block $H_\infty$ problem. In the latter, two $J$-spectral factorizations are performed instead of one in the two-block case [3].

**Overview.** In Section 3, we recall the method to solve two-block $H_\infty$ control problems using $J$-spectral factorization. In Section 4, we introduce the particular two-block problem that we will consider and we state the objectives. In Section 5, we analyze the modification on the matrix defining the central controller, induced by a weight modification. Using the result of Section 6, we give an approximation of the modification of the central controller in Section 5 and an approximation of the modified closed-loop transfer functions in Section 7. We finish by an illustration (Section 8) and some conclusions (Section 9). □

# 3 $J$-spectral factorization and $H_\infty$ control design

In this section, we first recall how we can solve a classical two-block $H_\infty$ control design problem (for a scalar model $G_{mod}$) using $J$-spectral factorization (see also [3, 4, 7] for more details). For this purpose let us first introduce the notion of homographic transformation.

**Definition 3.1 ([7]).** Consider a scalar transfer function $Q(s)$ and a transfer matrix $H(s)$ partitioned as follows and $H_{22}$ are scalars. Define the homographic transformation $HOM(H, Q)$ as

$$HOM(H,Q) \triangleq \frac{H_{11}Q + H_{12}}{H_{21}Q + H_{22}}. \tag{3.1}$$

□

[1]Used in the method of [2].

Having defined the notion of homographic transformation, we may now present the following result about the solution of a two-block $H_\infty$ control design problem.

**Proposition 3.2** **([7]).** *Consider the two-block $H_\infty$ control problem consisting of finding a stabilizing controller $C$ for the model $G_{mod}$ such that*

$$\left\| \begin{matrix} W_1(s)\ T_1(G_{mod}, C) \\ W_2(s)\ T_2(G_{mod}, C) \end{matrix} \right\|_\infty < 1 \tag{3.2}$$

*where $W_1$, $W_2$ are two stable and inversely stable weights and $T_1$, $T_2$ are two closed-loop transfer functions of the to-be-designed loop $[C\ G_{mod}]$. Suppose that this control design problem satisfies the general conditions of solvability. Then the central controller $C_c$ corresponding to this control design problem is given by*

$$C_c = HOM(\Theta, 0) \tag{3.3}$$

*where $\Theta$ is a $2 \times 2$ stable and inversely stable transfer matrix whose inverse $\Pi(s) \triangleq \Theta^{-1}$ is obtained via the J-spectral factorization of the augmented plant $H(s)$ i.e.,*

$$H^* \overbrace{\begin{pmatrix} I_2 & 0 \\ 0 & -1 \end{pmatrix}}^{\triangleq J_3} H = \Pi^* \overbrace{\begin{pmatrix} 1 & 0 \\ 0 & -1 \end{pmatrix}}^{\triangleq J_2} \Pi \tag{3.4}$$

*where $X^*(s)$ denotes the adjoint of $X(s)$ i.e., $X^T(-s)$. The augmented plant $H(s)$ is here defined as the transfer matrix such that*

$$HOM(H, C) = \begin{pmatrix} W_1(s)\ T_1(G_{mod}, C) \\ W_2(s)\ T_2(G_{mod}, C) \end{pmatrix}. \tag{3.5}$$

**Important remark.** The matrix $\Pi(s)$ delivered by the $J$-spectral factorization (3.4) of $H(s)$ is always defined to within left multiplication by a constant $J$-unitary matrix. In other words, there are different possible matrices $\Pi(s)$ solving (3.4). However, for some problems, there is a natural choice for $\Pi(j\infty)$, which thus pins down all of $\Pi(s)$. □

# 4 Considered problem

As stated in the introduction, the objective in this paper is to analyze the influence of a (small) modification in the weights on the obtained central controller $C_c$ and, more importantly, on the obtained closed-loop transfer functions $T_1(G_{mod}, C_c)$ and $T_2(G_{mod}, C_c)$. In the sequel, we will restrict attention to one particular (and classical) two-block problem. This problem is the one for which expression (3.2) is given by

$$\left\| \begin{matrix} W_1(s) \frac{C}{1+CG_{mod}} \\ W_2(s) \frac{1}{1+CG_{mod}} \end{matrix} \right\|_\infty < 1. \tag{4.1}$$

In other words, we consider the case where $T_1 = C/(1+CG_{mod})$ and $T_2 = 1/(1+CG_{mod})$.

For this particular two-block problem, we will analyze the following (classical) situation. We assume that, in a first step, a central controller $C_c$ has been designed using the criterion (4.1) with an appropriate constraint $W_2$ on the sensitivity function, but a very loose constraint $W_1$ on the other closed-loop transfer function. For instance, in this first step, the weight $W_1$ could have been chosen equal to a very small constant. However, this is of course not a requirement.

In a second step, the weight $W_1$ is adapted and becomes $W_{1,bis} = W_1 + \Delta W_1$, a second $H_\infty$ design procedure is achieved with the adapted weights $W_{1,bis} = W_1 + \Delta W_1$ and $W_2$, and a new central controller $C_{c,bis} = C_c + \Delta C$ is obtained delivering modified closed-loop transfer functions $T_{1,bis}$ and $T_{2,bis}$:

$$T_{1,bis} = \frac{C_c + \Delta C}{1 + (C_c + \Delta C)G_{mod}} \; (= T_1 + \Delta T_1), \tag{4.2}$$

$$T_{2,bis} = \frac{1}{1 + (C_c + \Delta C)G_{mod}} \; (= T_2 + \Delta T_2). \tag{4.3}$$

The final objective of our research is to fully understand the link existing between a change in the weights and the obtained closed-loop transfer functions. In this work, we will nevertheless restrict attention to the following tasks:

- find a frequency domain approximation $\Delta C_{appr}$ for $\Delta C$, as a function of $\Delta W_1$ and of the variables involved in the initial two-block problem (i.e., the one with $W_1$ and $W_2$).
- And then, knowing $\Delta C_{appr}$, to analyze the influence of this modification on the modified closed-loop transfer functions $T_{1,bis}$ and $T_{2,bis}$.

For this purpose, we will need to assume that the change $\Delta W_1$ is small (so we may use first order approximations) and since we are interested in frequency domain expressions, we will use the frequency domain expression (3.4) as basis of our analysis. It is to be noted that all approximations given in this paper will therefore only be relevant in the frequency domain.

In effect, we will construct a mapping $\Delta W_1 \longrightarrow \Delta C_{appr} \longrightarrow \Delta T_1$ and $\Delta T_2$. The smallness assumption means that the mapping is linear, but we will show that in general it is not memoryless, i.e., $\Delta C_{appr}$ evaluated at a frequency $\omega_1$ does not just depend on $\Delta W_1(j\omega_1)$, but on $\Delta W_1(j\omega)$ for, in principle, $\omega \in [0, \infty)$. In practice, $\Delta C_{appr}(j\omega_1)$ can nevertheless be expected to depend essentially on $\Delta W_1(j\omega)$ for $\omega$ confined to an interval around $\omega_1$.

# 5 Modification of the transfer matrix $\Theta$

In order to find a frequency domain approximation of $\Delta C$, we will first need to find an expression for the modification $\Delta\Theta$ in the transfer matrix $\Theta = \Pi^{-1}$ which defines the central controller as shown in (3.3). The expression for $\Delta\Theta$ will be given in Proposition 5.1, but beforehand let us introduce the expression of the augmented plant $H(s)$ for the particular two-block problem that we analyze in this paper. This augmented plant is indeed necessary in order to compute the transfer matrix $\Pi = \Theta^{-1}$ using expression (3.4). The matrix $H(s)$ is given by:

$$H(s) = \begin{pmatrix} \begin{pmatrix} W_1 \\ 0 \end{pmatrix} & \begin{pmatrix} 0 \\ W_2 \end{pmatrix} \\ G_{mod} & 1 \end{pmatrix}. \tag{5.1}$$

Expression (5.1) corresponds to the augmented plant of the considered two-block problem since

$$HOM(H, C) = \begin{pmatrix} W_1(s) \frac{C}{1+CG_{mod}} \\ W_2(s) \frac{1}{1+CG_{mod}} \end{pmatrix}. \tag{5.2}$$

**Proposition 5.1.** *Consider the two-block $H_\infty$ problem defined in (4.1), with $G_{mod}$ strictly proper. Let $\Theta$ denote the matrix defining the central controller obtained using the weights $W_1$ and $W_2$ i.e., $C_c = HOM(\Theta, 0)$, and force uniqueness of $\Theta$ by requiring $\Pi(j\infty)$ to be positive definite diagonal.*[2] *Consider now the modified two-block problem with the weights $W_{1,bis} = W_1 + \Delta W_1$ and $W_2$. Then, if $\Delta W_1$ is small, the matrix $\Theta_{bis}$ defining the modified central controller $C_{c,bis} = HOM(\Theta_{bis}, 0)$ can be approximated by*

$$\Theta_{bis} \approx \Theta - \Theta J_2 \Psi, \tag{5.3}$$

*where $\Psi$ is a stable transfer matrix such that*

$$\Psi + \Psi^* = (W_1^* \Delta W_1 + W_1 \Delta W_1^*) \begin{pmatrix} \theta_{11}^* \theta_{11} & \theta_{11}^* \theta_{12} \\ \theta_{12}^* \theta_{11} & \theta_{12}^* \theta_{12} \end{pmatrix}, \tag{5.4}$$

[2]This ensures that $C_c$ is strictly proper.

*and such that*

$$HOM(\Theta - \Theta J_2 \Psi, 0) \text{ is strictly proper.} \tag{5.5}$$

*$\theta_{ij}$ are the entries of the "initial" transfer matrix $\Theta$.*

**Proof.** In order to prove the result, we will make use of Expression (3.4) defining the inverse $\Pi$ of $\Theta$. For this purpose, denote first by $H$ and $H_{bis}$ the augmented plants corresponding to the two-block problems with $W_1$ (and $W_2$) and with $W_1 + \Delta W_1$ (and $W_2$), respectively. Using (5.1), we have then that

$$H^* J_3 H = \begin{pmatrix} W_1^* W_1 - G_{mod}^* G_{mod} & -G_{mod}^* \\ -G_{mod} & W_2^* W_2 - 1 \end{pmatrix} \quad \text{and} \tag{5.6}$$

$$\begin{aligned} H_{bis}^* J_3 H_{bis} &= H^* J_3 H + \begin{pmatrix} W_1^* \Delta W_1 + W_1 \Delta W_1^* & 0 \\ 0 & 0 \end{pmatrix} + \begin{pmatrix} \Delta W_1^* \Delta W_1 & 0 \\ 0 & 0 \end{pmatrix} \\ &\approx H^* J_3 H + \begin{pmatrix} W_1^* \Delta W_1 + W_1 \Delta W_1^* & 0 \\ 0 & 0 \end{pmatrix} \end{aligned} \tag{5.7}$$

where the last expression is a consequence of the assumption of a small $\Delta W_1$. Notice from (5.6) that at $s = j\infty$, $H^* J_3 H$ is diagonal, and we can take $\Pi(j\infty) = diag(|W_1(j\infty)|, \sqrt{1 - |W_2(j\infty)|^2})$. Expression (5.7) gives us a relation between the left-hand sides of (3.4) for the two considered two-block problems. Let us now consider the right-hand side. In order to find a relation between the right-hand sides in the two different design problems, let us denote the difference between $\Pi_{bis}$ and $\Pi$ by $\Delta\Pi$ (i.e., $\Pi_{bis} = \Pi + \Delta\Pi$). We have then the following:

$$\begin{aligned} \Pi_{bis}^* J_2 \Pi_{bis} &= \Pi^* J_2 \Pi + \Delta\Pi^* J_2 \Pi + \Pi^* J_2 \Delta\Pi + \Delta\Pi^* J_2 \Delta\Pi \\ &\approx \Pi^* J_2 \Pi + \Delta\Pi^* J_2 \Pi + \Pi^* J_2 \Delta\Pi. \end{aligned} \tag{5.8}$$

Since $\Pi$ and $\Pi_{bis}$ are the solutions of (3.4) for $H$ and $H_{bis}$, respectively, we have that $H^* J_3 H = \Pi^* J_2 \Pi$ and $H_{bis}^* J_3 H_{bis} = \Pi_{bis}^* J_2 \Pi_{bis}$. Consequently, from (5.7) and (5.8), we can deduce the following approximation of $\Delta\Pi$ as a function of $\Delta W_1$:

$$\begin{pmatrix} W_1^* \Delta W_1 + W_1 \Delta W_1^* & 0 \\ 0 & 0 \end{pmatrix} \approx \Delta\Pi^* J_2 \Pi + \Pi^* J_2 \Delta\Pi. \tag{5.9}$$

However, the previous expression is still an implicit expression of $\Delta\Pi$. Using some simple manipulations, we can deduce from (5.9) an explicit expression for $\Delta\Pi$ (and then for $\Delta\Theta$). For this purpose, let us pre-multiply (5.9) by $\Pi^{-*} = \Theta^*$ and post-multiply the same expression by $\Pi^{-1} = \Theta$:

$$\Theta^* \begin{pmatrix} W_1^* \Delta W_1 + W_1 \Delta W_1^* & 0 \\ 0 & 0 \end{pmatrix} \Theta \approx \Theta^* \Delta\Pi^* J_2 + J_2 \Delta\Pi\Theta. \tag{5.10}$$

Since $\Pi$ and $\Pi_{bis} = \Pi + \Delta\Pi$ are both stable and inversely stable, the right-hand side of (5.10) is the sum of a stable transfer matrix $J_2\Delta\Pi\Theta$ and its complex conjugate $\Theta^*\Delta\Pi^* J_2$. Let us now decompose the left-hand side of (5.10) as in (5.4).[3] From (5.4) and (5.10), it is then obvious that

$$J_2\Delta\Pi\Theta \approx \Psi. \tag{5.11}$$

From (5.11), we can write successively the following:

$$\Delta\Pi \approx J_2\Psi\Pi, \tag{5.12}$$

$$\Pi_{bis} = \Pi + \Delta\Pi \approx (I_2 + J_2\Psi)\Pi, \tag{5.13}$$

$$\Theta_{bis} = \Pi_{bis}^{-1} \approx \Pi^{-1}(I_2 - J_2\Psi) = \Theta - \Theta J_2\Psi. \tag{5.14}$$

Note that, in order to invert $\Pi_{bis}$, we have made use of the fact that the changes are small. □

# 6 Modification in the central controller

Proposition 5.1 gives us a (frequency domain) approximation of the modified transfer matrix $\Theta_{bis}$. This approximation is a function of $\Delta W_1$ and the variables involved in the initial two-block problem (i.e., the one with $W_1$ and $W_2$).

## 6.1 Approximation $\Delta C_{appr}$ of $\Delta C$

This result will now allow us to deduce an approximate expression for the modified central controller $C_{c,bis} = C_c + \Delta C$ using the relation between $\Theta_{bis}$ and the central controller. This expression is given in the following proposition.

**Proposition 6.1.** *Consider the matrix $\Theta$ involved in the two-block $H_\infty$ problem (4.1) with weights $W_1$ and $W_2$. Consider also the stable matrix $\Psi$ defined in (5.4) and (5.5). Then the modified controller $C_{c,bis}$ delivered by the two-block $H_\infty$ problem (4.1) with weights $W_{1,bis} = W_1 + \Delta W_1$ and $W_2$ can be approximated as*

$$C_{c,bis} = C_c + \Delta C \approx C_c \overbrace{- \frac{det(\Theta)}{\theta_{22}^2}\psi_{12}}^{\Delta C_{appr}}, \tag{6.1}$$

[3]The matrix $\Psi$ resulting from (5.4) is defined up to an additive constant skew matrix. This constant matrix is in our problem chosen to be zero and $\Theta_{bis}(j\infty)$ is again diagonal, with $C_{c,bis}$ again strictly proper.

*where $det(A)$ denotes the determinant of the matrix $A$ and $\psi_{ij}$ are the entries of $\Psi$.*

**Proof.** From Expression (3.3), we can write

$$C_{c,bis} = HOM(\Theta_{bis}, 0) = \frac{\theta_{12,bis}}{\theta_{22,bis}}, \tag{6.2}$$

where $\theta_{ij,bis}$ are the entries of $\Theta_{bis}$. Using (5.3), it is easy to find (approximate) expressions for $\theta_{12,bis}$ and $\theta_{22,bis}$ i.e.,

$$\theta_{12,bis} \approx \theta_{12} - \overbrace{\begin{pmatrix} \theta_{11} & \theta_{12} \end{pmatrix}}^{=\alpha^T} \overbrace{\begin{pmatrix} \psi_{12} \\ -\psi_{22} \end{pmatrix}}^{=\beta}, \tag{6.3}$$

$$\theta_{22,bis} \approx \theta_{22} - \overbrace{\begin{pmatrix} \theta_{21} & \theta_{22} \end{pmatrix}}^{=\gamma^T} \begin{pmatrix} \psi_{12} \\ -\psi_{22} \end{pmatrix}. \tag{6.4}$$

Using (6.3) and (6.4), the expression of the modified controller (i.e., (6.2)) can be approximated as

$$C_{c,bis} \approx \frac{\theta_{12} - \alpha^T\beta}{\theta_{22} - \gamma^T\beta}, \tag{6.5}$$

and using the first order Taylor expansion around $\beta = 0$ of the ratio between $\theta_{12} - \alpha^T\beta$ and $\theta_{22} - \gamma^T\beta$, we finally obtain

$$C_{c,bis} \approx \frac{\theta_{12} - \alpha^T\beta}{\theta_{22} - \gamma^T\beta} \approx \frac{\theta_{12}}{\theta_{22}} + \frac{\theta_{12}\gamma^T - \theta_{22}\alpha^T}{\theta_{22}^2}\beta = C_c - \frac{det(\Theta)}{\theta_{22}^2}\psi_{12}. \tag{6.6}$$

□

**Remark.** From the approximation of the controller $C_{c,bis}$ given in (6.1), we can deduce that the transfer function $\psi_{12}$ must be strictly proper i.e.,

$$\psi_{12}(j\infty) = 0 \tag{6.7}$$

in order that $\Delta C_{appr}(j\infty) = 0$ which is in accordance with the fact that $C_c(j\infty) = C_{c,bis}(j\infty) = 0$. This property is also in accordance with the definition and properties of the matrix $\Psi$ given in (5.4)–(5.5). Since (5.4) only defines $\Psi$ to within a skew matrix, $\psi_{12}(j\infty)$ can be adjusted. □

## 6.2 Link between $\psi_{12}$ and $\Delta W_1$

Proposition 6.1 delivers an approximation $\Delta C_{appr}$ of the modification in the central controller due to the deviation $\Delta W_1$. This approximation is a function of the matrix $\Theta$ of the initial two-block problem and of the second entry $\psi_{12}$ of the matrix $\Psi$ defined implicitly in (5.4). In order to be able to compute $\Delta C_{appr}$, we show in this subsection how to compute $\psi_{12}$ (explicitly) and moreover we analyze the relation between this transfer function $\psi_{12}$ and the deviation $\Delta W_1$.

Let us deduce the following from (5.4):

$$\psi_{12} + \psi_{21}^* = \overbrace{(W_1^* \Delta W_1 + W_1 \Delta W_1^*)\theta_{11}^* \theta_{12}}^{f(s)} \tag{6.8}$$

where $f(s)$ is a known transfer function, since it is a function of some entries of the matrix $\Theta$, the weight $W_1$ and the change of weight $\Delta W_1$. Notice also that $f$ is equal to 0 when $\Delta W_1 = 0$. An expression for $\psi_{12}$ can now be computed by partial fraction decomposition of $f(s)$ into its unstable part $\psi_{21}^*$ and its stable part $\psi_{12}$.

**Important comments.** The deviation $\Delta W_1$ will generally be a stable filter having a form similar to

$$\Delta W_1 = \frac{\pm K\ s}{(s+\omega_1)(s+\omega_2)}.$$

The modification of the weight due to $\Delta W_1$ is therefore mainly restricted to the frequency band $B_{\Delta W_1} = [\ \omega_1\ \omega_2\ ]$. Indeed, outside this band $B_{\Delta W_1}$, the amplitude of $\Delta W_1$ converges to zero ( i.e., $|\Delta W_1(j\omega)| \to 0$ when $\omega \to 0$ and when $\omega \to \infty$). The representation of such a filter $\Delta W_1$ can be found in Figure 8.4 (in Section 8). □

Due to the decomposition into partial fractions, the transfer function $\psi_{12}$ will generally have a different behaviour than $\Delta W_1$. The amplitude of $\psi_{12}$ will indeed only converge to 0 in high frequencies (see (6.7)). In low frequencies, this amplitude will generally converge to a non-zero constant and, in the band $B_{\Delta W_1} = [\ \omega_1\ \omega_2\ ]$, $\psi_{12}(j\omega)$ will have a non-constant dynamic behaviour. An example of such a transfer function $\psi_{12}$ can be found in Figure 8.4. Consequently, the region where $\psi_{12}(j\omega)$ will have a significant influence on $\Delta C_{appr}$ (i.e., $B_{\psi_{12}} = [\ 0, \omega_2\ ]$ is much more important than the region $B_{\Delta W_1}$ for $\Delta W_1$. This is in fact a well-known property of the decomposition into partial fractions.

### 6.3 Comments about $\Delta C_{appr}$

The approximation $\Delta C_{appr}$ (see (6.1)) can be computed *a-priori*, that is without having first to solve the second two-block $H_\infty$ problem. Indeed, $\Delta C_{appr}$ is only a function of the transfer matrix $\Theta$ (involved in the initial two-block $H_\infty$ problem) and of the transfer function $\psi_{12}$ which has been proved computable in the previous subsection.

Due to (6.1) and the properties of $\psi_{12}$ with respect to $\Delta W_1$ presented in the previous subsection, we can also state that the modification of the central controller due to $\Delta W_1$ will not only be restricted to the band $B_{\Delta W_1}$, but will also persist in low frequencies. In high frequencies, the amplitude $\Delta C_{appr}$ converges to 0 because of (6.7).

# 7 Consequences for the closed-loop transfer functions

The results presented in the previous section allow one to compute very easily an approximation of the modified central controller $C_{c,bis} = C_c + \Delta C$ due to the modification $\Delta W_1$, and this without having to perform the $H_\infty$ control design problem with the weights $W_1 + \Delta W_1$ and $W_2$ which would have given $C_{c,bis}$ as solution. According to Proposition 6.1, this approximation is given by

$$C_{c,bis} \approx C_c \overbrace{- \frac{det(\Theta)}{\theta_{22}^2} \psi_{12}}^{\Delta C_{appr}} . \tag{7.1}$$

¿From this approximation, it is then easy to compute the effects of the change in the weights on the obtained closed-loop transfer functions $T_{1,bis}$ and $T_{2,bis}$ (see (4.2) and (4.3)):

$$T_{1,bis} \approx \frac{C_c + \Delta C_{appr}}{1 + (C_c + \Delta C_{appr}) G_{mod}}, \tag{7.2}$$

$$T_{2,bis} \approx \frac{1}{1 + (C_c + \Delta C_{appr}) G_{mod}}. \tag{7.3}$$

Let us once again notice that the approximations given in (7.2) and (7.3) are both computable without having to perform the $H_\infty$ control design problem with the weights $W_1 + \Delta W_1$ and $W_2$ which would deliver $T_{1,bis}$ and $T_{2,bis}$ as closed-loop transfer functions. Consequently, the influence of the change in the weights can be evaluated *a-priori*.

Although the influence of a change of the weights will vary depending on the considered model, the (modified) weights[4] and so on, we can nevertheless deduce some general comments which will be based on the first order Taylor expansion around $\Delta C_{appr} = 0$ of the approximations (7.2) and (7.3):

[4]Indeed, $\Delta C_{appr}$ is a function of all these factors through $\Theta$ and $\psi_{12}$.

$$T_{1,bis} \approx \overbrace{\frac{C_c}{1+C_cG_{mod}}}^{T_1}\left(1+\frac{1}{1+C_cG_{mod}}\frac{\Delta C_{appr}}{C_c}\right)=\frac{C_c+\Delta C_{appr}}{1+C_cG_{mod}}, \quad (7.4)$$

$$T_{2,bis} \approx \overbrace{\frac{1}{1+C_cG_{mod}}}^{T_2}\left(1-\frac{G_{mod}}{1+C_cG_{mod}}\Delta C_{appr}\right); \quad (7.5)$$

and on the following classical behaviours of these closed-loop transfer functions in low frequencies (LF) and in high frequencies (HF):

$$LF: \quad T_1(j\omega) \approx T_{1,bis}(j\omega) \approx \frac{1}{G_{mod}(j\omega)}, \quad (7.6)$$

$$LF: \quad T_2(j\omega) \approx \frac{1}{G_{mod}(j\omega)C_c(j\omega)}, \qquad T_{2,bis}(j\omega) \approx \frac{1}{G_{mod}(j\omega)C_{c,bis}(j\omega)}, \quad (7.7)$$

$$HF: \quad T_1(j\omega) \approx C_c(j\omega), \qquad T_{1,bis}(j\omega) \approx C_{c,bis}(j\omega), \quad (7.8)$$

$$HF: \quad T_2(j\omega) \approx T_{2,bis}(j\omega) \approx 1. \quad (7.9)$$

**Comments on $T_{1,bis}$**

- At low frequencies, according to (7.6), the difference between $T_1$ and $T_{1,bis}$ will be negligible and this even though $\Delta C_{appr}(j\omega) \neq 0$ at those frequencies.
- At high frequencies, according to (7.8), $|T_{1,bis}(j\omega)| < |T_1(j\omega)|$ if $|C_{c,bis}(j\omega)| < |C_c(j\omega)|$, and vice versa. Moreover, since both controllers are strictly proper, $T_1(j\omega)$ and $T_{1,bis}(j\omega)$ generally converge to zero when $\omega \to \infty$
- According to (7.4), we will have the inequality $|T_{1,bis}(j\omega)| < |T_1(j\omega)|$ if $|C_{c,bis}(j\omega)| < |C_c(j\omega)|$ (and vice versa) around the peak of $T_1$. □

**Comments on $T_{2,bis}$**

- Since $\Delta C_{appr}(j\omega) \neq 0$ at low frequencies, we will have a difference between $T_2$ and $T_{2,bis}$ at those frequencies. According to (7.7) $|T_{2,bis}(j\omega)| < |T_2(j\omega)|$ if $|C_{c,bis}(j\omega)| > |C_c(j\omega)|$, and vice versa.
- At high frequencies, according to (7.9), $T_{2,bis}(j\omega) \approx T_2(j\omega) \approx 1$.

- A general comment about the difference between $T_{2,bis}$ and $T_2$ around the resonance peak is not obvious. □

**Further research to be achieved.** From the previous comments, we see that a next step in our research should be to understand the link that exists between $\Delta W_1$ and the fact that $|C_{c,bis}(j\omega)| < |C_c(j\omega)|$ (or conversely) at certain frequencies. For this purpose, a further investigation of the expression (6.1) will be necessary. □

# 8 Illustration

In this section, we will illustrate the results presented in this paper. We will consider the following system $G_{mod}$:

$$G_{mod} = \frac{10}{(s-1)(0.2s+1)}. \tag{8.1}$$

In the first $H_\infty$ control design problem (see (4.1)), we will as usual only constrain the sensitivity function and choose the constraint $W_1$ on the other transfer function as a small constant. The chosen weights are: $W_1(s) = 1/6$ and $W_2(s) = (0.1s+1)/(0.003(100s+1))$. The $H_\infty$ problem is solved with these elements and we obtain the following central controller:

$$C_c = \frac{50.2564(s+0.69)(s+5)}{(s+0.01)(s^2+30.85s+423.3)} \tag{8.2}$$

Figures 8.2 and 8.3 represent the amplitude of the closed-loop transfer functions $T_1$ and $T_2$ achieved by this controller $C_c$ with the system $G_{mod}$. Now in a second step, we want to decrease the resonance peak of $T_1(G_{mod}, C_c)$. So, we choose the following new weight $W_{1,bis}$:

$$W_{1,bis} = W_1 + \Delta W_1 = W_1 + \frac{1.8967s}{(s+17.78)(s+5.623)} \tag{8.3}$$

that can be considered as a small deviation of $W_1$. The $H_\infty$ problem is solved with this new weight and we obtain the following central controller:

$$C_{c,bis} = \frac{58.9691(s+17.78)(s+5.623)(s+5)(s+0.6782)}{(s+27.12)(s+5.628)(s+0.01)(s^2+32.49s+339.1)}. \tag{8.4}$$

The controller $C_{c,bis}$ is represented in Figure 8.1 and the new closed-loop transfer functions $T_{1,bis}$ and $T_{2,bis}$ in Figure 8.2 and 8.3, respectively. In these three last figures, the modified transfer functions are compared to the corresponding transfer function in the initial two-block problem.

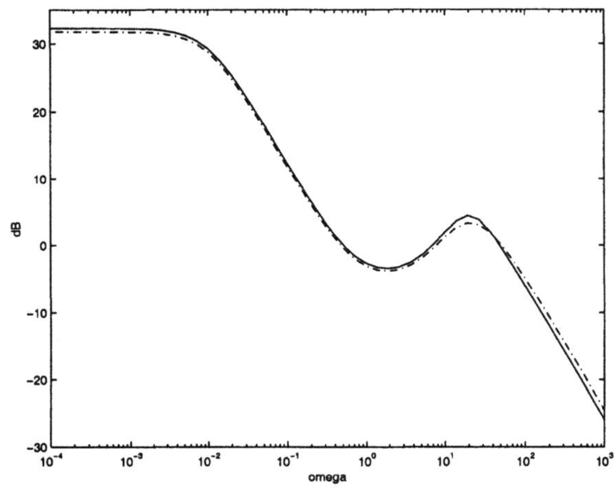


FIGURE 8.1. $|C_c(j\omega)|$ (solid) and $|C_{c,bis}(j\omega)|$ (dashdot)

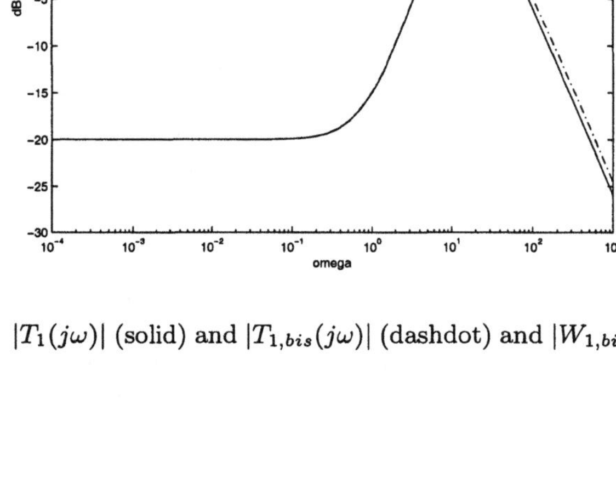


FIGURE 8.2. $|T_1(j\omega)|$ (solid) and $|T_{1,bis}(j\omega)|$ (dashdot) and $|W_{1,bis}(j\omega)^{-1}|$ (dotted)

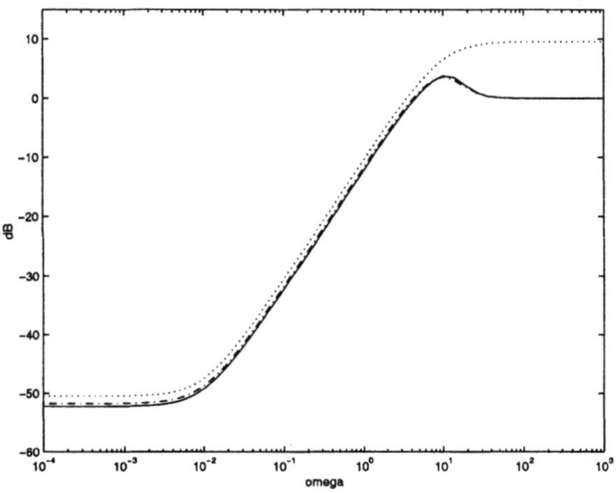


FIGURE 8.3. $|T_2(j\omega)|$ (solid) and $|T_{2,bis}(j\omega)|$ (dashdot) and $|W_2(j\omega)^{-1}|$ (dotted)

Now, we will show that the results presented in this paper would have allowed us to predict the modification caused by $\Delta W_1$ without having to perform the second $H_\infty$ design problem. For this purpose, let us first compute the transfer function $\psi_{12}$ that is necessary to approximate the change in the central controller according to (6.1). This function can be computed using the procedure presented in Section 6.2 and is represented in Figure 8.4. In this figure, we notice that $\psi_{12}$ converges to a non-zero constant in low frequencies as opposed to $\Delta W_1$ (which converges to 0). This is in accordance with our comments at the end of Section 6.2. From this function $\psi_{12}$ (and the matrix $\Theta$ of the first two-block problem), we can now compute $\Delta C_{appr}$. This last quantity is compared with the actual difference between $C_c$ and $C_{c,bis}$ in Figures 8.5 and 8.6. We observe that our $\Delta C_{appr}$ is a very good approximation of the actual difference between the two successive controllers. Moreover, we also observe that the change in the controller due to $\Delta W_1$ is not only located in the band where $\Delta W_1$ has a significant amplitude, but also persists at low frequencies. The change $\Delta C_{appr}$ converges to 0 when $\omega \rightarrow \infty$ as $\Delta W_1$ does.

In Figure 8.2 and 8.3, we observe behaviours that are also in accordance with our comments of Section 7. Indeed, at low frequencies, we have $|T_{1,bis}(j\omega)| \approx |T_1(j\omega)|$ and $|T_{2,bis}(j\omega)| > |T_2(j\omega)|$ since $|C_{c,bis}(j\omega)| < |C_c(j\omega)|$. At high frequencies, we have $T_{2,bis}(j\omega) \approx T_2(j\omega) \approx 1$ and $|T_{1,bis}(j\omega)| > |T_1(j\omega)|$ since $|C_{c,bis}(j\omega)| > |C_c(j\omega)|$ at those high frequencies, while both converging to 0.

Moreover, we see that the resonance peak of $T_{1,bis}$ has a smaller amplitude than the one of $T_1$. This could have been predicted from the fact that $|C_{c,bis}(j\omega)| < |C_c(j\omega)|$ around the frequency of this peak.

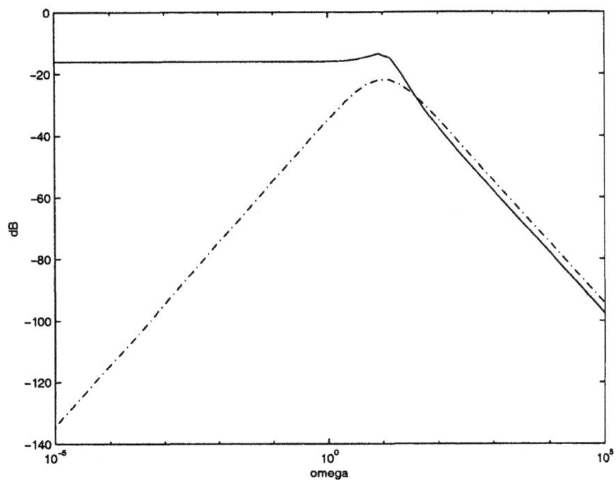


FIGURE 8.4. $|\psi_{12}(j\omega)|$ (solid) and $|\Delta W_1(j\omega)|$ (dashdot) at each frequency

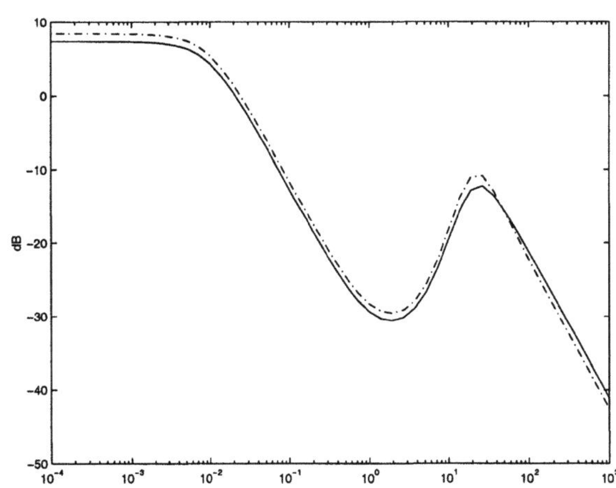


FIGURE 8.5. $|C_{c,bis}(j\omega) - C_c(j\omega)|$ (solid) and $|\Delta C_{appr}(j\omega)|$ (dashdot) at each frequency

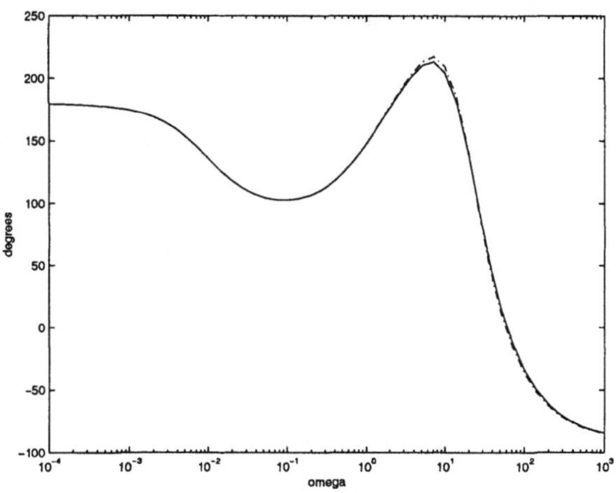


FIGURE 8.6. $arg(C_{c,bis}(j\omega) - C_c(j\omega))$ (solid) and $arg(\Delta C_{appr}(j\omega))$ (dashdot) at each frequency

# 9 Conclusions

In this work, we have analyzed the influence of a small weight modification in a classical two-block $H_\infty$ control design problem on the obtained central controller and on the obtained closed-loop transfer functions. The first contribution has been to give a frequency domain approximation of the change in the central controller induced by the (small) weight modification after a first $H_\infty$ control design step. This approximation is a function of the weight modification frequency response and of the variables involved in the initial control design problem, and is therefore computable before performing the new control design step (i.e., the one with the modified weights). We show also that the modification in the central controller persists outside the frequency band where the weight modification is mainly located. A last contribution is to analyze the influence of this modification of the central controller on the modified closed-loop transfer functions.

## References

[1] G. Ferreres and V. Fromion, "Computation of the robustness margin with the skewed $\mu$-tool," *Systems & Control Letters*, vol. 32, pp. 193–202, 1997.

[2] K. Glover and J. C. Doyle, "State-space formulae for all stabilizing controllers that satisfy an $H_\infty$-norm bound and relations to risk sensitivity," *Systems & Control Letters*, vol. 11, pp. 167–172, 1988.

[3] M. Green, K. Glover, D. Limebeer, and J. Doyle, "A J-spectral factorization approach to $H_\infty$ control," *SIAM Journal on Control and Optimization*, vol. 28, no. 6, pp. 1350–1371, 1990.

[4] V. Ionescu and R. Stefan, "Chain-scattering solution to the $\gamma$-DF problem: a Popov function approach," *IMA Journal of Mathematical Control and Information*, vol. 17, pp. 147–165, 2000.

[5] H. Kimura, "Directional interpolation approach to $H_\infty$ optimization and robust stabilization," *IEEE Transactions on Automatic Control*, vol. 32, pp. 1085–1093, 1987.

[6] H. Kimura, "Conjugation, interpolation and model-matching in $H_\infty$," *International Journal of Control*, vol. 49, pp. 269–307, 1989.

[7] H. Kimura, *Chain-Scattering Approach to $H_\infty$-Control.* Boston: Birkhäuser, 1997.

[8] G. Zames, "Feedback and optimal sensitivity: Model reference transformations, multiplicative seminorms, and approximate inverses," *IEEE Transactions on Automatic Control*, vol. AC-26, no. 2, pp. 301–320, April 1981.

Brian D. O. Anderson
Department of Systems Engineering
Research School of Information Sciences and Engineering
Australian National University
Canberra ACT 0200, Australia
Brian.Anderson@anu.edu.au

Xavier Bombois
Department of Applied Physics
Delft University of Technology
Lorentzweg 1, 2628 CJ
Delft, The Netherlands
x.j.a.bombois@tnw.tudelft.nl

# Performance Limits under Control Effort Constraints


## Jie Chen and Shinji Hara


*Dedicated to Professor Hidenori Kimura on the occasion of his 60th birthday, for a career of dedication and accomplishment*


ABSTRACT This paper presents some recent studies on performance limitations in optimal tracking and regulation control problems, in which objective functions of tracking error and regulated response, defined by integral square measures, are to be minimized jointly with the control effort, where the latter is measured by the plant input energy. The problems are solved explicitly by deriving analytical expressions for the best achievable performance. Other than the plant non-minimum phase zeros, time delays, and unstable poles, the results contain additional integral terms related to the gain characteristics of the plant and they reveal and quantify how the lightly damped poles, the anti-resonant zeros, and the bandwidth of the plant may all affect the performance. These effects are nonexistent when the control effort is not constrained, i.e., when it can be allowed to be infinite.



**Keywords:** Performance limitation, $\mathcal{H}_2$ control, Tracking and regulation, Unstable poles and zeros


# 1 Introduction

The progress of robust control theory in the last 25 years has been remarkable. This progress has led to a number of useful computer-aided design tools, and helped reduce the gap between control theory and control practice, making it possible to design high performance control systems for complex, real-world control problems. Professor Kimura, together with other leading control scientists, championed this course and has been recognized as one of the pioneers of robust control. Among his seminal contributions, we note his award winning works:

- Automatica Paper Prize Award in 1984, "Perfect and subperfect regulation in linear multivariable control systems," *Automatica*, vol. 18, no. 2, pp. 125–145, February 1982; and
- IEEE Control System Society Outstanding Paper Award, "Robust stabilizability for a class of transfer functions," *IEEE Transactions*

*on Automatic Control*, vol. AC-29, no. 9, pp. 788–793, September 1984.

Here in the first paper, Professor Kimura studied a tracking problem with respect to step reference commands and proposed design procedures which promise to achieve perfect tracking for minimum phase plants and sub-perfect tracking for non-minimum phase plants. In the second paper, Professor Kimura studied the robust stabilization problem for uncertain plants containing modeling errors. This work gave an analytical, constructive solution that can be used to synthesize a robustly stabilizing controller, and additionally, characterized the largest plant family that can be stabilized by a fixed controller in the $\mathcal{H}_\infty$ metric. In turn, it gives an indication on performance limitation that may result from plant uncertainty, and from plant unstable poles and non-minimum phase zeros. From a technical standpoint, the paper is one of the early attempts to apply the classical Nevanlinna–Pick interpolation theory to control synthesis problems.

Professor Kimura's work alluded to above has a direct bearing on the present paper, in which we also study the performance limitations in optimal $\mathcal{H}_2$ tracking and regulation control problems [3]. His work on the robust stabilization problem shares similarity with some of the recent studies on $\mathcal{H}_\infty$ type performance bounds; in particular, analytical characterizations have been obtained recently on the best achievable performance in sensitivity reduction, also based upon the Nevanlinna–Pick interpolation theory [2].

In recent years there has been growing attention devoted to the studies of intrinsic performance limits achievable by feedback control (see, e.g., [11, 1, 2] and the references therein). Two of the well-studied problems are the optimal reference tracking and optimal regulation problems [10, 5, 12, 9]. It has been known that the minimal tracking error depends upon the non-minimum phase zeros and time delays in the plant, while the minimal regulation energy depends upon the plant's unstable poles. In these studies, nevertheless, the performance criteria are highly idealistic, and thus serve more appropriately as an ideal, theoretical bound. Indeed, as an example, in the optimal tracking problems, in order to attain the minimal tracking error, the input to the plant is often required to have an infinite energy. This, of course, is seldom possible in practice. This consideration thus leads us to study the best achievable performances when only finite input energy is available.

More specifically, we consider an optimal tracking control problem in which not only the step error response, but also the plant input energy, both quantified under a square-integral or an $\mathcal{H}_2$ measure, is penalized. It is worth noting that such a performance objective, albeit a standard one in optimal control, has been investigated mainly in the so-called cheap control strategy [10, 12], that is, in the limit when the control penalty vanishes. In the present formulation, the control effort is *fixed* instead of being "cheap." This effectively restrains the *control effort*. Similarly, we also study an optimal output regulation problem which seeks to penalize jointly the output response and the control input energy. One should note that as some special cases of $\mathcal{H}_2$ optimal control, these problems can all be solved numerically. However, our aim here is to seek explicit, analytical expressions for the best achievable performance, rather than to provide a numerical solution.

Our primary motivation in this work is twofold. First, not only are the problems practically more relevant and generically more meaningful, but they in fact find rather pertinent applications, e.g., in the design of mechanical systems [7, 8]. Next, our investigation is driven by a deeper goal, in hope of discovering control constraints and limitations imposed by other sources than non-minimum phase zeros, unstable poles, and time delays. We maintain that this question may receive an answer only when more practical performance goals are taken into consideration, specifically when conflicting design objectives are to be considered jointly. This goal is met in the present paper. Generally, our results demonstrate in all cases how the best achievable performance may depend on the plant gain in the entire frequency range, and in particular help explain how the bandwidth, the lightly damped poles and the anti-resonant zeros of the plant may limit the achievable performance. The results thus unravel and quantify analytically yet another source of intrinsic feedback constraints, which may not be observed in the "single-objective" control design problems, such as the standard tracking and regulation problems, the Bode and Poisson integrals [12, 2], and the standard sensitivity and complementary sensitivity minimization problems [2].

The remainder of this paper is organized as follows. In Section 2, we provide a brief narrative of the technical background. We then proceed in Section 3 and Section 4 to formulate and solve the problems of optimal tracking and regulation under control effort constraint, respectively. The paper concludes in Section 5.

# 2 Preliminaries

We begin with a brief description of the notation. Denote the open right half plane by $\mathbf{C}_+$. For any complex number $z$, denote its complex conjugate by $\overline{z}$. For any signal $u(t)$, denote its Laplace transform by $\widehat{u}(s)$. The transpose and conjugate transpose of a matrix $A$ are denoted by $A^T$ and $A^H$, its largest singular values by $\overline{\sigma}(A)$, and its smallest eigenvalue by $\underline{\lambda}(A)$. For a pair of nonzero vectors $w$ and $v$, we define the principal angle $\angle(w,\ v)$ between their directions by $\cos\angle(w,\ v) := |w^H v|/(\|w\|\|v\|)$. Moreover, let $\|\cdot\|$ denote the Euclidean vector norm. The $\mathcal{L}_2$ norm is defined as

$$\|f\|_2 := \left(\frac{1}{2\pi}\int_{-\infty}^{\infty} f^H(j\omega)f(j\omega)d\omega\right)^{\frac{1}{2}}.$$

We suppress the dimensions of vectors and matrices, with the understanding that all the dimensions are compatible.

We shall consider the one-parameter unity feedback configuration of finite dimensional linear time-invariant systems as depicted in Figure 2.1. Here, $P$ represents the plant model and $K$ the compensator, whose transfer functions are denoted by $P(s)$ and $K(s)$, respectively. Throughout this paper, we omit the frequency variable $s$ in the transfer functions whenever no confusion occurs. The signals $r$, $e$, $u$, $d$, and $y$ are, respectively, the reference input, the error signal, the plant input, the disturbance input, and the system output. Furthermore, define the system sensitivity function by

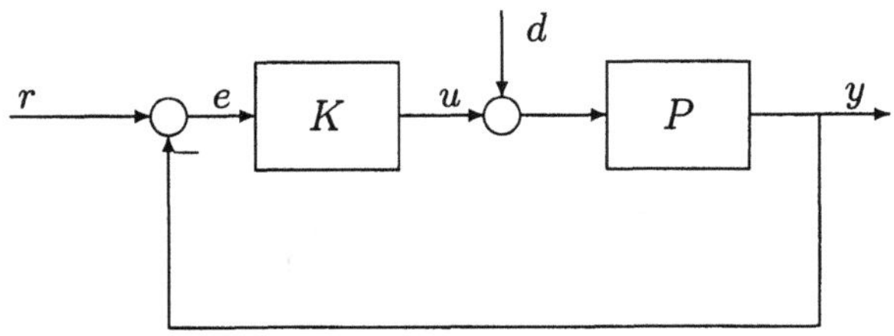


FIGURE 2.1. One-parameter control scheme

$$S(s) := (I + P(s)K(s))^{-1}.$$

For the rational transfer function matrix $P$, let its right and left coprime factorizations be given by

$$P = NM^{-1} = \tilde{M}^{-1}\tilde{N}, \tag{2.1}$$

where $N$, $M$, $\tilde{N}$, $\tilde{M} \in \mathbf{R}\mathcal{H}_\infty$ and satisfy the double Bezout identity for some $X$, $Y$, $\tilde{X}$, $\tilde{Y} \in \mathbf{R}\mathcal{H}_\infty$,

$$\begin{bmatrix} \tilde{X} & -\tilde{Y} \\ -\tilde{N} & \tilde{M} \end{bmatrix} \begin{bmatrix} M & Y \\ N & X \end{bmatrix} = I. \tag{2.2}$$

The set of all stabilizing compensators $K$ is characterized by the Youla parameterization [6]

$$\begin{aligned} \mathcal{K} := \Big\{K : \; K &= -(Y - MQ)(X - NQ)^{-1} \\ &= -(\tilde{X} - Q\tilde{N})^{-1}(\tilde{Y} - Q\tilde{M}), \; Q \in \mathbf{R}\mathcal{H}_\infty\Big\}. \end{aligned} \tag{2.3}$$

For a right-invertible $P$, it is well known (see, e.g., [11]) that each of its non-minimum phase zeros is also one for $N(s)$. Let $z_i \in \mathbf{C}_+$, $i = 1, \cdots, N_z$, be the non-minimum phase zeros of $P$. It is possible to factorize $N(s)$ as

$$N(s) = N_m(s) \left( \prod_{i=1}^{N_z} \begin{bmatrix} \eta_i & U_i \end{bmatrix} \begin{bmatrix} \frac{\bar{z}_i}{z_i} \frac{z_i - s}{\bar{z}_i + s} & 0 \\ 0 & I \end{bmatrix} \begin{bmatrix} \eta_i^H \\ U_i^H \end{bmatrix} \right), \tag{2.4}$$

where $N_m(s)$ represents the minimum phase part of $N(s)$, $\eta_i$ is a unitary direction vector associated with $z_i$, and the columns of $U_i$ form a unitary matrix together with $\eta_i$. Let $P$ have poles $p_i \in \mathbf{C}_+$, $i = 1, \cdots, N_p$. Note that for both right-invertible and left-invertible plants, it is always possible to factorize $M(s)$ as

$$M(s) = \left( \prod_{i=1}^{N_p} \begin{bmatrix} w_i & W_i \end{bmatrix} \begin{bmatrix} \frac{s - p_i}{s + \bar{p}_i} & 0 \\ 0 & I \end{bmatrix} \begin{bmatrix} w_i^H \\ W_i^H \end{bmatrix} \right) M_m(s), \tag{2.5}$$

where $M_m(s)$ is minimum phase, $w_i$ is a unitary direction vector associated with $p_i$, and the columns of $W_i$ form a unitary matrix together with $w_i$.

# 3 Optimal tracking performance

In the optimal tracking problem, we assume that $d = 0$ in Figure 2.1, and that the system is marginally stable, right-invertible, and initially at rest. We consider the step reference input

$$r(t) = \begin{cases} v & t \geq 0, \\ 0 & t < 0, \end{cases} \tag{3.1}$$

where the unitary vector $v$ specifies the input direction. We adopt the integral square criterion (for $0 \leq \epsilon \leq 1$)

$$J := (1 - \epsilon) \int_0^\infty \|y(t) - r(t)\|^2 dt + \epsilon \int_0^\infty \|u(t)\|^2 dt, \tag{3.2}$$

as our performance measure. Here, $\epsilon$ may be used to weigh the relative importance of tracking objective versus that of constraining the input energy. For a given $r$, we want to determine the optimal performance achievable by all stabilizing compensators,

$$J^* := \inf_{K \in \mathcal{K}} J.$$

In the limiting case, when $\epsilon = 0$, $J^*$ defines the minimal tracking error with no regard to input energy [5]. For $\epsilon = 1$, it reduces to an optimal energy regulation problem [9]. By invoking the Parseval identity, it readily leads to

$$J = (1 - \epsilon)\|S\widehat{r}\|_2^2 + \epsilon\|KS\widehat{r}\|_2^2. \tag{3.3}$$

To proceed, the following assumptions are enforced.

**Assumption 3.1.** $P(s)$ is right-invertible and it has no zero at $s = 0$. □

**Assumption 3.2.** $P(s)$ has a pole at $s = 0$, and $P(s)$ has no unstable poles except $s = 0$. □

We note that while Assumption 3.1 is standard in step reference tracking problems, it is clear from (3.3) that to maintain a finite energy cost precludes the possibility that $K$ may have an integrator. Note that Assumption 3.2 can be met in many cases of interest (e.g., mechanical systems [7, 8]), where the plant contains integrators.

Let us perform the inner-outer factorization

$$\begin{bmatrix} \sqrt{1-\epsilon}N_m \\ \sqrt{\epsilon}M \end{bmatrix} = \Theta_i \Theta_o, \tag{3.4}$$

where $\Theta_i, \Theta_o \in \mathbf{R}\mathcal{H}_\infty$ are inner and outer matrix functions, respectively. Next, define

$$f(s) := (1-\epsilon)v^T N_m(s)\Theta_o^{-1}(s)\Theta_o^{-T}(0)N_m^T(0)v \tag{3.5}$$

and factorize $f(s)$ as

$$f(s) = \left( \prod_{i=1}^{N_s} \frac{\bar{s}_i(s_i - s)}{s_i(\bar{s}_i + s)} \right) f_m(s),$$

where $s_i$ are the non-minimum phase zeros of $f(s)$ and $f_m(s)$ is minimum phase.

We are now ready to state our first main result which gives the performance limit for tracking.

**Theorem 3.3.** *Under Assumptions 3.1–3.2, let $z_i \in \mathbf{C}_+$, $i = 1, \cdots, N_z$, be the zeros of $P(s)$, which admit the decomposition (2.4). Then, the best achievable performance is given as follows:*

*1) SISO system:*

$$J^* = 2(1-\epsilon)\sum_{i=1}^{N_z} \frac{1}{z_i} + \frac{1-\epsilon}{\pi} \int_0^\infty \frac{1}{\omega^2} \log\left(1 + \frac{\epsilon}{(1-\epsilon)|P(j\omega)|^2}\right) d\omega. \tag{3.6}$$

*2) MIMO system:*

$$J^* = 2(1-\epsilon)\sum_{i=1}^{N_z} \frac{1}{z_i} \cos^2 \angle(\eta_i,\ v) + 2(1-\epsilon)\sum_{i=1}^{N_s} \frac{1}{s_i} + J_o, \tag{3.7}$$

*where*

$$J_o := -\frac{2(1-\epsilon)}{\pi} \int_0^\infty \frac{\log |f(j\omega)|}{\omega^2} d\omega. \tag{3.8}$$

*Furthermore,*

$$J_o \geq \frac{1-\epsilon}{\pi} \int_0^\infty \frac{1}{\omega^2} \log\left(1 + \frac{\epsilon}{(1-\epsilon)\overline{\sigma}^2(P(j\omega))}\right) d\omega. \tag{3.9}$$

Theorem 3.3 demonstrates that unlike in the standard tracking problem, the optimal tracking performance also depends on the minimum phase part of the plant, in addition to non-minimum phase zeros. This effect is captured by the the second term in (3.6), and the second and the third terms in (3.7), which are all nonnegative.

The implication of Theorem 3.3, in particular the result for SISO plants, is worth noting. First, while it has been long well known that non-minimum phase zeros impose fundamental performance limits, these results show that, with the objective under consideration herein, the achievable performance can be seriously constrained as well by those minimum phase zeros close to the imaginary axis, i.e., the anti-resonant zeros. This is seen by noting that in the vicinity of such zeros $|P(j\omega)|$ may be rather small, thus rendering the integral in (3.6) large. Evidently, the statement extends to multivariable systems as well, when the plant has such blocking zeros. Due to the weighting factor $1/\omega^2$, zeros close to the origin have more negative an effect. For the same reason, while in general the performance is affected by $\overline{\sigma}(P(j\omega))$ at all frequencies, the effect is particularly dominating in the low frequency range. This, of course, is expected, for to achieve good tracking performance requires a high loop gain, which, in the case of a low plant gain, requires high controller compensation. The latter is in direct conflict with the control effort constraint.

Based on Theorem 3.3, we consider a related constrained optimization problem instead of the unconstrained optimization problem with the performance index (3.2). This problem is defined as

$$\min_{K(s)\in\mathcal{K}_2} \quad \|\widehat{e}\|_2^2 := \int_0^\infty \|e(t)\|^2 dt,$$

$$\text{subject to} \quad \|\widehat{u}\|_2^2 := \int_0^\infty \|u(t)\|^2 dt \leq \gamma_u.$$

We denote the optimal cost by $J_e^*(\gamma_u)$.

**Theorem 3.4.** *Consider a SISO plant $P(s)$ and let a positive number $\gamma_u < \gamma_u^*$ be given, where*

$$\gamma_u^* := \frac{1}{\pi} \int_0^\infty \frac{1}{\omega^2 |P(j\omega)|^2} d\omega. \tag{3.10}$$

*Then, under Assumptions 3.1–3.2,*

$$J_e^*(\gamma_u) = 2\sum_{i=1}^{N_z} \frac{1}{z_i} - \alpha^* \gamma_u + \frac{1}{\pi} \int_0^\infty \frac{1}{\omega^2} \log\left(1 + \frac{\alpha^*}{|P(j\omega)|^2}\right) d\omega, \tag{3.11}$$

*where $\alpha^*$ is the unique positive solution of*

$$\int_0^\infty \frac{1}{\omega^2} \cdot \frac{1}{\alpha + |P(j\omega)|^2}\, d\omega = \pi \gamma_u. \tag{3.12}$$

*Moreover,*

$$J_e^*(\gamma_u) = J_e^*(\gamma_u^*) = 2\sum_{i=1}^{N_z} \frac{1}{z_i}\ , \quad \forall \gamma_u \geq \gamma_u^*. \tag{3.13}$$

Since $\alpha^*$ only depends on the plant gain, Theorem 3.4 also exhibits a nice separation between the effect due to the non-minimum phase zeros and that by the minimum phase part. Indeed, the first term results because of the plant non-minimum phase zeros and is independent of $\gamma_u$, while another two terms are completely determined by $|P(j\omega)|$ and $\gamma_u$.

In closing this section, we extend Theorem 3.3 to investigate how time delays may affect the tracking performance. For this purpose, we consider plants with measurement delays, that is,

$$P_d(s) = \Delta(s)P(s), \tag{3.14}$$

where $P(s)$ is a real rational transfer function matrix and

$$\Delta(s) := \operatorname{diag}\{e^{-\tau_1 s}, \cdots, e^{-\tau_m s}\}\ ;\ \tau_i > 0$$

collects the time-delay units in the $m$ channels of the plant. For the plant $P_d$, we define in exactly the same manner as in (3.2) the performance objective, and denote it by $J_d$. The optimal cost $J_d^*$ is the minimum $J_d$ achievable by all stabilizing compensators, which is obtained below.

**Theorem 3.5.** *Let $P_d(s)$ be given by (3.14) and suppose that $P(s)$ satisfies Assumptions 3.1–3.2. Then,*

$$J_d^* := \inf_{K \in \mathcal{K}_2} J_d = (1-\epsilon)\sum_{i=1}^{m} \tau_i |v_i|^2 + J^*, \tag{3.15}$$

*where $J^*$ is the optimal performance in the absence of delay, given in (3.7). In particular, if $P(s)$ is a scalar transfer function, then*

$$\begin{aligned} J_d^* =& (1-\epsilon)\tau + 2(1-\epsilon)\sum_{i=1}^{N_z}\frac{1}{z_i} \\ &+ \frac{1-\epsilon}{\pi}\int_0^\infty \frac{1}{\omega^2}\log\left(1 + \frac{\epsilon}{(1-\epsilon)|P(j\omega)|^2}\right) d\omega. \end{aligned} \tag{3.16}$$

# 4 Optimal regulation performance

In this section we formulate and solve an output regulation problem under finite control effort constraint. To this end, we consider the control scheme in Figure 2.1 with $r = 0$. Here we address minimum phase left-invertible plants and take $d$ to be the impulse signal

$$d(t) = \zeta\delta(t), \tag{4.1}$$

where $\zeta$ is a constant unitary vector. This signal may be interpreted either as a nonzero initial condition of the system or more generally as a disturbance signal entering at the plant input. The problem of interest is to regulate the output response $y$ to zero, by the design of the compensator $K$. We adopt the integral square energy criterion

$$E := (1-\epsilon)\int_0^\infty \|y(t)\|^2 dt + \epsilon\int_0^\infty \|u(t)\|^2 dt.$$

The best attainable performance then becomes

$$E^* := \inf_{K\in\mathcal{K}} E.$$

By means of the Parseval identity, it is easy to find that

$$E = (1-\epsilon)\|SP\widehat{d}\|_2^2 + \epsilon\|KSP\widehat{d}\|_2^2.$$

Furthermore, the following assumption is necessary.

**Assumption 4.1.** $P(s)$ is left-invertible and strictly proper, and it has no non-minimum phase zeros. □

This assumption that the plant be strictly proper implies that $y(0)$ is finite, a necessary condition for the output energy to be finite. We note that the same assumption is not required in the standard energy regulation problem [9] ($\epsilon = 1$).

Let $P(s)$ have unstable poles $p_i \in \mathbf{C}_+$, $i = 1, \cdots, N_p$, and $M(s)$ be factorized as in (2.5). Perform the inner and outer factorization

$$\begin{bmatrix} \sqrt{1-\epsilon}N \\ -\sqrt{\epsilon}M_m \end{bmatrix} = \Lambda_i\Lambda_o.$$

Moreover, define

$$g(s) := \epsilon\zeta^T M_m(s)\Lambda_o^{-1}(s)\Lambda_o^{-T}(\infty)M_m^T(\infty)\zeta \tag{4.2}$$

and factorize $g(s)$ as

$$g(s) = \left(\prod_{i=1}^{N_s}\frac{s-s_i}{s+\overline{s}_i}\right) g_m(s),$$

where $s_i$ are the non-minimum phase zeros of $g(s)$ and $g_m(s)$ is the minimum phase.

**Theorem 4.2.** *Under Assumption 4.1, let $p_i \in \mathbf{C}_+$, $i = 1, \cdots, N_p$, be the poles of $P(s)$, which admit the decomposition (2.5). Then, the best achievable performance is given as follows:*

*1) SISO system:*

$$E^* = 2\epsilon \sum_{i=1}^{N_p} p_i + \frac{\epsilon}{\pi} \int_0^\infty \log\left(1 + \frac{1-\epsilon}{\epsilon}|P(j\omega)|^2\right) d\omega. \tag{4.3}$$

*2) MIMO system:*

$$E^* = 2\epsilon \sum_{i=1}^{N_p} p_i \cos^2 \angle(w_i, \zeta) + 2\epsilon \sum_{i=1}^{N_s} s_i + E_o, \tag{4.4}$$

*where*

$$E_o := -\frac{2\epsilon}{\pi} \int_0^\infty \log|g(j\omega)| d\omega. \tag{4.5}$$

*Furthermore,*

$$E_o \geq \frac{\epsilon}{\pi} \int_0^\infty \log\left(1 + \frac{1-\epsilon}{\epsilon} \underline{\lambda}\left[P^H(j\omega)P(j\omega)\right]\right) d\omega. \tag{4.6}$$

Theorems 4.2 shows that the regulation performance also hinges closely on the plant gain and bandwidth, other than its unstable poles. However, a clear distinction exists between the tracking and the regulation problems. Unlike in the tracking problem, a large plant gain is seen to be undesirable for regulation. This is not surprising, for a larger plant gain will necessarily amplify more the disturbance, making it more difficult to return the response to zero. Similarly, the lightly damped poles in the plant are now problematic factors. This too is consistent with one's intuition, since such poles tend to cause more unsettling response and thus require more control effort.

Consider next the constrained optimization problem

$$\inf_{K \in \mathcal{K}} \quad \|\widehat{u}\|_2^2 := \int_0^\infty \|u(t)\|^2 dt,$$

$$\text{subject to} \quad \|\widehat{y}\|_2^2 := \int_0^\infty \|y(t)\|^2 dt < \gamma_y.$$

Let the optimal cost be denoted by $E_e^*(\gamma_y)$. This problem is similar to that of tracking with the input energy constraint. The question being asked is "What is the minimum control effort required to achieve a given regulation performance level, for minimum phase unstable SISO plants?" The following closed-form solution of the problem is the dual result to Theorem 3.4.

**Theorem 4.3.** *Suppose that $P(s)$ is a strictly proper scalar transfer function satisfying Assumption 4.1. Let a positive number $\gamma_y < \gamma_y^*$ be given, where*

$$\gamma_y^* := \frac{1}{\pi}\int_0^\infty |P(j\omega)|^2\, d\omega. \tag{4.7}$$

*Then,*

$$J_e^*(\gamma_y) = 2\sum_{i=1}^{N_p} p_i + \frac{1}{\pi}\int_0^\infty \log(1+\beta^*|P(j\omega)|^2)d\omega - \beta^*\gamma_y, \tag{4.8}$$

*where $\beta^*$ is the unique positive solution of*

$$\int_0^\infty \frac{|P(j\omega)|^2}{1+\beta|P(j\omega)|^2}\, d\omega = \pi\gamma_y. \tag{4.9}$$

*Moreover,*

$$E_e^*(\gamma_y) = E_e^*(\gamma_y^*) = 2\sum_{i=1}^{N_p} p_i \ , \ \forall \gamma_y > \gamma_y^*. \tag{4.10}$$

*In other words, for any $\gamma_y > \gamma_y^*$, the minimum control effort for the prescribed regulation performance is $\|\widehat{u}\|_2^2 = E_e^*(\gamma_y^*) = 2\sum_{i=1}^{N_p} p_i$.*

# 5 Conclusion

In this paper we have presented a number of recent results on $\mathcal{H}_2$-type optimal tracking and regulation control problems, which attempt to minimize jointly the tracking error and plant input energy, and the output response along with input energy, respectively. The results enable us to conclude that under finite control effort, the best tracking and regulation performance both depend on the plant gain in the entire frequency range, and consequently both can be significantly affected by the plant bandwidth. Furthermore, the results help clarify the roles of lightly damped poles and anti-resonant zeros, which have been seen to have a direct pertinence on the performance.

The results can be extended to two-parameter control systems. A two-parameter control system offers a number of distinctive advantages. For the tracking problem, it can cope with unstable plants while attaining the same best tracking performance. For the regulation problem, it allows non-minimum phase plants and achieves the same regulation performance solely determined by the plant unstable poles. These results are presented in [4]. Since two-parameter controllers constitute the most general control structure, the results in the present paper in fact provide the fundamental tracking and regulation performance limits, whenever the plant input energy must be so constrained.

# Acknowledgement

This research was supported in part by the NSF/USA under the Grant ECS-9912533, and in part by The Grant-in-Aid for COE Research Project of Super Mechano-Systems by The Ministry of Education, Science, Sport and Culture, Japan.


## References


[1] K. J. Åström, "Limitations on control system performance," *European J. of Control*, vol. 6, no. 1, pp. 2–20, 2000.

[2] J. Chen, "Logarithmic integrals, interpolation bounds, and performance limitations in MIMO systems," *IEEE Trans. on Automatic Control*, vol. 45, no. 6, pp. 1098–1115, June 2000.

[3] J. Chen, S. Hara, and G. Chen, "Best tracking and regulation performance under control effort constraint," in *Proc. 40th IEEE Conf. Decision Contr.*, Orlando, FL, Dec. 2001.

[4] J. Chen, S. Hara, and G. Chen, "Best tracking and regulation performance under control effort constraint: two-parameter controller case," in *Proc. IFAC World Congress*, Barcelona, Spain, July 2002.

[5] J. Chen, O. Toker, and L. Qiu, "Limitations on maximal tracking accuracy," *IEEE Trans. on Automatic Control*, vol. 45, no. 2, pp. 326–331, Feb. 2000.

[6] B. A. Francis, *A Course in $H_\infty$ Control Theory.* Lecture Notes in Control and Information Sciences, Berlin: Springer-Verlag, 1987.

[7] S. Hara and N. Naito, "Control performance limitation for electromagnetically levitated mechanical systems," in *Proc. 3rd MOVIC*, Zurich, 1998, pp. 147–150.

[8] T. Iwasaki, S. Hara, and Y. Yamauchi, "Structure/control design integration with finite frequency positive real property," in *Proc. 2000 Amer. Contr. Conf.*, Chicago, IL, June 2000, pp. 549–553.

[9] L. Qiu and J. Chen, "Time domain performance limitations of feedback control," in *Mathematical Theory of Networks and Systems*, A. Beghi, L. Finesso, and G. Picci, Eds. Il Poligrafo, 1998, pp. 369–372.

[10] L. Qiu and E. J. Davison, "Performance limitations of non-minimum phase systems in the servomechanism problem," *Automatica*, vol. 29, no. 2, pp. 337–349, Feb. 1993.

[11] M. M. Seron, J. H. Braslavsky, and G. C. Goodwin, *Fundamental Limitations in Filtering and Control.* London: Springer-Verlag, 1997.

[12] M. M. Seron, J. H. Braslavsky, P. V. Kokotovic, and D. Q. Mayne, "Feedback limitations in nonlinear systems: From Bode integrals to cheap control," *IEEE Trans. Automatic Control*, vol. 44, no. 4, pp. 829–833, April 1999.

[13] M. Vidyasagar, *Control System Synthesis: A Factorization Approach.* Cambridge, MA: MIT Press, 1985.

Jie Chen
Department of Electrical Engineering
University of California, Riverside
CA 92521, USA
jchen@ee.ucr.edu

Shinji Hara
Department of Information Physics and Computing
Graduate School of Information Science and Technology
The University of Tokyo
Hongo 7–3–1, Bunkyo-ku, Tokyo 113–8656, Japan
hara@crux.t.u-tokyo.ac.jp

# Complexity of Systems and Controllers

## Koji Tsumura

*Dedicated to Professor Hidenori Kimura on the occasion of his 60th birthday*

ABSTRACT In this article, we introduce some recent results on the complexity of the model and controller using a unified probabilistic approach to model estimation/selection and controller design. The objective systems are assumed to include unknown random parameters with probability distributions. The first issue is what evaluation function, for model estimation, is reasonable with respect to the controller design. Second, we analyse the effects of the complexity of the parameter distribution model and the class of controller on the expectation of the evaluation functions for model estimation. Finally, we discuss the distribution of systems with a result on a metric structure of a set of analytic functions.

**Keywords:** Model estimation, Model selection, Complexity, Information criterion, Robust control, $\epsilon$-entropy

## 1 Introduction

In this article we explain the author's recent results for the following simple questions:

1 What is a good model for the control systems?

2 What is the meaning of the complexity of the model and controller in control system design?

The first issue has attracted the interest of researchers in the area of control theory and system identification in the last decade and there are several interesting results. The author also has considered this problem with a more probabilistic approach in order to improve the conservatism of the ordinary deterministic robust control. The author also found that not only the complexity of the model but also that of the controller plays an important role in the first issue. The answer for the second issue is given from that result.

Normally there are two processes in control system design, modelling, or estimation of the system, and controller design. There is a long history and many results on the modelling/estimation problem and, in particular a universal criterion, e.g., the Kullback–Leibler information [12], or the likelihood, is usually used to evaluate an estimated model. The basic idea of this criterion is to treat every model, system parameter and data item equally, and this is, of course, reasonable in some sense. If the modelling/estimation gives perfect information about systems, then those processes and controller design will be divisible. However, of course, we only receive imperfect information and a model for the plant to be controlled cannot be evaluated without regard to the performance attained by the closed loop system, which is designed from that model, because the most important purpose of control is the performance of the controlled systems. This indivisibility of modelling/estimation and controller design has attracted much interest in the field of model estimation. Coupled with robust control, this has been discussed in the literature [6, 11, 8, 16, 4, 3, 27, 2]. On the other hand, the author has also considered this problem using a more probabilistic approach. The control design policy considered is to improve the probabilistic control performance, e.g., expectation of performance. In order to design a controller for that purpose we need the model of the distribution of systems, which should be estimated from the given data and be efficient for the design of the controller.

Alternatively, for the model selection problem itself, there is a well-known result in statistical estimation that we should select a simple model among the models at the same level as the estimator. This fact is explained by Akaike's information criterion (AIC) [1] or any other criteria [20, 10, 19, 18] with a keyword of *complexity of model*, and they reveal deep relations between the data and the model. Of course, their studies have been purely focused on the model estimation and selection problem. Therefore, when we consider model selection with respect to the control systems it is interesting to see the difference from that given by the normal model selection methods. Moreover, we also prefer simple controllers, as is the case with the model. There may be several reasons for this, but there has been insufficient discussion. The author has considered this problem and has shown that the complexity of the controller also has considerable influence on the model selection and the selection of the class of controllers. The results are given in [22, 26, 24, 25].

Finally, we define a type of a unified distribution on a metric space of analytic functions related to the above problems. We show that it does not necessarily give a natural distribution of systems and discuss the meanings of the results.

The results on the above discussions are explained in Section 2 – Section 4, and moreover, the result on the structure of the distribution of systems is shown in Section 5.

# 2 Preliminaries

First, we assume a plant $P$ contains some unknown elements to be considered. For example, assume that a feedback system is given by

$$\begin{aligned}\ddot{z}(t) &= xz(t) + a_2\dot{z}(t) + u(t) + w(t),\\ u(t) &= k_1 z(t) + k_2\dot{z}(t), \quad y(t) = c_1 z(t) + c_2\dot{z}(t), \end{aligned} \tag{2.1}$$

where $x$ is an unknown parameter. Hereafter denote the plant indexed by $x$ as $P_x$. In order to derive the information on $x$, system identification is a possibility. The estimated values may differ for each identification experiment due to observation noise or perturbation of the plant arising from changes in the system's environment. As a result a set of data from the estimation may be obtained as

$$\{x_1, x_2, \ldots, x_n\}, \tag{2.2}$$

where $n$ is the number of the experiments.

**Remark 2.1.** In the literature [26] similar results to this article are given in the case of system identification from I/O data of systems by employing a Bayesian model. □

The policy of ordinary robust control is to design a controller to improve the worst case performance for the range of the data (2.2). Such design policy is reasonable in some sense, however, we note that the result is, of course, generally conservative. Another possible design policy is to use the stochastic property of the data in order to reduce the conservatism. For example, define its control performance $\gamma$ as a regulation problem by

$$\gamma = \int_0^\infty y^2(t) + u^2(t)\,dt, \; w(t) = 0, z(0) = 0, \dot{z}(0) = 1, \tag{2.3}$$

then we obtain

$$\gamma(\phi = (k_1, k_2), x) = \frac{c_1^2 + k_1^2 - (x + k_1)c_2^2 - (x + k_1)k_2^2}{2(x + k_1)(a_2 + k_2)}. \tag{2.4}$$

When $x$ is a random variable, $\gamma$ is also so and a possible design policy is

$$\min_\phi \mathbf{E}_x\left(\gamma(\phi, x)\right). \tag{2.5}$$

In order to realize such a controller design, we need a stochastic property of $x$, such as distribution density $f(x)$, which should be estimated from the data (2.2). This process is a standard stochastic model estimation of $f(x)$ from the data of $x$ and there is a long history in this area and many results are known. The policy of the standard model estimation/selection problem is to regard all data equally, and simply select the model that best explains the distribution of $\{x\}$, and this discussion is independent of the design of the closed loop system, which is composed of $P_x$ and a controller $\phi = (k_1, k_2)$. Of course this is justified in the sense of pure model estimation. However, when we use the estimated model for the design of the closed loop systems, the problem of "which model is optimal" is not straightforward. Suppose a model $f_1(x)$ shows the best similarity with the distribution of $\{x_i\}$ in some region $X_1 \subset X$, whereas another model $f_2(x)$ shows a different similarity in another region $X_2 \subset X$. If the attained control performance $\gamma(\phi, x)$ with some controller $\phi$ is poor in $X_2$, and the purpose of the control is to compensate comparably poor performance, then the model $f_2(x)$ will be preferable and this selection may not coincide with the standard estimation model.

This situation means that if the final purpose of control is the improvement of the control performance, then it is reasonable to select a model with respect to that. In the standard model estimation/selection, a penalized maximum likelihood method [5, 7] is known to adjust some types of specified problems. However, controller design is complicated by the fact that the estimated model is used for designing the controller, and therefore, as a result, the penalty will change as the model itself changes.

A key idea to deal with this problem is to consider the model estimation not on the space of the system index $x$, but on the space of attained performance $\gamma$ of the closed loop system. From (2.4), we know that the controller $\phi$ can be considered as a mapping $S_\phi$ from $x$ on $X$ to $\gamma$ on $\Gamma$ as

$$S_\phi : X \to \Gamma.$$

Corresponding to this mapping, the probability density $f(x)$ on $X$ is transformed to that on a space of the performance index $\Gamma$ as

$$S_\phi : f(x) \to h(\gamma).$$

For the analysis in this discussion, several formulae on variable transformations on measurable spaces are necessary, and next we show the basic notation [13, 9]. Let $(\mathcal{A}, X, \mu)$ and $(\mathcal{B}, Y, \nu)$ be two measurable spaces, where $\mathcal{A}$ and $\mathcal{B}$ are $\sigma$-algebras composed of all the subsets of $X$ and $Y$, $\mu$ and $\nu$ are measures on $X$ and $Y$, and let $f(x)$ be a probability density on $(\mathcal{A}, X, \mu)$. Moreover, let $S : X \to Y$ be a transformation. Then, by employing the Radon–Nikodym theorem, we obtain the following proposition.

**Proposition 2.2 (e.g., [13]).** *For any non-singular transformation $S : X \to Y$ on measurable spaces $(\mathcal{A}, X, \mu)$, $(\mathcal{B}, Y, \nu)$, there exists a unique operator $P : \mathcal{L}^1 \to \mathcal{L}^1$ for all $B \in \mathcal{B}$, which satisfies*

$$\int_{B \in \mathcal{B}} Pf(y)\nu(dy) = \int_{S^{-1}(B)} f(x)\mu(dx). \tag{2.6}$$

**Definition 2.3.** Let $(\mathcal{A}, X, \mu)$, $(\mathcal{B}, Y, \nu)$ be measurable spaces. For a non-singular measurable transformation $S : X \to Y$, the unique operator $P : \mathcal{L}^1 \to \mathcal{L}^1$, which satisfies (2.6) is called a Frobenius–Perron operator. □

If the transformation $S$ is invertible, the analytical representation of the Frobenius–Perron operator is given as:

**Proposition 2.4 (e.g., [13]).** *For a non-singular invertible measurable transformation $S : X \to Y$ on measurable spaces $(\mathcal{A}, X, \mu)$, $(\mathcal{B}, Y, \nu)$,*

$$\begin{aligned}\int_{S^{-1}(B)} f(x)\mu(dx) &= \int_B f(S^{-1}(y))\mu S^{-1}(dy)\\ &= \int_B f(S^{-1}(y))J^{-1}(y)\nu(dy),\end{aligned} \tag{2.7}$$

*where $\mu S^{-1}$, $J^{-1}$ are defined by*

$$\mu S^{-1}(B) = \mu(S^{-1}(B)) = \int_B J^{-1}(y)\nu(dy),\ B \in \mathcal{B}.$$

*Therefore,*

$$Pf(y) = f(S^{-1}(y))J^{-1}(y).$$

*Moreover, when $S(y)^{-1}$ is differentiable, $J^{-1}(y)$ is given by*

$$J^{-1}(y) = \left|\det\left(\frac{\partial S^{-1}(y)}{\partial y^{\mathrm{T}}}\right)\right|.$$

If $S$ is not invertible, we can calculate $Pf(y)$ analytically or numerically with some modification of the above, but we omit the details here, see [25] for more information. In the remainder of this article measures such as $\mu$ or $\nu$ are considered to be the normal Lebesgue measures.

# 3 Simultaneous model estimation and controller design

In this section, we consider a simultaneous model estimation and controller design problem. Let $x \in \mathcal{R}^p$ be a random index vector of systems, $g(x)$ be the true multi-dimensional probability density of $x$, and assume it is unknown, $f_\theta(x)$ be a model of $g(x)$, which is parameterized by $\theta \in \mathcal{R}^{\dim\theta}$, $x^{(n)} := \{x_1, x_2, \ldots, x_n\}$ be a set of sampled data of $x$, and $\gamma$ be a control performance vector of a closed loop system.

For model estimation, the maximum likelihood method is essential. The log-likelihood is defined by

$$\frac{1}{n}\sum_{i=1}^{n}\log f_\theta(x_i), \tag{3.1}$$

and its maximizer $f_{\theta_o}$ can be considered a good estimator of $g$ in a sense of maximum likelihood. The quantity (3.1) is also an approximation of

$$\int g(x) \log f_\theta(x) dx, \tag{3.2}$$

by substitution of the corresponding empirical density $\hat{g}(x) := \sum_i \delta(x - x_i)$ for the true density $g(x)$. The maximization of (3.2) is equivalent to the minimization of the Kullback–Leibler information between $g$ and $f_\theta$ [12],

$$\int g(x) \left(\log g(x) - \log f_\theta(x)\right) dx. \tag{3.3}$$

Now we consider the consistency between the cost function for model estimation and controller design. A key idea is a variable transformation of $x$ as described in Section 2. In the usual case of model estimation of the distribution of $x^{(n)}$, the modelling of $f_\theta(x)$, a probability density of the system index $x$, is considered directly. However, following the discussion in Section 2, it is reasonable to estimate the distribution of the performance index $\gamma$ substituted for that of $x$ from the viewpoint of the control problem. Note that the control performance $\gamma$ is a function of a system index $x$ and a controller $\phi$, and when $\phi$ is fixed, $\gamma$ is given by a mapping

$$S_\phi(x) = \gamma \tag{3.4}$$

from $x$ to $\gamma$. In this sense, a controller $\phi$ is regarded as deciding a mapping from a system index $x$ to a performance index $\gamma$. Corresponding to the variable transformation (3.4), the probability density $f_\theta(x)$ or $g(x)$ is transformed to $P_\phi f_\theta(\gamma)$ or $P_\phi g(\gamma)$, respectively, where the Frobenius–Perron operator $P_\phi$ is defined by (3.4).

In the case of the Kullback–Leibler information, (3.3) is itself constant for the variable transformation because

$$\begin{aligned} &\int P_\phi g(\gamma) \log P_\phi g(\gamma) - P_\phi g(\gamma) \log P_\phi f_\theta(\gamma) d\gamma \\ &\quad = \int g(x) \log g(x) J^{-1}(S_\phi(x)) - g(x) \log f_\theta(x) J^{-1}(S_\phi(x)) dx \\ &\quad = \int g(x) \log g(x) - g(x) \log f_\theta(x) dx, \end{aligned}$$

and the transformation is meaningless in this case (the Kullback–Leibler information is defined originally to have such a property). However, the variable transformation of (3.2),

$$\int P_\phi g(\gamma) \log P_\phi f_\theta(\gamma) d\gamma, \tag{3.5}$$

has an interesting property: Note that (3.5) can be decomposed as

$$\int P_\phi g(\gamma)\log P_\phi f_\theta(\gamma)d\gamma = \int P_\phi g(\gamma)\log P_\phi g(\gamma)d\gamma - \int g(x)\log\frac{g(x)}{f_\theta(x)}dx. \tag{3.6}$$

The first term on the right-hand side of (3.6) is minus the entropy of the distribution of the control performance, and the second term is minus the normal Kullback–Leibler information between $g(x)$ and $f_\theta(x)$. Therefore, the maximization of (3.6) can be interpreted as a combination of restricting the dispersion of the resultant performance, and approximating $g(x)$ by $f_\theta(x)$. In particular, to restrict the dispersion of the performance means a robustness of the controlled system against parameter uncertainty and this is the purpose of robust control. For these reasons (3.5) is sufficiently meaningful.

Now consider the problem:

$$\max_\theta \sum_{i=1} L_{\phi,\theta}(\gamma_i), \tag{3.7}$$

$$L_{\phi,\theta}(\gamma^{(n)}) := \frac{1}{n}\sum_{i=1}^{n}\log P_\phi f_\theta(\gamma_i), \tag{3.8}$$

where $\gamma_i := S_\phi(x_i)$ and $\gamma^{(n)} := \{\gamma_1, \gamma_2, \ldots, \gamma_n\}$, which is an extension for the maximization of (3.1). We call this problem a "penalized maximum likelihood problem".

By employing the formulae for variable transformations introduced in Section 2, $L_{\phi,\theta}(\gamma^{(n)})$ is written as

$$\begin{aligned} L_{\phi,\theta}(\gamma^{(n)}) &= \frac{1}{n}\sum_{i=1}^{n}\log P_\phi f_\theta(\gamma_i) \\ &= \frac{1}{n}\sum_{i=1}^{n}\log f_\theta(x_i) + \frac{1}{n}\sum_{i=1}^{n}\log K_\phi(x_i), \qquad (3.9) \\ K_\phi(x) &:= J_\phi^{-1}\left(S_\phi(x)\right). \qquad (3.10) \end{aligned}$$

The equations (3.7) and (3.9) show that when $\phi$ depends on $\theta$, that is, the controller $\phi$ is designed with regard to the model of the probability distribution $f_\theta(x)$ of the system, the maximization of the penalized likelihood (3.7) is essentially different from that of (3.1). In the remainder of this subsection, we consider how $\phi$ is designed from $f_\theta$.

The paradigm of ordinary robust control is that the worst case performance of the closed loop system for a set of models is considered for improvement, and the optimal controller $\phi_{\mathrm{WC}}$ is given by

$$\phi_{\mathrm{WC}} := \arg\min_\phi \max_{x\in X} S_\phi(x). \tag{3.11}$$

Such a policy is reasonable for the absolute guarantee of stability or performance for control systems and for the situation where only the range of uncertainty is known. However in general, ordinary robust control gives conservative results. If some stochastic property of the uncertainty of the system is known and its information can be utilized in the design of the controller, then a less conservative performance of the resultant control system can be expected, in a probabilistic sense. One of the possible evaluations of the controller $\phi$ is the expectation of the attained performance as $\int \gamma P_\phi f_\theta(\gamma) d\gamma$, and the optimizer is given as

$$\phi(\theta) = \arg\min_{\phi} \int \gamma \, P_\phi f_\theta(\gamma) d\gamma. \tag{3.12}$$

A variety of combinations of (3.7) (or (3.5)) and (3.12) are considered, such as

$$\theta := \arg\max_{\theta} \int \log P_{\phi(\theta)} f_\theta(\gamma) P_{\phi(\theta)} \hat{g}(\gamma) d\gamma \tag{3.13}$$

$$\text{where} \quad \phi(\theta) := \arg\min_{\phi} \int \gamma P_\phi f_\theta(\gamma) d\gamma. \tag{3.14}$$

**Remark 3.1.** There are other possible cost functions for model estimation [14, 10], such as the Cramer–von Mises discrepancy or the Gauss discrepancy, which may be preferable with respect to controller design. The other cost functions for controller design and the combinations of these cost functions such as (3.13) and (3.14) are also possible. The following results in this article are also applicable for such a general situation, however, we omit these. See [24, 25] for the details. □

**Remark 3.2.** One of the important issues for the combined optimization problem such as (3.13) and (3.14) is the well-posedness of the problems, or the consistency of the estimated model and the true system. Those issues are considered in [24, 25]. In the remainder of this article, such properties are assumed to be satisfied. □

# 4 Estimation of cost functions

## 4.1 Model selection

In this section, we consider the estimation of the cost function (3.5) at $\theta = \hat{\theta}$ and consider the model selection problem. Solving (3.13) and (3.14), its optimizer $f_{\hat{\theta}}(x)$ is given among a class of models $\{f_\theta(x), \, \theta \in \Theta\}$. The model selection problem is to determine which class of model is best if we have several classes of model for the distribution of the data $\{x_i\}$, e.g., $f_\theta(x)$, $h_\psi(x)$, ... , and solve for the optimizers $f_{\hat{\theta}}(x)$, $h_{\hat{\psi}}(x)$, ... , respectively, by solving (3.13) and (3.14). The comparison of the values (3.5) at $f_\theta(x) = f_{\hat{\theta}}(x)$, $h_\psi(x) = h_{\hat{\psi}}(x)$, ... , that is,

$$\int P_\phi g(\gamma) \log P_\phi f_{\hat{\theta}}(\gamma) d\gamma, \int P_\phi g(\gamma) \log P_\phi h_{\hat{\psi}}(\gamma) d\gamma, \ldots \tag{4.1}$$

is reasonable for this model selection because (3.5) is the original cost function for the model estimation. Of course, these values also cannot be calculated because the true density, $g(x)$, is unknown and we estimate it by using the empirical data. In the case of the normal Kullback–Leibler information, Akaike's information criterion (AIC) [1], Takeuchi's information criterion (TIC) [20, 10], etc. are known to be the estimation of (3.2) when $f_{\hat{\theta}}$ is the maximum likelihood model. Following the procedure used for deriving these the estimation of (3.5) at $\theta = \hat{\theta}$ is given by:

**Theorem 4.1 ([22, 24, 25]).** *Assume that $\theta_o$ exists and $\hat{\theta}$ is its consistent estimation given by solving (3.7). Define the following:*

$$\begin{aligned}
\mathrm{IC}(\theta_1, \theta_2) &:= \frac{1}{n}\sum_{i=1}^{n} \log P_{\phi(\theta_1)} f_{\theta_1}(\gamma_i) - \frac{1}{n} b_1(\phi, f_{\theta_2}), \qquad (4.2)\\
b_1(\phi, f_\theta) &:= \mathrm{tr}\, J(\theta)^{-1} I(\theta),\\
J(\theta) &:= -\int \frac{\partial^2}{\partial\theta\partial\theta^{\mathrm{T}}} \log\left(f_\theta(x) K_{\phi(\theta)}(x)\right) g(x) dx,\\
I(\theta) &:= \int \frac{\partial}{\partial\theta} \log\left(f_\theta(x) K_{\phi(\theta)}(x)\right) \frac{\partial}{\partial\theta^{\mathrm{T}}} \log\left(f_\theta(x) K_{\phi(\theta)}(x)\right) g(x) dx;
\end{aligned}$$

*then, the following holds.*

$$\mathbf{E}_g\left[\mathrm{IC}(\hat{\theta}, \theta_o) - \int P_{\phi(\hat{\theta})} g(\gamma) \log P_{\phi(\hat{\theta})} f_{\hat{\theta}}(\gamma) d\gamma\right] = o\left(n^{-1}\right). \tag{4.3}$$

**Remark 4.2.** The result (4.3) shows that $\mathrm{IC}(\hat{\theta}, \theta_o)$ is a good estimation of the value of (3.5) at $f_\theta = f_{\hat{\theta}}$ when $n$ is sufficiently large and it can be used for the model selection with respect to the control performance.

However, $\theta_o$ and $g$ in $\mathrm{IC}(\hat{\theta}, \theta_o)$ are unknown and $\mathrm{IC}(\hat{\theta}, \hat{\theta})$ with the substitution $\hat{g}$ for $g$ is considered a possible estimation. Further approximation is also applicable, but we omit it here. □

**Remark 4.3.** In a similar manner to AIC and TIC, the bias term $\frac{1}{n}\mathrm{tr}\, J^{-1} I$ is approximately proportional to the dimension of $\theta$. That is, too complex a model should be avoided because it tends to be biased towards the given data and lacks generality. The difference from the normal AIC/TIC, except for the definition of the cost function, is that IC depends not only on $f_\theta(x)$, but also on the design method used for the controller $S_\phi(x)$. This suggests that when the model is selected based on IC, the selected model or its estimation value will change, depending on the class of controller and the design method. □

## 4.2 Controller design selection

The equation (3.12) is difficult to solve in a general sense because it contains the integral of the performance. One reasonable solution is the application of the Monte Carlo method; that is, at first we generate sampled points $\tilde{x}_i$ ($i = 1 \sim m$) by using the probability density $f_{\hat{\theta}}(x)$ and transform them to the performances $\{\tilde{\gamma}_i\}$. Then, the controller is selected as the optimizer of the sum of the sampled data as

$$\hat{\phi}(\hat{\theta}) = \arg\min_{\phi} \frac{1}{m} \sum_{i=1}^{m} S_\phi(\tilde{x}_i) =: \Phi(\hat{f}_{\hat{\theta}}), \tag{4.4}$$

where $\hat{f}_{\hat{\theta}}$ represents the empirical distribution density of the sampled data of $f_{\hat{\theta}}$ and $\phi \in \mathcal{R}^r$. Note that IC should be evaluated at $\phi_o = \Phi(f_{\hat{\theta}})$; however, in this case only $\hat{\phi} = \Phi(\hat{f}_{\hat{\theta}})$ can be practically calculated. This means that $\hat{\phi}$ is given from the data $\{x_i\}$ and $\{\tilde{x}_i\}$, and therefore it is a random variable, as is IC. From this viewpoint, we consider giving the expectation of IC at $\phi = \phi_o$ and derive the following.

**Theorem 4.4 ([22, 24, 25]).** *Suppose that $\Phi$ is second-order compact differentiable. Then*

$$\mathbf{E}_{f_{\hat{\theta}}} \left[ \frac{1}{n} \sum_{i=1}^{n} \log P_{\hat{\phi}} f_{\hat{\theta}}(\gamma_i) - \frac{1}{n} \sum_{i=1}^{n} \log P_{\phi_o} f_{\hat{\theta}}(\gamma_i) \right] = \frac{1}{m} b_2(\phi_o, f_{\hat{\theta}}) + o(m^{-1}), \tag{4.5}$$

$$\begin{aligned} b_2(\phi_o, f_{\hat{\theta}}) &:= C(\phi_o, \hat{\theta})^{\mathrm{T}} D(\phi_o, \hat{\theta}) + \operatorname{tr} E(\phi_o, \hat{\theta}) F(\phi_o, \hat{\theta}), \\ C(\phi, \theta) &:= \int \Phi^{(2)}(\gamma, \gamma; f_\theta) P_\phi f_\theta d\gamma, \quad D(\phi, \theta) := \frac{1}{2} \sum \frac{\partial}{\partial \phi} \log K_\phi(x_i), \\ E(\phi, \theta) &:= \int \Phi^{(1)}(\gamma; f_\theta) \Phi^{(1)}(\gamma; f_\theta)^{\mathrm{T}} P_\phi f_\theta d\gamma, \\ F(\phi, \theta) &:= \frac{1}{2} \sum \frac{\partial^2}{\partial \phi \partial \phi^{\mathrm{T}}} \log K_\phi(x_i), \end{aligned} \tag{4.6}$$

*where $\Phi^{(i)}$ denotes the i-th derivative of the functional $\Phi$.*

**Remark 4.5.** From (4.3) and (4.5), we can derive a criterion for the estimation/selection of the system model and also the selection of the controller design method as

$$\mathrm{ICC} := \frac{1}{n} \sum_{i=1}^{n} \log P_{\hat{\phi}} f_{\hat{\theta}}(\gamma_i) - \frac{1}{n} b_1(\phi_o, f_{\theta_o}) - \frac{1}{m} b_2(\phi_o, f_{\hat{\theta}}). \tag{4.7}$$

The formulae (4.2) and (4.7) show that ICC adds a new bias term to IC, which shows it is approximately proportional to the number of parameters of the controller. This suggests that when the structure of the controller is complex, the error between the original cost function

$$\int P_{\phi_o} f_{\hat{\theta}}(\gamma) \log P_{\phi_o} f_{\hat{\theta}}(\gamma) d\gamma$$

and the actual $\frac{1}{n}\sum_{i=1}^{n} \log P_{\hat{\phi}} f_{\hat{\theta}}(\gamma_i)$ becomes large, and it makes it difficult to find the optimal estimation parameter and controller. The new bias term is the compensation for this error and this criterion is used for the selection of the model and also a class of controller. In this sense, this result answers the above-mentioned question in Section 1 on the selection of a simple controller in practical design and implementation. The difference from the bias term $b_1$ of IC is that the bias term $b_2$ of ICC does not necessarily deteriorate. Therefore, a complex controller does not necessarily spoil the estimation of the discrepancy. □

# 5 Metric structure of a set of systems

In the previous sections, we assumed the probability distribution of the systems and considered the expectation of the performance of the control systems. In this paradigm we suppose implicitly a natural distribution of systems, such as appropriate dispersion in a system space. Such a preconception is reasonable if we restrict the class of systems to be finite dimensional parametrized systems and assume an appropriate distribution, such as the normal distribution or a uniform distribution on the space of their parameters. However, if we consider the distribution of systems with respect to a distance defined on their normed space, a question arises as to whether such a preconception on the distribution of systems is reasonable.

In this section, we deal with a normed space of analytic functions, which correspond to transfer functions of systems, and a set within the ball of size $M$ is considered. In order to investigate the metric structure of the ball, we employ the notion of $\epsilon$-entropy [15, 17, 28, 21], which was defined by Kolmogorov in the field of function approximation and later was introduced by Zames in control theory [29, 30]. In this section, we define a notion of dimension of a set of analytic functions by employing the $\epsilon$-covering, which is a fundamental notion for the definition of the $\epsilon$-entropy. Moreover, we discuss the meaning of the dimension in control theory with respect to the results in the former sections.

## 5.1 Metric dimension

First let $X$ be a metric space and $A$ a subset of $X$. A family $U_1, U_2, \ldots, U_n$ of subsets of $X$ is called an $\epsilon$-covering of $A$ if the diameter of each $U_k$ does not exceed $2\epsilon$ and if the sets $\cup_1^n U_k$ cover $A$, $A \subset \cup_1^n U_k$. For a given

$\epsilon > 0$, define the minimum number $n$ of sets $U_k$ that cover $A$ by

$$N_\epsilon(A) := \min n, \tag{5.1}$$

and its logarithm

$$H_\epsilon(A) = \log N_\epsilon(A), \tag{5.2}$$

which is called the $\epsilon$-entropy (or metric entropy) of the set $A$.

Now define a bounded subset of the $s$-dimensional complex space $G$, and $z = (z_1, z_2, \ldots, z_s) \in G$ a point in $G$. Let $A^s(G)$ be the set of all analytic functions in $G$ with real coefficients. We define $A = A^s(F, G; M)$ the subset of $A^s(G)$ where $f(z) \in A^s(F, G; M)$ satisfies $|f(z)| \leq M$ for $z \in G$ with the infinity norm on a compact subset $F \subset G$. Let $r = (r_1, \ldots, r_s)$, $k = (k_1, \ldots, k_s)$ ($r_j > 0$, $k_j = 0, 1, \ldots, j = 1, \ldots, s$) be the points in the $s$-dimensional Euclidean space, "$x > 0$" or "$x < r$" means $x_j > 0$ or $x_j < r_j$ ($j = 1, \ldots, s$), respectively, $|z| := (|z_1|, \ldots, |z_s|)$, $z^k = \prod_{j=1}^s z_j^{k_j}$, $z^k = \prod_{j=1}^s z_j^k$, and $(x, y) = \sum_{j=1}^s x_j y_j$.

Let $D$ be a polycylinder $|z| < r$, then the analytic functions $f(z)$ have Taylor expansions

$$f(z) = \sum_k a_k z^k = \sum_{k_1, \ldots, k_s = 0}^{\infty} a_{k_1 \ldots k_s} z_1^{k_1} \ldots z_s^{k_s}, \tag{5.3}$$

where $a_k$ is a real coefficient.

When $F := \{z; |z| \leq r\}$, $G := \{z; |z| < r'\}$, and $r_j < r_j'$ ($j = 1, \ldots, s$), for simplicity we denote $A^s(F, G; M)$ as $A^s(r, r'; M)$. Moreover, for convenience we define a number $q_j = \frac{r_j'}{r_j}$.

Vituskin gave the $\epsilon$-entropy of $A = A^s(r, r'; M)$ [28] and we can extend the result to the more general case [23]. For the extension, we define a set of functions;

$$F_n := \left\{ f_n(z) := \sum_{k \in T(n)} a_k z^k \right\}, \tag{5.4}$$

where $a_k \in \mathcal{R}$ and

$$T(n) := \{k, (h, k) = h_1 k_1 + \cdots + h_s k_s \leq n, h_j = \log q_j\}, \tag{5.5}$$

and denote

$$A^s(r, r'; M, n) := F_n \cap A^s(r, r'; M). \tag{5.6}$$

Here we obtain the following result by employing a similar process to the proof of [15].

**Proposition 5.1 ([23]).** *For the set* $A = A^s(r, r'; M, n)$,

$$H_\epsilon(A) = \frac{1}{(s+1)! \prod_1^s \log q_j} \log^s \frac{M}{\epsilon + E} \left( \log \frac{M}{\epsilon + E} - (s+1) \log \frac{\epsilon}{\epsilon + E} \right) + O\left( \log^s \frac{M}{\epsilon + E} \log\log \frac{M}{\epsilon + E} \right), \tag{5.7}$$

*for an appropriate real number* $E > 0$.

Proposition 5.1 gives the number of $\epsilon$-covering of a ball of given size $M$, therefore it represents a type of volume of the ball. In order to estimate the change of the volume of the ball for its size, we give the following corollary to this proposition.

**Corollary 5.2 ([23]).** *For the two sets* $A_M = A^1(r, r'; M, n)$ *and* $A_m = A^1(r, r'; m, n)$, $M \geq m$, *when* $\epsilon$ *and* $E$ *are sufficiently small, the following is satisfied:*

$$\frac{N_\epsilon(A_m)}{N_\epsilon(A_M)} \simeq \left( \frac{m}{M} \right)^{\frac{1}{2a} \log \frac{mM}{\epsilon 2}}, \tag{5.8}$$

*where* $a := \log q_1$.

From (5.8), when $\epsilon \to 0$,

$$\frac{N_\epsilon(A_m)}{N_\epsilon(A_M)} \to \left( \frac{m}{M} \right)^{\log \frac{1}{\epsilon^{1/a}}}, \tag{5.9}$$

and, moreover,

$$\left( \frac{m}{M} \right)^{\log \frac{1}{\epsilon^{1/a}}} \overset{\epsilon \to 0}{\longrightarrow} \begin{cases} 0 & \text{if } m < M, \\ 1 & \text{if } m = M. \end{cases} \tag{5.10}$$

Equation (5.10) shows that when $\epsilon$ tends to 0, then almost all the points of $\epsilon$-covering converge to the region that is apart from the origin by $M$. By analogy with the ball in Euclid space, we call

$$\log \frac{1}{\epsilon^{1/a}} \tag{5.11}$$

a *metric dimension* of the balls with respect to the $\epsilon$-covering.

## 5.2 Discussion

We discuss the meaning of Corollary 5.2 in relation to the results of the former sections. As described above, objective systems to be controlled are supposed to be distributed appropriately in a set of systems, however, Corollary 5.2 says that such a situation may not be normal. It is probable that the distribution of the real objective systems behaves with respect to a distance on the space of systems and, therefore, a notion of the distribution of systems derived from the $\epsilon$-covering is reasonable. In this sense a pseudo-uniform distribution is given as follows: first, consider a ball in the space of systems, and generate an $\epsilon$-covering for the objective set, and then, randomly sample the centres of each $\epsilon$-ball, which is a composition of the $\epsilon$-covering. However, in that case, almost all the sampling points are near the sphere of the objective ball of size $M$ when $\epsilon \to 0$ and therefore the gaps from the centre are unevenly distributed.

If the distribution of actual systems is not such a case, the systems are not distributed obeying the "uniform distribution" based on the $\epsilon$-covering and there may be another mechanism generating the systems. Unfortunately, the analysis on this point in control theory is underdeveloped, and more research activity with experimental data is desired in future.

# 6 Conclusion

In this article, we present some resent research results on the complexity of model and controller related to the first two questions imposed in Section 1. A unified model selection/estimation and controller design process is considered and the criteria for model selection and that of the class of controller introduced. These explain why too complex a controller, or model, should be avoided. Finally, the distribution of systems in the normed space is analysed and discussed.

There are many open problems, such as the relation between the complexity of model and controller and adaptive control, the computational complexity in solving the unified optimization problem or the distribution of actual systems discussed in Section 5. These are left for future work.

## References

[1] H. Akaike, "A new look at the statistical model identification," *IEEE Trans. Automatic Control*, vol. AC-19, no. 6, pp. 716–723, 1974.

[2] T. Asai, S. Hara, and T. Iwasaki, "Simultaneous modeling and synthesis for robust control by LFT scaling," in *Preprints of 13th IFAC World Congress*, 1996, pp. 309–314.

[3] J. Chen, G. Gu, and C. N. Nett, "Worst case identification of continuous time systems via interpolation," *Automatica*, vol. 30, no. 12, pp. 1825–1837, 1994.

[4] L. Giarre, B. Z. Kacewicz, and M. Milanese, "Model quality evaluation in set membership identification," *Automatica*, vol. 33, no. 6, pp. 1133–1139, 1997.

[5] I. J. Good and R. A. Gaskins, "Nonparametric roughness penalties for probability densities," *Biometrika*, vol. 58, no. 2, pp. 255–277, 1971.

[6] G. C. Goodwin, M. Gevers, and B. Ninness, "Quantifying the error in estimated transfer functions with application to model order selection," *IEEE Trans. Automatic Control*, vol. AC-37, pp. 913–928, 1992.

[7] P. J. Green, "Penalized likelihood for general semi-parametric regression models," *International Statistical Review*, vol. 55, no. 3, pp. 245–259, 1987.

[8] A. J. Helmicki, C. A. Jacobson, and C. N. Nett, "Control oriented system identification: A worst-case/deterministic approach in $H_\infty$," *IEEE Trans. Automatic Control*, vol. AC-36, no. 10, pp. 1163–1176, 1991.

[9] E. Hewitt and K. Stromberg, *Real and Abstract Analysis.* New York: Springer-Verlag, 1975.

[10] S. Konishi and G. Kitagawa, "Generalized information criteria in model selection," *Biometrika*, vol. 83, no. 4, no. 875–890, 1996.

[11] R. L. Kosut, M. K. Lau, and S. P. Boyd, "Set-membership identification of systems with parametric and nonparametric uncertainty," *IEEE Trans. Automatic Control*, vol. AC-37, no. 7, pp. 929–941, 1992.

[12] S. Kullback, *Information Theory and Statistics.* Mathematical Statistics, New York: John Wiley & Sons, Inc., 1959.

[13] A. Lasota and M. C. Mackey, *Chaos, Fractals, and Noise: Stochastic Aspects of Dynamics*, 2nd ed. Applied Mathematical Sciences, vol. 97, New York: Springer-Verlag, 1995.

[14] H. Linhart and W. Zucchini, *Model Selection.* New York: John Wiley & Sons, Inc., 1986.

[15] G. G. Lorentz, *Approximation of Functions.* New York: Chelsea Publishing Company, 1986.

[16] M. Milanese and A. Vicino, "Optimal estimation theory for dynamic systems with set membership uncertainty: An overview," *Automatica*, vol. 27, pp. 997–1009, 1991.

[17] A. Pinkus, *n-Widths in Approximation Theory.* Berlin: Springer-Verlag, 1985.

[18] J. Rissanen, "Universal coding, information, prediction and estimation," *IEEE Trans. Information Theory*, vol. IT-30, pp. 629–636, 1984.

[19] G. Schwarz, "Estimating the dimension of a model," *The Annals of Statistics*, vol. 6, no. 2, pp. 461–464, 1978.

[20] K. Takeuchi, "Distribution of information statistics and criteria for adequacy of models" (in Japanese), *Mathematical Sciences*, vol. 153, pp. 12–18, 1976.

[21] V. M. Tichomirov, "Diameters of sets in function spaces," *Uspehi Mat. Nauk, Russian Math Surveys*, vol. 15, pp. 75–111, 1960.

[22] K. Tsumura, "Information criterion for model selection with controller design," in *Proceedings of the 12th IFAC Symposium on System Identification*, Santa Barbara, 2000.

[23] K. Tsumura, "Metric structure of a set of analytic functions," in *Proceedings of the 30th SICE Symposium on Control Theory*, 2001.

[24] K. Tsumura, "Unification of modeling, estimation and controller design," in *Proceedings of the 40th Conference on Decision and Control*, 2001, pp. 1995–2000.

[25] K. Tsumura, "Unification of modeling, estimation and controller design," submitted for publication.

[26] K. Tsumura and H. Kimura, "Criterion for selection of model and controller design based on I/O data," in *Proceedings of the 39th Conference on Decision and Control*, 2000, pp. 2837–2842.

[27] K. Tsumura and S. Shin, "Simultaneous modeling and data-distribution-dependent robust control system design," in *Preprints of the 11th IFAC Symposium on System Identification*, Kitakyushu, 1997, pp. 129–134.

[28] A. G. Vitushkin, *Theory of the Transmission and Processing of Information.* New York: Pergamon, 1961.

[29] G. Zames, "On the metric complexity of causal linear systems: $\epsilon$-entropy and $\epsilon$-dimension for continuous time," *IEEE Trans. Automatic Control*, vol. AC-24, no. 2, pp. 222–230, 1979.

[30] G. Zames and J. G. Owen, "A note on metric dimension and feedback in distance time," *IEEE Trans. Automatic Control*, vol. AC-38, no. 4, pp. 664–667, 1993.

Koji Tsumura
Department of Information Physics and Computing
The University of Tokyo
Hongo 7-3-1, Bunkyo, Tokyo 113-8656, Japan
tsumura@crux.t.u-tokyo.ac.jp

# Off-Line Reference Shaping for Systems with State and Control Constraints

**Toshiharu Sugie**

*Dedicated to Professor Hidenori Kimura on the occasion of his 60th birthday*

ABSTRACT This paper shows a method to shape the given reference signals for closed loop systems so as not to violate the input and state constraints of the plants. In particular, improvement of the output response is taken into account explicitly subject to the given constraints, and the method is computationally efficient because of LMI formulae. The effectiveness of the proposed method is evaluated by simulation and experiment.

**Keywords:** constrained systems, saturation, servo systems, LMI, optimization

## 1 Introduction

In real plants, there always exist constraints on state and/or input such as actuator saturation and amplitude limitation of certain states. If we ignore these constraints, we may have wind-up phenomena or sometimes the control system may become unstable. Therefore, a lot of work has been done to cope with such constraints. One way to do this is to construct a feedback compensator which is robust against these constraints. Various types of anti-windup controllers (e.g., [1, 11, 9, 12]) fall into this category. Another way is to modify the reference signal in such a way that the given closed loop system does not violate the state/input constraints. This type of approach is sometimes called *reference governor* or *reference management* [10, 4, 5, 6, 2, 3, 7]. Modifying the reference signal is relatively simple, and it could be easily adopted to any given control systems. Therefore this paper focuses on this type of approach.

Unfortunately, as for the latter type approaches, most of the existing methods suffer from several drawbacks: That is, (i) the original output response can not be improved, (ii) the state measurement is required, and (iii) the constraints are satisfied often in a conservative way. These shortcomings are mainly due to the problem setting which allows arbitrary time varying reference signals and requires on-line computation of the reference.

However, in many practical cases, the reference signals are often given in advance. This implies that it is not necessary to modify the reference signals on-line. Therefore it is possible to overcome the defects mentioned above by simplifying the problem setting, and it is worth doing so.

Based on the above observation, Sugie and Yamamoto [14] have developed a method to shape the original reference signals in an off-line manner for constrained linear systems without the above drawbacks. The method improves the tracking property subject to the given state/input constraints, which has not been fully addressed so far, although it is very important in many practical cases. This work is followed by [8] and [13]. The purpose of this paper is to review the off-line shaping method and to demonstrate its effectiveness through simulation and experiment based on [14, 13].

# 2 Motivating example and system description

First, we show a motivating example. Then the system considered here is described.

## 2.1 Motivating example

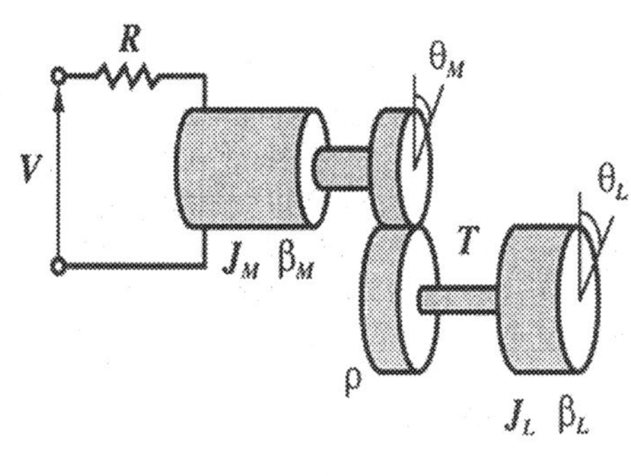


FIGURE 2.1. Positioning servo system

We consider the plant depicted in Figure 2.1, which was originally given by [3]. The plant consists of a DC servo motor, reduction gear and a load connected by a flexible shaft. In the figure, $\theta_M$ and $\theta_L$ denote the position of the motor and the load, respectively, and $V$ is the input voltage. The symbol $T$ denotes the torsional torque of the flexible shaft. The control purpose is to make the load position $\theta_L$ track the given reference signal $r$ without any steady-state error.

The plant is described by

$$\begin{aligned} \dot{x}_p &= A_p x_p + B_p u, \\ y &= [\,1 \;\; 0 \;\; 0 \;\; 0\,]\, x_p, \end{aligned}$$

where $x_p := [\theta_L \; \dot{\theta}_L \; \theta_M \; \dot{\theta}_M]^T$ , $u = V$, $y = \theta_L$ and $(A_p, B_p, C_p)$ are appropriate constant matrices.

The torque $T$ is given by

$$T = \begin{bmatrix} k_\theta & 0 & -\dfrac{k_\theta}{\rho} & 0 \end{bmatrix} x_p.$$

It is assumed that the plant has input saturation and the shaft torque $T$ is desired to be less than a certain number from the safety viewpoint, which are described by

$$|T| \leq T_{max}, \qquad |V| \leq V_{max},$$

where $T_{max} = 78.54$[Nm] and $V_{max} = 220$[V] in the simulation.

We discretize the plant by zero-order hold with the sampling period $T_s = 0.1$[s], and adopt an appropriate stabilizing controller of unity feedback type. The initial conditions of the plant and controller are zero. Now, suppose the desired output is given by $r_0 = 30$[deg].

In Figure 2.2 (a), (b) and (c), we show the closed loop system responses of $y(=\theta_L)$, $T$ and $V$ by broken lines when we inject the step reference $r(t) = r_0$. In Figure 2.2 (c), the broken line shows the input before saturation and the dotted broken lines show the limit of the constraints. Therefore, the actual input is saturated at the constraint boundary. The figure makes several points: (i) The output response is poor (partly because of input saturation). (ii) The torque and input constraints are not satisfied. However, if we use a new reference $r(t)$ shown in Figure 2.2 (d) instead of the step signal $r(t) = r_0$, we obtain the responses shown in Figure 2.2 (a), (b) and (c) by thick lines. Though the new reference trajectory looks strange at first glance, the output response is much better than the original one. Furthermore, the constraints on $T$ and $V$ are satisfied as shown in Figure 2.2 (b) and (c). This is surprising. How can we find such a nice reference? Is it possible to find it without a heavy computational burden while taking account of more general constraints? This is the main issue of the paper.

## 2.2 System description

Given the plant and its stabilizing controller, we consider the closed loop system $\Sigma_{cl}$ that they comprise. The system $\Sigma_{cl}$ is described by

$$\Sigma_{cl}: \quad \begin{aligned} x(t+1) &= A_{cl}x(t) + B_{cl}r(t), & (2.1) \\ y(t) &= C_{ycl}x(t) + D_{ycl}r(t), & (2.2) \\ z(t) &= C_{zcl}x(t) + D_{zcl}r(t) & (2.3) \end{aligned}$$

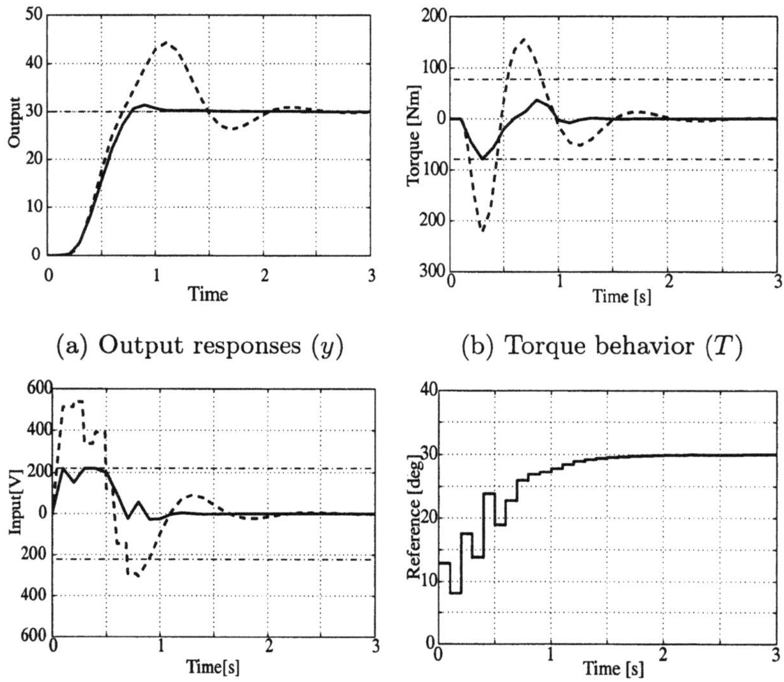


(a) Output responses ($y$) (b) Torque behavior ($T$)

(c) Input voltage ($V$) (d) New reference ($r$)

FIGURE 2.2. Step responses subject to input saturation

where $x(t) \in \mathbf{R}^n$ is the closed loop state, $y(t) \in \mathbf{R}^m$ is the output to be controlled, and $r(t) \in \mathbf{R}^m$ is the shaped reference to find. The vector $z(t) \in \mathbf{R}^\ell$ denotes the constraint variable.

We make several assumptions:

[A1] The initial condition is set to be zero.

[A2] If $r(t) = r_0$ (constant), then $y(t)$ converges to $r_0$ as $t \to \infty$.

[A3] The vector $z(t)$ should satisfy the following constraint;

$$z(t) \in \mathcal{Z} = \{z \mid |z_j| < 1 \ \ \forall j\}, \qquad \forall\, t \geq 0 \tag{2.4}$$

where $z_j(t)$ denotes the $j$-th element of $z(t)$.

The goal is to find a modified reference sequence $\{r(t)\}_{t=0,1,\ldots}$ which makes $y(t)$ track the given step reference $r_0$ as quickly as possible subject to the constraint (2.4).

# 3 Off-line reference shaping

In this section, we describe how to obtain the desired reference sequence which takes account of both the tracking property and the constraints.

### 3.1 Finite horizon case

First, we consider the following finite horizon optimization problem.

**Problem 3.1.** Given the step reference $r_0$ for the closed loop system $\Sigma_{cl}$, find the modified reference signal sequence $r(t)(t = 0, 1, \dots, N)$ which minimizes

$$J = J_1 + J_2, \tag{3.1}$$

$$J_1 := \sum_{t=0}^{N} \{\|r_0 - y(t)\|^2\}, \qquad J_2 := w \sum_{t=0}^{N} \{\|r_0 - r(t)\|^2\}, \tag{3.2}$$

subject to

$$z(t) \in \mathcal{Z}, \qquad t = 0, 1, \dots, N \tag{3.3}$$

where $w > 0$ is the weighting coefficient and $N$ corresponds to the terminal time, both of which are determined by the designer in advance. □

The important point here is that we try to find the sequence $r(t)(t = 0, 1, \dots, N)$ by taking account of the tracking property (which corresponds to minimization of $J_1$) explicitly, while most of the former research concentrates on minimization of $J_2$. This makes a great difference as will be shown later.

In what follows, we will show how to solve the above problem via LMIs. From (2.1) – (2.3), $z(t)$ and $y(t)$ are given by

$$z(t) = \sum_{j=0}^{t-1} C_{zcl} A_{cl}^{t-j-1} B_{cl} r(j) + D_{zcl} r(t), \tag{3.4}$$

$$y(t) = \sum_{j=0}^{t-1} C_{ycl} A_{cl}^{t-j-1} B_{cl} r(j) + D_{ycl} r(t). \tag{3.5}$$

Define the vectors $Z$, $Y$ and $R$ by

$$\begin{aligned} Z &= \left[z^T(0),\ z^T(1),\ \dots,\ z^T(N)\right]^T, \\ Y &= \left[y^T(0),\ y^T(1),\ \dots,\ y^T(N)\right]^T, \\ R &= \left[r^T(0),\ r^T(1),\ \dots,\ r^T(N)\right]^T. \end{aligned}$$

Then Equations (3.4), (3.5) can be rewritten as

$$\begin{aligned} Z &= M_z R, \\ Y &= M_y R \end{aligned}$$

where $M_z$ is the block lower triangular Toeplitz matrix whose $(i,j)$ blocks are given by

$$[M_z]_{ii} = D_{zcl}, \qquad (i = 1, \ldots, N+1),$$

$$[M_z]_{ij} = C_{zcl} A_{cl}^{j-i-1} B_{cl}, \qquad (N+1 \geq j > i \geq 1).$$

The matrix $M_y$ is similarly defined from (3.5).

Using these equations, we can represent the criterion $J$ by

$$\begin{aligned} J &= (R_0 - M_y R)^T (R_0 - M_y R) + (R_0 - R)^T W (R_0 - R) \\ &= R_0^T (I + W) R_0 - R^T (M_y^T R_0 + W R_0) \\ &\quad -(R_0^T M_y + R_0^T W) R + R^T (M_y^T M_y + W) R \end{aligned}$$

where $W = wI$ and $R_0 = [r_0^T, \ r_0^T, \ \ldots, \ r_0^T]^T$.

Let $\gamma$ be an upper bound of $J$, then we have the matrix inequality

$$\begin{aligned} \gamma \ - \ & R_0^T (I + W) R_0 + R^T (M_y^T R_0 + W R_0) \\ & + (R_0^T M_y + R_0^T W) R - R^T (M_y^T M_y + W) R > 0. \end{aligned}$$

Since $M_y^T M_y + W > 0$ holds, the above inequality is reduced to the following LMI via Schur Complement.

$$F_1(\gamma, R) := \begin{bmatrix} \Theta_1 & R \\ R^T & \gamma - \Theta_2 + 2\Theta_3 R \end{bmatrix} > 0 \qquad (3.6)$$

where $\Theta_i$'s are defined by $\Theta_1 = (M_y^T M_y + W)^{-1}$, $\Theta_2 = R_0^T (I + W) R_0$, and $\Theta_3 = (R_0^T M_y + R_0^T W)$.

As for the constraints, (3.3) is described by

$$-1 < (M_z R)_i < 1, \qquad \forall i \in [1, (N+1)\ell] \qquad (3.7)$$

from $Z = M_z R$, where $(M_z R)_i$ means the $i$-th element of $M_z R$. These are nothing but LMIs with respect to $R$.

Consequently, Problem 3.1 can be reduced to the following LMI optimization problem.

**Problem 3.2.** For the system $\Sigma_{cl}$ described by (2.1), (2.2) and (2.3), find an $R$ which minimizes $\gamma$ subject to (3.6) and (3.7). □

Once the solution $R (= [r^T(0), r^T(1), \ldots, r^T(N)]^T)$ is obtained, we adopt the modified reference sequence $r(t)$ $(t = 0, \ldots, N)$ to the closed loop. It is well known that LMI problems are computationally efficient to solve.

### 3.2 Infinite horizon case

It is easy to extend the above result to the infinite horizon case. We will show how to find the sequence $r(t)$ which minimizes

$$J' = \sum_{t=0}^{\infty} \{\|r_0 - y(t)\|^2 + w\|r_0 - r(t)\|^2\} \tag{3.8}$$

subject to the constraint (2.4). Here we assume

$$r(t) = r_0, \qquad (t \geq N+1) \tag{3.9}$$

where $N$ is the parameter chosen by the designer in advance. From this assumption, we have

$$J' = J_1 + J_2 + J_3, \qquad J_3 = \sum_{t=N+1}^{\infty} \{\|r_0 - y(t)\|^2\}$$

where $J_1$ and $J_2$ are given by (3.2). Since we have already discussed $J_1$ and $J_2$ in the previous subsection, we focus on $J_3$ here.

Let $x_{ss}$ and $z_{ss}$ be the steady state values of $\Sigma_{cl}$ for the step reference $r(t) = r_0$, and define the error vector by

$$e(t) = x(t) - x_{ss}.$$

Then noting that

$$e(N+k) = A_{cl}^{k-1} e(N+1) \qquad k \geq 1$$

holds, we have

$$\begin{aligned} J_3 &= \sum_{k=1}^{\infty} \|C_{ycl} e(N+k)\|^2 \\ &= e(N+1)^T V e(N+1) \end{aligned}$$

where $V$ is the observability grammian defined by

$$V = \sum_{k=1}^{\infty} (A_{cl}^T)^{k-1} C_{ycl}^T C_{ycl} A_{cl}^{k-1}.$$

It is well known that $V$ is computed by solving

$$V - A_{cl}^T V A_{cl} = C_{ycl}^T C_{ycl}.$$

As for $e(N+1)$, we have

$$\begin{aligned} e(N+1) &= x(N+1) - x_{ss} \\ &= GR - x_{ss} \end{aligned}$$

where $G$ is given by

$$G = [\ A_{cl}^N B_{cl},\ A_{cl}^{N-1} B_{cl},\ \ldots, A_{cl} B_{cl}, B_{cl}\ ].$$

Therefore, we obtain

$$J_3 = (GR - x_{ss})^T V (GR - x_{ss}).$$

Now, letting $\gamma_3$ be an upper bound of $J_3$, we have

$$\begin{bmatrix} -\gamma_3 & (GR - x_{ss})^T V \\ V(GR - x_{ss}) & -V \end{bmatrix} < 0. \tag{3.10}$$

Concerning the constraints, we get

$$\begin{aligned} z(N+k) &= C_{zcl} e(N+k) + z_{ss} \\ &= C_{zcl} A_{cl}^{k-1}(GR - x_{ss}) + z_{ss}, \qquad k \geq 1. \end{aligned}$$

Therefore, the constraint

$$-1 < z(N+k) < 1, \qquad \forall k \geq 1 \tag{3.11}$$

is an LMI with respect to $R$.

To summarize, the infinite horizon optimization problem can be cast as the following problem:

**Problem 3.3.** For the system $\Sigma_{cl}$ given by (2.1) – (2.3), find an $R$ which minimizes $\gamma + \gamma_3$ subject to (3.6), (3.7), (3.10), and (3.11). □

The above problem is computationally tractable except (3.11). One practical way to avoid the difficulty is as follows: (i) Initialize the counter; $k_0 = 1$. (ii) Solve the above problem to obtain $R$ by ignoring (3.11) for $k \geq k_0$. (iii) If the obtained $R$ satisfies (3.11), then $R$ is a solution. (iv) Otherwise, replace $k_0$ by $k_0 + 1$, and go back to (ii).

This procedure is guaranteed to stop in a finite number of steps.

## 3.3 Remarks on the proposed method

Several remarks may be in order.

### *3.3.1 Extension to other constraints*

Though we have discussed only the constraint given by (2.4) so far, it is obvious that any constraint that is described by LMIs can be treated. For example, as for the output, we can take account of overshoot/undershoot constraint. Also, as for the input, we can deal with rate constraint as well as amplitude constraint.

### 3.3.2 Comparison with reference governors

We compare our method with the exiting reference governor type methods [4, 5, 6, 3]. The essential difference lies how the reference signal is modified. Typically, the reference governor is looking for the modified signal $r(t)$ subject to

$$r(t) = r(t-1) + \beta(t)(r_0(t) - r(t-1)). \tag{3.12}$$

Here, $\beta$ is the parameter to be determined within $(0 \leq \beta \leq 1)$, and $r_0(t)$ is the original (possibly time-varying) reference signal. The largest $\beta$ is sought subject to the state/input constraints. The basic idea is to find $r(t)$ which is as close to $r_0(t)$ as possible. In terms of our criterion $J$ given in (3.1), they consider $J_2$ only.

Basically, it is much more reasonable to adopt $J_1$ rather than $J_2$ in the cost function from the tracking viewpoint. The following simulation results demonstrate this point clearly. We compare our method with the original one [3].

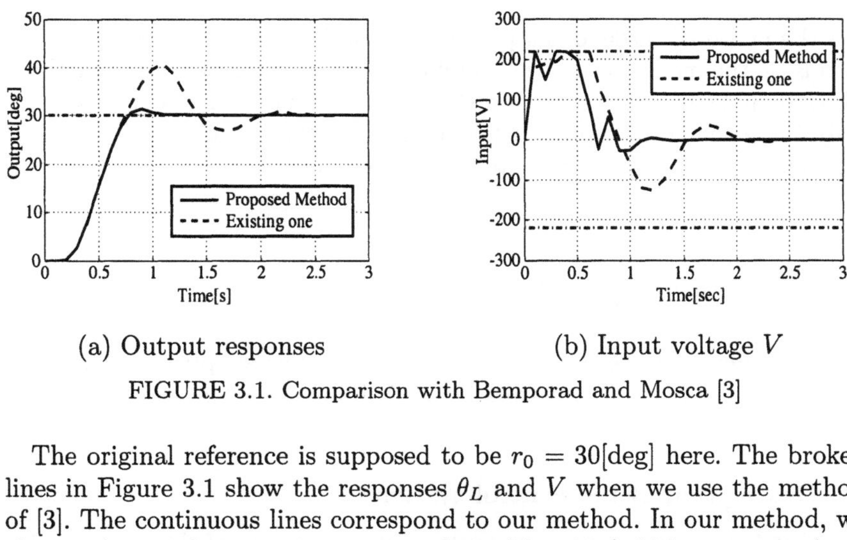


(a) Output responses (b) Input voltage $V$

FIGURE 3.1. Comparison with Bemporad and Mosca [3]

The original reference is supposed to be $r_0 = 30$[deg] here. The broken lines in Figure 3.1 show the responses $\theta_L$ and $V$ when we use the method of [3]. The continuous lines correspond to our method. In our method, we choose the weighting as $w = 0.1$, and set $N = 30$ (which corresponds to 3[s] in real time). The modified reference signal generated by our method is already shown in Figure 2.2 (d) in Section 2. Both methods succeed in satisfying the constraints, and they use the input aggressively (i.e., in a less conservative way). The constraint on $T$ is also satisfied though we omit the figure here. As for the output response, the former method gives some overshoot, while ours does not. This benefit comes from the criterion which takes account of tracking error (i.e., $J_1$).

In addition, if the given reference signal does not violate the state/input constraint, the reference governor does not work. Therefore, if the original control system has poor tracking property (e.g., big overshoot or slow response), there is no way to improve it in this case. On the contrary, our method does improve it, which will be shown later.

On the other hand, we assume that the reference signal is given in advance, which is restrictive when compared to the existing methods. This is the price to pay. However, at this cheap cost, we are able to improve the output responses via simple and computationally efficient LMI optimization.

#### *3.3.3 Role of $J_2$*

From the arguments in Section 3.3.2, one may think that the term $J_2$ is not necessary at all. However, this is not true. Through various numerical examples, we have experienced that we often get very badly behaved reference signal $r(t)$ if we ignore the term $J_2$. An example to show this point is reported in [14].

#### *3.3.4 Robustness against the modeling error*

Since the proposed method is a pure feedforward control, the performance depends on the model accuracy. Therefore, it is important to robustify the off-line shaping method in the presence of modeling errors. One way to satisfy the given state/input constraints against the model uncertainty is developed by [13], though the detail is omitted here.

# 4 Experimental evaluation

This section evaluates the effectiveness of the off-line shaping method by experiment. The method employed here is a robustified version [13].

## 4.1 Description of the control system

The plant is the two-mass-spring system which consists of the motor and the disc connected by a flexible joint. The plant model in continuous-time is described by

$$\begin{aligned} \dot{x}(t) &= Ax(t) + Bu(t), && (4.1) \\ y(t) &= Cx(t), && (4.2) \end{aligned}$$

$$x(t) = \left[\theta_d,\ \dot{\theta}_d,\ \theta_m,\ \dot{\theta}_m\right]^T, \quad u(t) = \tau, \quad y(t) = \theta_d,$$

$$
\begin{aligned}
A &= \begin{bmatrix} 0 & 1 & 0 & 0 \\ -k/j_d & -d_d/j_d & k/j_d & 0 \\ 0 & 0 & 0 & 1 \\ k/j_m & 0 & -k/j_m & -d_m/j_m \end{bmatrix}, \\
B &= [0,\ 0,\ 0,\ 1/j_m]^T, C = [1,\ 0,\ 0,\ 0]
\end{aligned}
$$

where $\theta_d$[rad] and $\theta_m$[rad] denote the rotation angles, and $\tau$ [Nm] is the input torque. The subscripts $d$ and $m$ mean the disc and motor, respectively. The plant parameters are given as follows. The moments of inertia are given by $j_d = 2.76 \times 10^{-3}$ [Nms$^2$/rad] and $j_m = 2.50 \times 10^{-3}$ [Nms$^2$/rad]. The viscous coefficients are given by $d_d = 2.76 \times 10^{-3}$[Nms/rad] and $d_m = 1.83 \times 10^{-1}$[Nms/rad]. The spring constant of the flexible joint is $k = 6.16$.

Then, we obtain the discrete-time plant model

$$
\begin{aligned}
x(t+1) &= A_p x(t) + B_p u(t), \\
y(t+1) &= C_p x(t)
\end{aligned}
$$

by sampling the continuous-time system (4.1),(4.2) with the sampling time $T_s = 0.001$[s] and zero-order hold. The input torque $\tau$ has the saturation constraint given by

$$
-5.5 \leq \tau \leq 5.5.
$$

Then, for this system, we construct the servo system by state feedback with an integrator.

$$
\begin{aligned}
u(t) &= Fx(t) + G\sum_{k=0}^{t}(r(k) - y(k)), \\
F &= \left[-5.74 \times 10^1 \quad -4.85 \times 10^{-1} \quad 3.96 \quad -1.32\right], \\
G &= 7.07 \times 10^{-1}.
\end{aligned}
$$

These parameters are chosen based on the LQ optimal control.

We generate the modified reference signal $r(t)$ for $r_0 = 90$[deg] and $r_0 = 60$[deg]. From the step response by simulation, we choose $N = 150$ (which corresponds to $t = 0.15$ [s]).

## 4.2 Experimental results and discussions

The experimental results of $r_0 = 60$[deg] are shown in Figure 4.1 (a) (b). Figure 4.1 (a) shows the plant output, where the lines and broken lines correspond to our method and the step reference (i.e., $r(t) = r_0$) cases, respectively. The modified reference is shown in Figure 4.1 (b). Though the inputs are not shown here due to space limitation, the inputs stay inside the unsaturated area ($|u| < 5.5$) in both cases. Compared to the

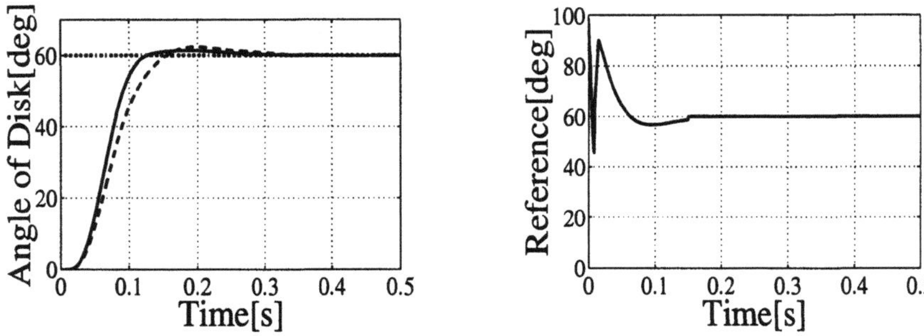


(a) Plant output $y$ (b) Modified reference $r$

FIGURE 4.1. Experimental results ($r_0 = 60$)

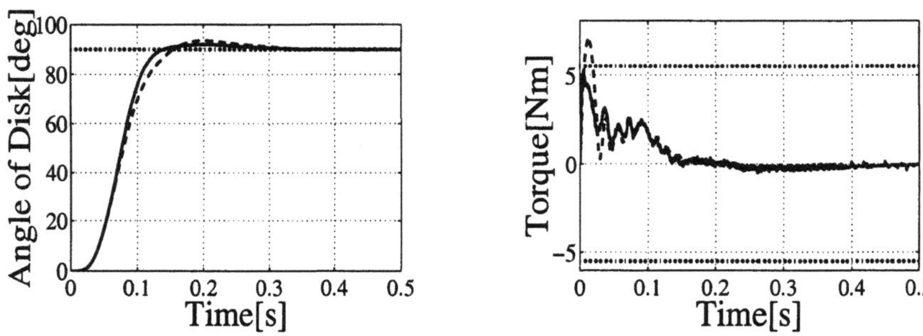


(a) Plant output $y$ (b) Input torque $u$

FIGURE 4.2. Experimental results ($r_0 = 90$)

step reference, our modified reference signal improves the response. The improvement has never been achieved by the existing reference governors.

We also performed the experiment with $r_0 = 90$ [deg]. The results are shown in Figure 4.2 (a) (b). When we use the step reference $r(t) = r_0$, the input $u$ goes beyond the saturation bound as shown in Figure 4.2 (b) by the broken line. On the other hand, by applying the modified reference, the input stays within the unsaturated area all the time, as shown by the thick line. Nevertheless, it improves the output a little bit as shown in Figure 4.2 (a), where the thick line corresponds to our modified reference and the broken one exhibits the step response.

These experimental results demonstrate the validity of our method.

# 5 Conclusion

In this paper, we have shown a method to shape the original step reference signals for closed loop systems having state/input constraints in such a way that the plant output tracks the reference quickly without violating the state/input constraints. In particular, improvement of the output response is taken into account *explicitly* subject to the given constraints. Therefore, contrary to the existing reference governors, the output responses can be improved even when the constraints are already satisfied for the original reference signals. Furthermore, since the proposed method utilizes the LMI optimization, it is computationally efficient. The effectiveness of the proposed method has been demonstrated by simulation and experiments.

There are many future topics to be investigated; for instance, (i) to develop a reference shaping method which is robust against modeling errors (see Section 3.3.4), (ii) to shape the periodic references, and (iii) to extend the result to sampled-data systems, and so on. Concerning (i) and (ii), we already have partial answers which will be reported elsewhere. Since the off-line reference shaping could be a powerful tool for control systems with constraints, it is interesting to challenge one of these problems.

# Acknowledgment

This paper is based on the joint work with my students, Mr. Hiroyuki Yamamoto and Mr. Takayuki Fukui. The author thanks them for their valuable help.

Finally the author would like to express his sincere gratitude to Professor Hidenori Kimura. He has been a charismatic leader who shows us the way to go and tells us what is important to do essential work. Without his strong leadership and his attractive personality, more than a half of what I did so far would have been impossible to achieve.

## References

[1] K. J. Åström and L. Runqwist, "Integrator windup and how to avoid it," in *Proc. American Control Conf.*, 1989, pp. 1693–1698.

[2] A. Bemporad, A. Casavola, and E. Mosca, "Nonlinear control of constrained linear systems via predictive reference management," *IEEE Trans. on Automatic Control*, vol. 42, no. 3, pp. 340–349, 1997.

[3] A. Bemporad and E. Mosca, "Fulfilling hard constraints in uncertain linear systems by reference managing," *Automatica*, vol. 34, no. 4, pp. 451–461, 1998.

[4] E. G. Gilbert and K. T. Tan, "Linear systems with state and control constraints: The theory and application of maximal output admissible sets," *IEEE Trans. on Automatic Control*, vol. 36, pp. 1008–1019, 1991.

[5] E. G. Gilbert and K. T. Tan, "Nonlinear control of discrete-time linear systems with state and control constraints: A reference governor with global convergence properties," in *Proc. IEEE Conf. on Decision and Control*, 1994, pp. 144–149.

[6] E. G. Gilbert and K. T. Tan, "Discrete-time reference governors and the nonlinear control of systems with state and control constraints," *Int. Journal of Robust and Nonlinear Control*, vol. 5, pp. 487–504, 1995.

[7] K. Hirata and M. Fujita, "Set of admissible reference signals and control of systems with state and control constraints," in *Proc. IEEE Conf. on Decision and Control*, 1999, pp. 1427–1432.

[8] K. Hirata and K. Kogiso, "A performance improving off-line reference management for systems with state and control constraints," in *Proc. 40th IEEE Conf. on Decision and Control*, Orlando, 2001, pp. 4645–4650.

[9] P. Kapasouris and M. Athans, "Control systems with rate and magnitude saturation for neutrally open loop systems," in *Proc. IEEE Conf. Decision and Control*, 1990, pp. 3404–3409.

[10] P. Kapasouris, M. Athans, and G. Stein, "Design of feedback control systems for stable plants with saturating actions," in *Proc. IEEE Conf. Decision and Control*, 1988, pp. 469–479.

[11] M. V. Kothare, P. J. Campo, M. Morari, and C. N. Nett, "A unified framework for the study of anti-windup designs," *Automatica*, vol. 30, no. 12, pp. 1869–1883, 1994.

[12] M. V. Kothare and M. Morari, "Multiplier theory for stability analysis of anti-windup control systems," *Automatica*, vol. 35, no. 5, pp. 917–928, 1999.

[13] T. Sugie and T. Fukui, "Robust reference management for constrained systems and its experimental evaluation," submitted.

[14] T. Sugie and H. Yamamoto, "Reference management for closed loop systems with state and control constraints," in *Proc. American Control Conf.*, Arlington, 2001, pp. 1426–1431.

Toshiharu Sugie
Department of Systems Science
Kyoto University
Uji, Kyoto 611-0011, Japan
sugie@i.kyoto-u.ac.jp

# Finite Frequency Characterization of Easily Controllable Plant toward Structure/Control Design Integration

**Shinji Hara and Tetsuya Iwasaki**

*Dedicated to Professor Hidenori Kimura on the occasion of his 60th birthday*

ABSTRACT This paper summarizes the authors' recent results on finite frequency characterization of easily controllable plants under control effort constraint, the aim being the development of a new approach for plant/control design integration. We first show by a motivating example that the closed-loop bandwidth achievable with a reasonable control effort is closely related to the frequency range for which the plant is high-gain and exhibits positive-realness. We then present an LMI characterization of the finite frequency Kalman–Yakubovich–Popov (KYP) lemma and derive several related conditions. Finally, the conditions for the finite frequency positive-real (FFPR) and the finite frequency high-gain (FFHG) properties are shown.

**Keywords:** KYP lemma, Structure/control design integration, Finite frequency property, Passivity, LMI

## 1 Introduction

The control theory aiming at designing a best (optimal or robust) controller for a given plant or a specified set of plants to be controlled has progressed to a high degree of control performance in practical applications. Professor Kimura has made many fundamental contributions in this field and is recognized as one of the pioneer researchers in robust control. His thought was always deep and he has continued to seek the essentials of controlled dynamical systems in order to reduce the gap between control theory and its applications in the real world. His research style has encouraged young Japanese researchers, including the authors, to do high level theoretical work in control.

This paper is not about robust control but is about plant/control design integration. However, the flavor of the research is similar to Kimura's. We believe that the theoretical foundation for characterizing easily controllable plants in practice is essential to develop a new systematic way of design integration. This paper is concerned with finite frequency characterization of easily controllable plants under control effort constraint, aiming at the development of a new approach for integrated design of controlled mechanical systems.

Given a set of performance specifications, the mechanical control system design mainly consists of the following two steps: the mechanical structure design and the controller design. Conventionally, these two steps are followed *sequentially*; the structure is first designed to meet requirements involving stiffness, strength, weight, etc., which are not directly related to the closed-loop (i.e., controlled) dynamic performance, and a controller is then designed for a *given* mechanical system. Thus, the conventional method does not provide optimal solutions in general, and hence there is room for sensible design strategies to improve the overall performance.

There are many research results along this direction. Recently, simultaneous design methods have been proposed [12, 16, 5, 4, 14] via numerical optimization of the parameters of both structure and controller. Although these methods may improve the performance of a given mechanical control system used as the initial condition of the optimization algorithm, the resulting dynamical system depends heavily on the initial design and can be far from the global optimum due to the non-convexity of the problem. To have a sensible initial design, we have to go back to the conventional sequential design. Thus we need a fundamental design principle that essentially eliminates the limitations of the two-step design.

Also note that numerical optimizations do not give any information on the relation between the control performance and the structure. Thus, such methods do not indicate the crucial properties of the structure that lead to good controlled performance.

The objective of this paper is to summarize the authors' new approach to structure/control design integration, which is still a two-step design but is essentially different from conventional methods in that the controlled performance is explicitly taken into account in the mechanical design. Our approach is to identify and characterize properties of mechanical systems that are critical to achieving good controlled dynamic performance when an appropriate feedback loop is closed. Once we design a mechanical system with such properties, standard optimal control methods can be applied to complete the whole design process to yield controlled mechanical systems with near optimal performance.

It is widely accepted as a fact in the mechanical design community that a good controlled performance can be expected if a mechanical system is designed such that the flexible modes are "in-phase" with respect to the rigid-body mode. There is not much theoretical justification for this claim but empirical evidence is rather convincing [13]. Our research is initially motivated by this claim, and we have investigated implications of the in-phase property in terms of the language of the control community. Our aim here is to summarize the authors' recent results on characterizations of easily controllable plants or mechanical systems under control effort constraint, leading toward a new approach for integrated design.

In Section 2, we investigate the limits of a servo tracking performance for a typical lightly damped flexible mechanical system. The investigation shows that the closed-loop bandwidth achievable with a reasonable control effort is closely related to the frequency range for which the plant is high-gain and exhibits positive-realness. In Section 3, we first present an LMI characterization of the finite frequency Kalman–Yakubovich–Popov (KYP) lemma and derive several related conditions. The conditions for the finite frequency positive-real (FFPR) and the finite frequency high-gain (FFHG) properties are discussed in Sections 4 and 5, respectively. Section 6 concludes the paper.

We use the following notation. For a matrix $A$, $A^{\mathsf{T}}$, $A^*$, $\overline{\sigma}(A)$ and $\underline{\sigma}(A)$ denote its transpose, complex conjugate transpose, maximum singular value and minimum singular value, respectively. For a symmetric or Hermitian matrix, $A > (\geq)0$ and $A < (\leq)0$ denote positive (semi-)definiteness and negative (semi-)definiteness.

## 2 Motivating example

We here consider a simple servo-tracking control design problem and answer the following basic question: What property of the plant to be controlled limits the closed-loop performance, in particular, the control bandwidth?

We use the control system setup depicted in Figure 2.1 where $K(s)$ is the controller, $r$ is the unit step command input, $y$ is the plant output which we force to follow the step command, $e$ is the tracking error, and $u$ is the control input. The class of target plants is lightly damped flexible systems expressed as

$$P(s) = \frac{1}{s^2} + \frac{k_1}{s^2 + 2\zeta_1\omega_1 s + \omega_1^2} + \frac{k_2}{s^2 + 2\zeta_2\omega_2 s + \omega_2^2}$$

where $(\zeta_1, \omega_1) = (0.001, 10), (\zeta_2, \omega_2) = (0.001, 100)$. For fixed values of $k_1$ and $k_2$, define the optimal $\mathcal{H}_2$ tracking performance with control level $\gamma_u$ by

$$\gamma_e(k_1, k_2, \gamma_u) := \min_{K(s)} \sqrt{\int_0^\infty e(t)^2 dt},$$

$$\text{subject to} \quad \sqrt{\int_0^\infty u(t)^2 dt} < \gamma_u.$$

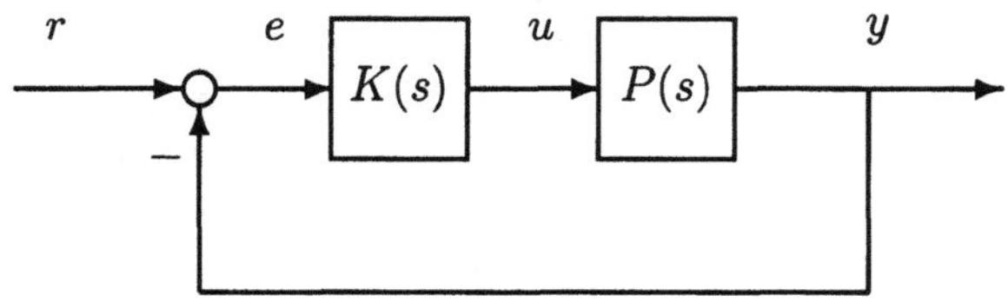


FIGURE 2.1. Servo tracking control system setup

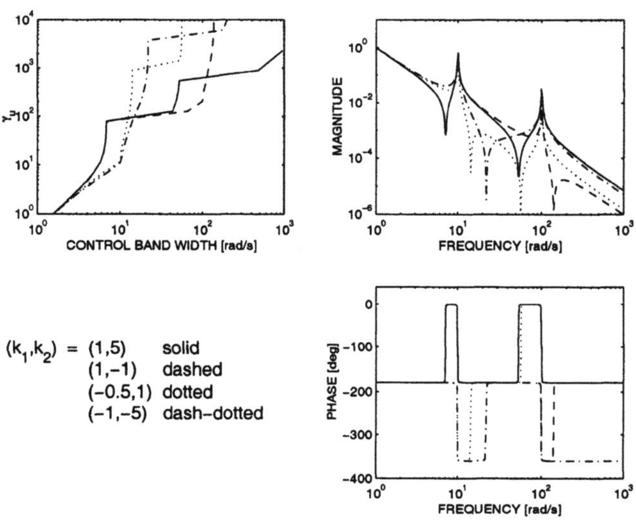


FIGURE 2.2. What limits the control bandwidth?

Figure 2.2 (upper left) shows, for the four cases of the plant parameters, the relations between $\gamma_u$ and the control bandwidth of the optimal closed-loop system where the cross-over frequency of the open-loop transfer function $K(s)P(s)$ is used as an estimate of the control bandwidth. Shown to the right is the frequency response of each plant. From the figure, we see that:

- We have to exert lots of control effort ($\gamma_u$) to make the control bandwidth go over the frequencies at which the phase or the gain of the plant drops, and the phase drop effect is dominant.

- The amount of the required control effort is large if the phase shifts to $-360$ [deg] instead of $-180$ [deg].

- If the allowable control effort is neither too small nor too large, then the control bandwidth coincides roughly with the smallest frequency at which the phase of the open-loop transfer function goes below −180 [deg].

From these observations, we conclude that we should design the mechanical system such that its phase does not go below −180 [deg] and its gain is larger than a certain value within the desired control bandwidth. In other words, for $G(s) := sP(s)$, we should require the following finite frequency positive-real (FFPR) property:

$$G(j\omega) + G(j\omega)^* \geq 0, \quad {}^\forall\omega \text{ such that } |\omega| \leq \varpi, \tag{2.1}$$

where $\varpi$ is the desired control bandwidth. For $P(s)$ itself, the following finite frequency high-gain (FFHG) property

$$\underline{\sigma}(P(j\omega)) \geq \gamma, \quad {}^\forall\omega \text{ such that } |\omega| \leq \varpi \tag{2.2}$$

is required.

# 3 Finite frequency KYP lemma

In this section, we first give an LMI characterization of a general finite frequency condition, or the finite frequency Kalman–Yakubovich–Popov lemma, and then specialize it to obtain FFPR and FFHG conditions.

**Theorem 3.1.** *Let complex matrices $A \in \mathbb{C}^{n\times n}$, $B \in \mathbb{C}^{n\times m}$, a Hermitian matrix $\Theta \in \mathbb{C}^{(n+m)\times(n+m)}$, and real scalars $\omega_1 \leq \omega_2$ be given. Suppose $(A, B)$ is a controllable pair, i.e.,*

$$\operatorname{rank}\begin{bmatrix} A - \lambda I_n & B \end{bmatrix} = n, \quad \forall\,\lambda \in \mathbb{C}$$

*holds. Then the following statements are equivalent.*

*(i) With $\boldsymbol{\Omega} := \{\, \omega \in \mathbb{R} : \det(j\omega I - A) \neq 0,\ \omega_1 \leq \omega \leq \omega_2 \,\}$, we have*

$$\begin{bmatrix} (j\omega I - A)^{-1}B \\ I_m \end{bmatrix}^* \Theta \begin{bmatrix} (j\omega I - A)^{-1}B \\ I_m \end{bmatrix} \leq 0, \quad \forall\omega \in \boldsymbol{\Omega}. \tag{3.1}$$

*(ii) There exist $n \times n$ Hermitian matrices $P$ and $Q \geq 0$ satisfying*

$$\begin{bmatrix} A & B \\ I_n & 0 \end{bmatrix}^* \begin{bmatrix} -Q & P + j\omega_c Q \\ P - j\omega_c Q & -\omega_1\omega_2 Q \end{bmatrix} \begin{bmatrix} A & B \\ I_n & 0 \end{bmatrix} + \Theta \leq 0, \tag{3.2}$$

*where $\omega_c = (\omega_1 + \omega_2)/2$.*

*(See [9] for the proof.)*

LMI characterizations of the FFPR and FFBR (Finite Frequency Bounded Real) conditions can be derived by setting

$$\Theta = -\begin{bmatrix} 0 & C^{\mathsf{T}} \\ C & D + D^{\mathsf{T}} \end{bmatrix} \tag{3.3}$$

and

$$\Theta = \begin{bmatrix} C^{\mathsf{T}}C & C^{\mathsf{T}}D \\ D^{\mathsf{T}}C & D^{\mathsf{T}}D - I \end{bmatrix} \tag{3.4}$$

in Theorem 3.1, respectively.

Set $-\omega_1 = \omega_2 = \varpi$ in the above theorem, then we can get an LMI condition for the low frequency case.

**Theorem 3.2.** *Let real matrices $A \in \mathbb{R}^{n\times n}$, $B \in \mathbb{R}^{n\times m}$, a symmetric matrix $\Theta \in \mathbb{R}^{(n+m)\times(n+m)}$, and a positive scalar $\varpi$ be given. Suppose $(A, B)$ is a controllable pair. Then the following statements are equivalent.*

*(i) With $\mathbf{\Omega}_L := \{\, \omega \in \mathbb{R} : \ \det(j\omega I - A) \neq 0, \ |\omega| \leq \varpi \,\}$, we have*

$$\begin{bmatrix} (j\omega I - A)^{-1}B \\ I_m \end{bmatrix}^* \Theta \begin{bmatrix} (j\omega I - A)^{-1}B \\ I_m \end{bmatrix} \leq 0, \qquad \forall \omega \in \mathbf{\Omega}_L. \tag{3.5}$$

*(ii) There exist $n \times n$ real symmetric matrices $P$ and $Q \geq 0$ satisfying*

$$\begin{bmatrix} A & B \\ I_n & 0 \end{bmatrix}^{\mathsf{T}} \begin{bmatrix} -Q & P \\ P & \varpi^2 Q \end{bmatrix} \begin{bmatrix} A & B \\ I_n & 0 \end{bmatrix} + \Theta \leq 0. \tag{3.6}$$

The equivalence holds true without assuming controllability of $(A, B)$ when non-strict inequalities in (3.5), (3.6), and $Q \geq 0$, are replaced by strict inequalities and $A$ has no eigenvalues $\lambda \in \mathbb{C}$ such that $\lambda = j\omega$ and $|\omega| \leq \varpi$. It can also be shown that when $\varpi$ approaches infinity, the parameter $Q$ must approach zero and we recover the standard Kalman–Yakubovich–Popov lemma [1, 15, 17].

A condition for the high frequency property can be derived based on the result in [10].

**Theorem 3.3.** *Let a positive scalar $\varpi$, real matrices $A \in \mathbb{R}^{n\times n}$, $B \in \mathbb{R}^{n\times m}$, a Hermitian matrix $\Theta \in \mathbb{C}^{(n+m)\times(n+m)}$ be given. Suppose $A$ has no eigenvalues on the imaginary axis. Then the following statements are equivalent.*

*(i) The high frequency KYP condition holds:*

$$\begin{bmatrix} (j\omega I - A)^{-1}B \\ I_m \end{bmatrix}^* \Theta \begin{bmatrix} (j\omega I - A)^{-1}B \\ I_m \end{bmatrix} < 0, \qquad \forall\, |\omega| \geq \varpi. \tag{3.7}$$

*(ii) There exist Hermitian matrices P and $Q > 0$ such that*

$$\begin{bmatrix} A & B \\ I_n & 0 \end{bmatrix}^* \begin{bmatrix} Q & P \\ P & -\varpi^2 Q \end{bmatrix} \begin{bmatrix} A & B \\ I_n & 0 \end{bmatrix} + \Theta < 0. \tag{3.8}$$

Before closing this section, we will provide the finite frequency KYP lemma for discrete-time systems, which can be derived by using a standard bilinear transform.

**Theorem 3.4.** *Let complex matrices $A \in \mathbb{C}^{n\times n}$, $B \in \mathbb{C}^{n\times m}$, a Hermitian matrix $\Theta \in \mathbb{C}^{(n+m)\times(n+m)}$, scalars $\theta_1, \theta_2 \in \mathbb{R}$ which satisfy $\theta_2 - \theta_1 \in (0, \pi)$ be given. Suppose $(A, B)$ is a controllable pair, i.e.,*

$$\operatorname{rank} \begin{bmatrix} A - \lambda I_n & B \end{bmatrix} = n, \qquad \forall\, \lambda \in \mathbb{C}$$

*holds. Then the following statements are equivalent.*

*(i) With $\boldsymbol{\Omega}_D := \{\ \theta \in \mathbb{R} :\ \det(e^{j\theta} I - A) \neq 0,\ \theta_1 \leq \theta \leq \theta_2\ \}$, we have*

$$\begin{bmatrix} (e^{j\theta} I - A)^{-1} B \\ I \end{bmatrix}^* \Theta \begin{bmatrix} (e^{j\theta} I - A)^{-1} B \\ I \end{bmatrix} \leq 0, \qquad \forall \omega \in \boldsymbol{\Omega}_D. \tag{3.9}$$

*(ii) There exist Hermitian matrices $X, Y \in \mathbb{C}^{n\times n}$ such that $Y \geq 0$ and*

$$\begin{bmatrix} A & B \\ I & 0 \end{bmatrix}^* \begin{bmatrix} -X & e^{j\theta_c} Y \\ e^{-j\theta_c} Y & X - \gamma Y \end{bmatrix} \begin{bmatrix} A & B \\ I & 0 \end{bmatrix} + \Theta \leq 0 \tag{3.10}$$

*where*

$$\theta_c := \frac{\theta_1 + \theta_2}{2}, \qquad \gamma := 2\cos\left(\frac{|\theta_1 - \theta_2|}{2}\right).$$

# 4 Finite frequency positive-real property

## 4.1 LMI characterizations of FFPR property

An LMI characterization of the FFPR condition can be derived by setting $\Theta$ as in (3.3) in Theorem 3.2, and the following theorem represents the dual version [7, 8, 10], which is obtained by replacing $(A, B, C, D)$ by $(A^{\mathsf{T}}, C^{\mathsf{T}}, B^{\mathsf{T}}, D^{\mathsf{T}})$.

**Theorem 4.1.** *Consider a real rational, square transfer function $G(s)$ with minimal realization*

$$G(s) = C(sI - A)^{-1} B + D.$$

*Let a positive scalar $\varpi$ be given. Then, $G(s)$ is FFPR with bandwidth $\varpi$, if and only if there exist real symmetric matrices $P$ and $Q \geq 0$ such that*

$$\begin{bmatrix} A & I \\ C & 0 \end{bmatrix} \begin{bmatrix} -Q & P \\ P & \varpi^2 Q \end{bmatrix} \begin{bmatrix} A & I \\ C & 0 \end{bmatrix}^{\mathsf{T}} \leq \begin{bmatrix} 0 & B \\ B^{\mathsf{T}} & D^{\mathsf{T}} + D \end{bmatrix}. \tag{4.1}$$

The next topic is to the robustification of the finite frequency condition in Theorem 4.1.

Consider a linear time-invariant uncertain system

$$\begin{bmatrix} \dot{x} \\ z \\ y \end{bmatrix} = \begin{bmatrix} A & B_1 & B_2 \\ C_1 & D_{11} & D_{12} \\ C_2 & D_{21} & D_{22} \end{bmatrix} \begin{bmatrix} x \\ w \\ u \end{bmatrix}, \qquad w = \Delta z \tag{4.2}$$

where all the coefficient matrices $A$, $B_i$, $C_i$ and $D_{ij}$ $(i = 1, 2)$ are real, and $\Delta$ is an uncertain matrix belonging to a known subset of real matrices $\mathbf{\Delta}$. Let the transfer function from $u$ to $y$ be denoted by $G_\Delta(s)$. We shall give a sufficient condition for the robust FFPR property:

$$G_\Delta(j\omega) + G_\Delta(j\omega)^* \geq 0, \quad \forall\, \omega \in \mathbf{\Omega}_L, \quad \Delta \in \mathbf{\Delta}. \tag{4.3}$$

**Theorem 4.2.** *Consider the uncertain system (4.2) with $\Delta \in \mathbf{\Delta}$ where $\mathbf{\Delta}$ is a subset of real matrices. Assume that $D_{21} = 0$ and $\det(I - D_{11}\Delta) \neq 0$ holds for all $\Delta \in \mathbf{\Delta}$. Denote by $G_\Delta(s)$ the transfer function from $u$ to $y$. Let a positive scalar $\varpi$ be given. Then, the robust FFPR condition holds if there exist real symmetric matrices $Q$, $Y$, $\Phi \in \mathbf{\Phi}$, and $\Lambda \in \mathbf{\Lambda}$ such that*

$$\begin{bmatrix} \mathbf{A} & I \\ \mathbf{C} & 0 \end{bmatrix} \begin{bmatrix} -Q & P \\ P & \varpi^2 Q \end{bmatrix} \begin{bmatrix} \mathbf{A} & I \\ \mathbf{C} & 0 \end{bmatrix}^{\mathsf{T}} \leq \begin{bmatrix} \mathbf{B} & 0 \\ \mathbf{D} & \mathbf{J} \end{bmatrix} \Phi \begin{bmatrix} \mathbf{B} & 0 \\ \mathbf{D} & \mathbf{J} \end{bmatrix}^{\mathsf{T}}, \tag{4.4}$$

$$Q \geq \mathbf{B}_1 \Lambda \mathbf{B}_1^{\mathsf{T}} \tag{4.5}$$

*where*

$$\mathbf{J} := [\, I \;\; 0 \,]^{\mathsf{T}},$$

$$\mathbf{\Phi} := \left\{ \begin{bmatrix} \Psi & 0 \\ 0 & \Pi \end{bmatrix} : \begin{bmatrix} \Gamma & 0 \\ 0 & \Gamma \end{bmatrix}^{\mathsf{T}} \Psi \begin{bmatrix} \Gamma & 0 \\ 0 & \Gamma \end{bmatrix} \leq 0, \forall\, \Gamma \in \mathbf{\Gamma} \right\},$$

$$\Pi := \begin{bmatrix} 0 & I \\ I & 0 \end{bmatrix},$$

$$\mathbf{\Lambda} := \{\, \Lambda : \; \Gamma^{\mathsf{T}} \Lambda \Gamma \geq 0, \quad \forall\, \Gamma \in \mathbf{\Gamma} \,\},$$

$$\mathbf{\Gamma} := \left\{ \begin{bmatrix} \Delta^{\mathsf{T}} \\ I \end{bmatrix} : \; \Delta \in \mathbf{\Delta} \right\},$$

*and*

$$\begin{bmatrix} \mathbf{A} & \mathbf{B}_1 & \mathbf{B}_2 \\ \mathbf{C} & 0 & \mathbf{D}_2 \end{bmatrix} := \left[\begin{array}{cc|cc|c} A & 0 & 0 & B_1 & B_2 \\ C_1 & 0 & I & D_{11} & D_{12} \\ \hline C_2 & 0 & 0 & 0 & D_{22} \\ 0 & I & 0 & 0 & 0 \end{array}\right], \qquad \begin{bmatrix} \mathbf{B} \\ \mathbf{D} \end{bmatrix} := \left[\begin{array}{ccc} \mathbf{B}_1 & \mathbf{A}\mathbf{B}_1 & \mathbf{B}_2 \\ \hline 0 & \mathbf{C}\mathbf{B}_1 & \mathbf{D}_2 \end{array}\right].$$

Conditions (4.4) and (4.5) are both LMIs in the variables $P$, $Q$, $\Phi$, and $\Lambda$, and hence are suitable for numerical computation. However, constraints $\Phi \in \boldsymbol{\Phi}$ and $\Lambda \in \boldsymbol{\Lambda}$ are characterized by infinitely many inequalities and make the numerical verification of robust finite frequency condition (4.3) difficult. To make it tractable, inner approximations of the sets $\boldsymbol{\Phi}$ and $\boldsymbol{\Lambda}$ may be used. In particular, when $\boldsymbol{\Delta}$ is the set of diagonal matrices with real uncertain parameters of bounded magnitude on the diagonal, such inner approximations can be given by the $D$–$G$ scaling [3] or the LFT scaling [2]. The former results in a sufficient condition for (4.3) that can be checked in polynomial time, while the latter improves conservatism of the former at the expense of more computation. See [11] for the details.

We consider the following situation in order to apply the above matrix inequality condition to the structure/control design integration. If $B_2$, $D_{12}$ and $D_{22}$, or $\mathbf{B}_2$ and $\mathbf{D}_2$, only depend affinely on the mechanical design parameter $\theta$, then conditions (4.4) and (4.5) are LMIs in terms of $P$, $Q$, $\Phi$, $\Lambda$, and $\theta$. Therefore, the design parameter $\theta$ can be found efficiently. The situation includes actuator location problems (the dual situation relates to sensor location problems).

## 4.2 Time domain interpretation of FFPR property

The following theorem provides a physical interpretation of being FFPR in terms of the passivity property with respect to a certain class of input signals. This time domain characterization of the FFPR property will be useful for extending the idea of our integrated design for nonlinear systems.

**Theorem 4.3.** *Consider a stable square transfer function $G(s)$. Let $u$, $y$, and $x$ be the input, the output, and the state of any minimal realization of $G(s)$. Then the following statements are equivalent.*

*(i) $G(s)$ is FFPR with bandwidth $\varpi$.*

*(ii) The passivity property*

$$\int_{-\infty}^{\infty} y(t)^{\mathsf{T}} u(t)\, dt \geq 0 \tag{4.6}$$

*holds for all square integrable inputs $u$ such that*

$$\int_{-\infty}^{\infty} \dot{x}(t)\dot{x}(t)^{\mathsf{T}} dt \leq \varpi^2 \int_{-\infty}^{\infty} x(t)x(t)^{\mathsf{T}} dt. \tag{4.7}$$

*(See [9] for the proof.)*

Roughly speaking, Theorem 4.3 states that the FFPR condition means that the system possesses the passivity property (4.6) with respect to the input signal $u$ that does not drive the states too quickly where the bound on the "quickness" is given by $\varpi$ in the sense of (4.7). It should be noted that condition (4.7) is coordinate free, i.e., if (4.7) holds for some minimal realization, then it holds for all other minimal realizations.

In the frequency domain, it can be shown that inequality (4.7) can be written as

$$\mathrm{Re} \int_0^{\varpi} (\varpi^2 - \omega^2)\hat{x}\hat{x}^* d\omega \geq \mathrm{Re} \int_{\varpi}^{\infty} (\omega^2 - \varpi^2)\hat{x}\hat{x}^* d\omega$$

where $\hat{x}$ is the Fourier transform of $x$. This inequality roughly means that the frequency component of $\hat{x}$ within the range $|\omega| \leq \varpi$ is "larger" than that outside of the range. Thus, Theorem 4.3 also shows that the FFPR condition requires the system to possess the passivity property for the input signal $u$ that mainly excites the system modes with natural frequency below $\varpi$.

# 5 Finite frequency high-gain property

This section is devoted to matrix inequality characterizations for finite frequency high-gain (FFHG) property. We first note that the FFHG condition (2.2) is equivalent to

$$\bar{\sigma}(P^{-1}(j\omega)) \leq 1/\gamma, \quad {}^{\forall}\omega \text{ such that } |\omega| \leq \varpi. \tag{5.1}$$

Hence, we need to introduce the descriptor form expression for the inverse of a square transfer function

$$P(s) = C_p(sI - A_p)^{-1}B_p + D_p$$

which is given by

$$P(s)^{-1} = \hat{P}(s) := \hat{C}(s\hat{E} - \hat{A})^{-1}\hat{B} + \hat{D}$$

where

$$\hat{A} := \begin{bmatrix} A_p & B_p \\ C_p & D_p \end{bmatrix}, \quad \hat{B} := \begin{bmatrix} 0 \\ -I \end{bmatrix},$$

$$\hat{C} := \begin{bmatrix} 0 & I \end{bmatrix}, \quad \hat{D} := 0, \quad \hat{E} := \begin{bmatrix} I & 0 \\ 0 & 0 \end{bmatrix}.$$

Using the following lemma, we can get necessary and sufficient conditions for the FFHG [6].

**Lemma 5.1.** *A transfer function $\tilde{G}(s)$ with descriptor representation*

$$\tilde{G}(s) = \tilde{C}(s\tilde{E} - \tilde{A})^{-1}\tilde{B} + \tilde{D}$$

*and* $\varpi > 0$ *are given. Suppose*

$$\det(\lambda\tilde{E}-\tilde{A}) \neq 0, \quad {}^{\forall}\lambda \in \mathbf{C}, \ \lambda+\bar{\lambda}=0$$

*holds. Then, the following two conditions are equivalent.*

*(i)* $\overline{\sigma}(\tilde{G}(j\omega)) \leq \tau, \quad {}^{\forall}\omega$ *such that* $|\omega| \leq \varpi$.

*(ii)* ${}^{\exists}P = P^{\mathsf{T}}, \ Q = Q^{\mathsf{T}} \geq 0$ *s.t.,*

$$\begin{bmatrix} \tilde{A} & \tilde{E} \\ \tilde{C} & 0 \end{bmatrix} \begin{bmatrix} -Q & P \\ P & \varpi^2 Q \end{bmatrix} \begin{bmatrix} \tilde{A} & \tilde{E} \\ \tilde{C} & 0 \end{bmatrix}^{\mathsf{T}} + \begin{bmatrix} \tilde{B}\tilde{B}^{\mathsf{T}} & \tilde{B}\tilde{D}^{\mathsf{T}} \\ \tilde{D}\tilde{B}^{\mathsf{T}} & \tilde{D}\tilde{D}^{\mathsf{T}} - \tau^2 I \end{bmatrix} \leq 0. \tag{5.2}$$

**Theorem 5.2.** *Consider a real rational, square transfer function* $P(s)$ *with minimal realization* $C_p(sI-A_p)^{-1}B_p$ *and suppose that* $A_p$ *has no imaginary axis eigenvalues. Let a positive scalar* $\varpi$ *be given. Then,* $P(s)$ *is FFHG with bandwidth* $\varpi$*, if and only if one of the following two conditions holds.*

*(i)* ${}^{\exists}P_{11} = P_{11}^{\mathsf{T}}, \ P_{12}, \ Q := \begin{bmatrix} Q_{11} & Q_{12} \\ Q_{12}^{\mathsf{T}} & Q_{22} \end{bmatrix} = Q^{\mathsf{T}} \geq 0$ *s.t.,*

$$\begin{bmatrix} A_p & B_p & I \\ C_p & 0 & 0 \\ 0 & I & 0 \end{bmatrix} \begin{bmatrix} -Q_{11} & -Q_{12} & P_{11} \\ -Q_{12}^{\mathsf{T}} & -Q_{22} & P_{12}^{\mathsf{T}} \\ P_{11} & P_{12} & \varpi^2 Q_{11} \end{bmatrix} \begin{bmatrix} A_p & B_p & I \\ C_p & 0 & 0 \\ 0 & I & 0 \end{bmatrix}^{\mathsf{T}} + \begin{bmatrix} 0 & 0 & 0 \\ 0 & I & 0 \\ 0 & 0 & -(1/\gamma^2)I \end{bmatrix} \leq 0. \tag{5.3}$$

*(ii)* ${}^{\exists}P_{11} = P_{11}^{\mathsf{T}}, \ S_{23}, \ Q := \begin{bmatrix} Q_{11} & Q_{12} \\ Q_{12}^{\mathsf{T}} & Q_{22} \end{bmatrix} = Q^{\mathsf{T}} \geq 0$ *s.t.,*

$$\begin{bmatrix} -Q_{22} & -Q_{22}B^{\mathsf{T}} & 0 & 0 \\ * & \begin{matrix} -A_pQ_{11}A_p^{\mathsf{T}}+P_{11}A_p^{\mathsf{T}}+A_pP_{11} \\ -B_pQ_{12}^{\mathsf{T}}A_p^{\mathsf{T}}-A_pQ_{12}B_p^{\mathsf{T}} \\ +S_{23}^{\mathsf{T}}B_p^{\mathsf{T}}+BS_{23}+\varpi^2Q_{11} \end{matrix} & \begin{matrix} -A_pQ_{11}C_p^{\mathsf{T}}+P_{11}C_p^{\mathsf{T}} \\ -B_pQ_{12}^{\mathsf{T}}C_p^{\mathsf{T}} \end{matrix} & -A_pQ_{12}+S_{23}^{\mathsf{T}} \\ * & * & -C_pQ_{11}C_p^{\mathsf{T}}+I & -C_pQ_{12} \\ * & * & * & -Q_{22}-(1/\gamma^2)I \end{bmatrix} \leq 0. \tag{5.4}$$

The conditions (i) and (ii) are both LMI in terms of variables $P_{11}, P_{12}$ and $Q$ if the plant parameters $A_p, B_p$ and $C_p$ are given. Hence, the FFHG condition can be readily checked. Moreover, the maximal frequency for the plant having the FFPR property can be computed by solving a generalized eigenvalue problem.

# 6 Conclusion

We have summarized the authors' recent results on finite frequency characterization for easily controllable plants or mechanical structures in practice. We have first shown an LMI characterization of the finite frequency KYP lemma and derived the conditions for the FFPR and the FFHG properties.

The theoretical results in this paper can be used to develop a systematic method for plant/control design integration. Especially for designing mechanical systems, the validity of the method of integrated design based on the finite frequency properties was confirmed by applying it to the shape design of a swing-arm for magnetic storage devices [8] and the smart structure design using piezo-electric films in order to reduce the higher mode oscillation [6].

# Acknowledgments

This research was supported in part by The Grant-in Aid for COE Research Project of Super Mechano-Systems by The Ministry of Education, Science, Sport and Culture in Japan.

## References

[1] B. D. O. Anderson and S. Vongpanitlerd, *Network Analysis and Synthesis.* Prentice Hall, 1973.

[2] T. Asai, S. Hara, and T. Iwasaki, "Simultaneous parametric uncertainty modeling and robust control synthesis by LFT scaling," *Automatica*, vol. 36, pp. 1457–1467, 2000.

[3] M. Fan, A. Tits, and J. Doyle, "Robustness in the presence of mixed parametric uncertainty and unmodeled dynamics," *IEEE Trans. Auto. Contr.*, vol. 36, no. 1, pp. 25–38, 1991.

[4] K. M. Grigoriadis and F. Wu, "Integrated $H_\infty$ plant/controller design via linear matrix inequalities," in *Proc. IEEE Conf. Decision Contr.*, 1997.

[5] K. M. Grigoriadis, G. Zhu, and R. E. Skelton, "Optimal redesign of linear systems," *ASME J. Dyn. Syst. Meas. Contr.*, vol. 118, pp. 596–605, 1996.

[6] S. Hara, T. Iwasaki, and F. Shimizu, "Finite frequency characterization of easily controllable mechanical systems under control effort constraint," in *Proc. IFAC World Congress*, 2002.

[7] T. Iwasaki and S. Hara, "Integrated design of dynamical systems: Requirements for easily controllable structures," in *Pre-print of TITech COE/Super Mechano-Systems Workshop'99*, 1999, pp. 68–72.

[8] T. Iwasaki, S. Hara, and H. Yamauchi, "Structure/control design integration with finite frequency positive real property," in *Proc. American Contr. Conf.*, 2000.

[9] T. Iwasaki, S. Hara, and H. Yamauchi, "Dynamical system design from a control perspective: Finite frequency positive-realness approach," submitted to *IEEE Trans. Auto. Contr.*, 2002.

[10] T. Iwasaki, G. Meinsma, and M. Fu, "Generalized $S$-procedure and finite frequency KYP lemma," *Mathematical Problems in Engineering*, vol. 6, pp. 305–320, 2000.

[11] T. Iwasaki and G. Shibata, "LPV system analysis via quadratic separator for uncertain implicit systems," *IEEE Trans. Auto. Contr.*, vol. 46, no. 8, pp. 1195–1208, 2001.

[12] I. Kajiwara and A. Nagamatsu, "An approach of optimal design for simultaneous optimization of structure and control systems using sensitivity analysis," *J. SICE*, vol. 26, no. 10, pp. 1140–1147, 1990.

[13] K. Ono and T. Teramoto, "Design methodology to stabilize the natural modes of vibration of a swing-arm positioning mechanism," *ASME Adv. Info. Storage Syst.*, vol. 4, pp. 343–359, 1992.

[14] J. Onoda and R. Haftka, "An approach to structure/control simultaneous optimization for large flexible spacecraft," *AIAA Journal*, vol. 25, no. 8, pp. 1133–1138, 1987.

[15] A. Rantzer, "On the Kalman–Yakubovich–Popov lemma," *Sys. Contr. Lett.*, vol. 28, no. 1, 1996.

[16] C. Sultan and R. E. Skelton, "Integrated design of controllable tensegrity structures," in *Proc. Int. Mech. Eng. Congress*, Dallas, TX, 1997.

[17] J. C. Willems, "Least squares stationary optimal control and the algebraic Riccati equation," *IEEE Trans. Auto. Contr.*, vol. 16, pp. 621–634, 1971.

Shinji Hara
Department of Information Physics and Computing
Graduate School of Information Science and Technology
The University of Tokyo
Hongo 7–3–1, Bunkyo-ku, Tokyo 113–8656, Japan
hara@crux.t.u-tokyo.ac.jp

Tetsuya Iwasaki
Department of Mechanical and Aerospace Engineering
University of Virginia, 122 Engineer's Way, P.O.Box 400746,
Charlottesville, VA 22904-4746, USA
iwasaki@virginia.edu

# Part IV

# Hybrid, Chaotic, and Nonlinear Systems

# On Stabilization of Switched Linear Systems

## Lei Guo

*Dedicated to Professor Hidenori Kimura on the occasion of his 60th birthday*

ABSTRACT This paper presents several theoretical results on stabilization of switched linear systems in both continuous- and discrete-time cases. It will be shown that in the continuous-time case, stabilizing feedback laws can be designed under the natural condition that the system switches (arbitrarily) among a finite set of controllable linear systems with finite frequency of switching. The situation in the discrete-time case is somewhat different, and the adaptive stabilization problem will be solved for the case where the switching is modelled as a finite state hidden Markov chain. Some necessary and sufficient conditions will be given to characterize the feedback stabilizability of the switched linear systems.

**Keywords:** Switching, Stabilization, Controllability, Estimation, Adaptation, Markov chain

## 1 Introduction

I knew Professor Kimura from the literature a long time ago, but it was not until the 1999 IEEE-CDC held in Phoenix that we had a chance to meet each other, although we are "neighbors" geographically. At that conference, I gave a talk on the capability of feedback in dealing with uncertainties [17], and since then, my research in this direction has always been encouraged by Professor Kimura. The original motivation of this work was also to understand the capability of feedback in dealing with uncertainties in complex control systems, and the results that will be presented in the paper may be regarded as a preliminary step towards this goal from the switching perspective.

It is a well-known fact that switched models and switched control laws have the capability to deal with complex nonlinear systems, when their operation conditions change with time (e.g., by tuning the parameters in the proportional-integral-derivative (PID) controller as in process control, or using gain scheduling techniques as in flight control). Moreover, when a nonlinear system is working around the "equilibrium" point, it is oftentimes treated effectively by linear models or linear feedbacks (e.g., by the PID controllers). Mathematically, switched linear models have a nice structure, in the sense that they are "simple" enough so that they may be analyzed with the help of the rich theory of linear systems, and at the same time, they are also "flexible" enough so that complicated nonlinear dynamics may be treated or approximated based on such models.

Switched systems are indeed frequently encountered in practice, for example, in robot manipulators, traffic management, power systems, and even in human behaviors, etc. It is for this reason, switched systems have attracted considerable research efforts from researchers in the area of control in either a deterministic setting, see, e.g., [8],[12], [4], [15], or in a stochastic framework with the switching laws modelled as a finite state Markov chain, see, e.g, [2], [7], [16],[5] and [6], among many others. In particular, a nice condition for the stabilization problem has been obtained in [10] for the nonadaptive case where the Markovian switchings are observable. In the adaptive case where the Markovian switchings are not observable, however, the problem is more complicated.

When the switching law has no given mode (or is arbitrary), one way to investigate the stability and stabilization problems is to find a common Lyapunov function to dominate all the switching models, see, e.g., [14], [11], [3], [1]. However, it is known that the common Lyapunov function method usually leads to unnecessary and restrictive conditions for stability or stabilization problems of switched linear systems. Another commonly used approach is to assume that the system would remain un-switched for a period long enough to absorb the overshot effect in the transient phases, see [13]. Finally, a related but different stabilization problem for switched linear systems is to assume that the switching rule is a control variable. With this additional freedom of control, one can show that even if each particular model is not stabilizable, the switched system with manipulated switching may still be stabilizable [15].

In this paper, we are concerned with the common situation where the switching laws are not known *a priori*, which have to be observed or estimated on-line. The main results are based on some recent developments in [19],[18] and [9], and the rest of the paper is organized as follows. In Section 2, we will study the stabilization problem of continuous-time switched linear systems with arbitrary switching laws, under the natural condition that the system switches among a finite set of controllable linear systems with finite switching frequency. We will take a new and essentially direct approach, which is based on pole-placement and on estimation of the growth rate of matrix exponential functions. In Section 3 of the paper, we will study the adaptive stabilization problem of discrete-time linear time-varying (LTV) systems with hidden Markovian switchings, and provide a general necessary and sufficient condition to characterize the adaptive stabilizability. Such a condition depends on the existence of a set of algebraic coupled Riccati-like equations that are different from those already known for the non-adaptive case.

## 2 Continuous-time linear systems

In this section, we consider the following continuous-time switched linear control system:

$$\dot{x}(t) = A_{\sigma(t)}x(t) + B_{\sigma(t)}u(t), \quad x(t) \in \mathcal{R}^n, \ u(t) \in \mathcal{R}^m, \tag{2.1}$$

where the switching law $\sigma(t) : [0, \infty) \longrightarrow \Lambda$ is a right-continuous piecewise constant mapping, and where $\Lambda = \{1, 2, \dots, N\}$ for some integer $N$.

We need to introduce the definition of the frequency of switching first.

**Definition 2.1.** The frequency of switching of a switching process $\sigma(t)$ : $[0, \infty) \longrightarrow \Lambda$, where $\Lambda = \{1, 2, \dots, N\}$, denoted by $f$, is defined as

$$f = \limsup_{t \longrightarrow \infty} \frac{\{\text{Number of switches of } \sigma(\cdot) \text{ in } [0, t]\}}{t}. \tag{2.2}$$

Similarly, the uniform frequency of switching is defined by

$$f^* = \limsup_{t-s \longrightarrow \infty} \frac{\{\text{Number of switches of } \sigma(\cdot) \text{ in } [s, t]\}}{t-s}. \tag{2.3}$$

□

Obviously, most practical systems (if not all) have finite frequency of switching, i.e., $f < \infty$.

**Definition 2.2.** The switching moments $0 < t_1 < t_2 < \cdots$ of the switching process $\sigma(t)$ are defined recursively by

$$t_{k+1} = \inf\{t > t_k : \sigma(t) \neq \sigma(t_k)\}, \ t_0 = 0.$$

Also, the *dwell time* of $\sigma(t)$ is defined by

$$\tau = \inf_k \delta_k$$

where $\delta_k$ is the switching duration of $\sigma(t)$ on an interval $[t_{k-1}, t_k)$, defined by

$$\delta_k = t_k - t_{k-1} .$$

□

To see how the switching duration $\delta_k$ relates to the frequency of switching, we denote

$$\delta = \liminf_{k \longrightarrow \infty} \bar{\delta}_k , \tag{2.4}$$

where $\bar{\delta}_k$ is the averaged switching duration

$$\bar{\delta}_k = \frac{1}{k} \sum_{i=1}^{k} \delta_i. \tag{2.5}$$

**Remark 2.3.** There is a nice relationship between the frequency of switching $f$ and the long-run averaged values of the switching duration $\delta$, i.e., $f < \infty$ if and only if $\delta > 0$. Furthermore, $\delta = \frac{1}{f}$. □

Throughout the paper, we will need the following standard assumption:

**H1).** The models $(A_i, B_i)$, $i = 1, \ldots, N$, are controllable.

The following theorem provides a basic result on stabilization of switched linear systems [9].

**Theorem 2.4.** *Let Assumption H1) hold, and let the switching process $\sigma(t)$ be observable with finite frequency of switching $f$. If an upper bound of $f$ is known* a priori, *then a finite set of gain matrices $\{K_i, i = 1, ..., N\}$ can be found such that the switched linear system (2.1) under the state feedback law $u(t) = K_{\sigma(t)}x(t)$:*

$$\dot{x}(t) = \left(A_{\sigma(t)} + B_{\sigma(t)}K_{\sigma(t)}\right) x(t) \tag{2.6}$$

*is exponentially stable. Furthermore, if the uniform frequency of switching $f^*$ is finite with a known upper bound, then (2.6) is uniformly exponentially stable.*

**Remark 2.5.** The controllability condition H1) can not be weakened to stabilizability in general. This is because there are examples (see, e.g., [12]) showing that switching between even two stable linear systems can indeed give rise to instability of the switched linear systems. Thus, in general, we cannot guarantee the stability of the uncontrollable part of the switched linear systems, if we only assume stabilizability of each linear model and allow the switching law to be arbitrary. □

**Remark 2.6.** If $\{\sigma(t)\}$ is a continuous-time homogeneous Markov chain with standard transition function

$$\lim_{t \longrightarrow 0} p_{ij}(t) = \delta_{ij}, \forall i, j,$$

$$p_{ij}(t) = P\{\sigma(t) = j | \sigma(0) = i\},$$

then the frequency of switching is finite almost surely. □

When the upper bound on $f$ is unavailable, we first need to construct an on-line estimate of the upper bound on the frequency of switching. By Remark 2.3, we need only to consider the estimation of the lower bound to the long-run averaged switching duration defined by (2.4).

Let us introduce the following recursively defined sequence:

$$\hat{\delta}_{k+1} = \min\{\hat{\delta}_k, \bar{\delta}_{k+1}\}, \quad \hat{\delta}_0 = 1, \qquad k = 0, 1, 2, \ldots, \tag{2.7}$$

where $\bar{\delta}_{k+1}$ is defined as in (2.5).

Since $\hat{\delta}_k$ is a monotonically decreasing nonnegative sequence, it has a nonnegative limit, denoted as $\hat{\delta}$. Moreover, this limit must be positive, because by Remark 2.3, $\delta$ defined by (2.4) is positive. Furthermore, it is easy to see that $0 < \hat{\delta} \leq \delta$. Consequently, by Remark 2.3, we have the following upper bound on $f$:

$$f \leq \frac{1}{\hat{\delta}}. \tag{2.8}$$

With the help of $\hat{\delta}_k$, we can still design stabilizing linear feedback laws in the case where the upper bound on $f$ is not directly available [9].

**Theorem 2.7.** *Under conditions of Theorem 2.4, if an upper bound of the frequency of switching of $\sigma(t)$ is not available* a priori, *then a linear state feedback control can also be constructed such that the switched linear system (2.1) is exponentially stable.*

Finally, we consider the case where the switching process $\sigma(t)$ itself is not observable. Intuitively, when the dwell time $\tau$ as defined in Definition 2.2 is positive, one would naturally try to first identify the switching signals at the beginning of each time interval $[t_k, t_{k+1})$ using a short time period (say, $<< \tau$) of the observed signals, and then to control the identified system in the rest of the time interval.

To identify the unobservable switching signal, let us consider a short time period $t \in [t_k, t_k + h)$, with $h < \tau$, and introduce the following "filtered" signals $y(t), \eta(t)$ and $\phi(t)$:

$$y(t) = x(t) - \eta(t), \tag{2.9}$$

$$\dot{\eta}(t) + \eta(t) = x(t), \quad \eta(t_k) = x(t_k), \tag{2.10}$$

$$\dot{\phi}(t) + \phi(t) = \begin{pmatrix} x(t) \\ u(t) \end{pmatrix}, \quad \phi(t_k) = 0. \tag{2.11}$$

It is obvious that $y(t), \eta(t)$ and $\phi(t)$ are all available signals given the observations $\{x(s), u(s), t_k \leq s \leq t\}$.

Set

$$Y_k = \int_{t_k}^{t_k+h} \phi(t) y(t)^T dt; \quad \Phi_k = \int_{t_k}^{t_k+h} \phi(t)\phi(t)^T dt. \tag{2.12}$$

Then, we can define an estimate for $\sigma(t_k)$ by minimizing the "prediction errors":

$$\hat{\sigma}(t_k) = \underset{1 \leq i \leq N}{\operatorname{argmin}} \left\| Y_k - \Phi_k \begin{pmatrix} A_i^T \\ B_i^T \end{pmatrix} \right\|. \tag{2.13}$$

It can be shown that if the "information" matrix $\Phi_k$ is of full rank, then $\hat{\sigma}(t_k)$ can indeed correctly estimate the unknown $\sigma(t_k)$. We remark that $\hat{\sigma}(t_k)$ can also do so even if there are bounded noises in the systems, provided that $\Phi_k$ has a certain level of excitation.

The following theorem states that when the switchings are not directly observable, a stabilizing control law based on adaptive estimation can also be constructed (see [9] for details).

**Theorem 2.8.** *Let the Condition H1) be satisfied for the switched linear system (2.1). If the switching process is not directly observable, but the dwell time is positive, then a control law can also be constructed to make the switched linear system exponentially stable.*

# 3 Discrete-time linear systems

We now consider the following discrete-time linear model:

$$x_{t+1} = A(\sigma_t)x_t + B(\sigma_t)u_t + w_{t+1}, \ t \geq 1, \tag{3.1}$$

where $x_t \in R^n$, $u_t \in R^m$ and $w_{t+1} \in R^n$ are the state, input and noise vectors respectively.

We need the following assumptions:

**A1).** $\{\sigma_t\}$ is an unobservable Markov chain which is homogeneous, irreducible and aperiodic, and takes values in a finite set $\{1, 2, \ldots, N\}$ with transition matrix denoted by $P = (p_{ij})_{NN}$, where by definition

$$p_{ij} = P\{\sigma_t = j | \sigma_{t-1} = i\}.$$

**A2).** There exists some $m \times n$ matrix $L$ such that

$$\det\left[(A_i - A_j) - (B_i - B_j)L\right] \neq 0, \quad \forall i \neq j, \tag{3.2}$$

where $1 \leq i, j \leq N$, and $A_i \triangleq A(i) \in R^{nn}$, $B_i \triangleq B(i) \in R^{nm}$ are the system matrices.

**A3).** $\{w_t\}$ is a martingale difference sequence which is independent of $\{\sigma_t\}$, and satisfies

$$\alpha I \leq Ew_t w_t', \quad Ew_t' w_t \leq \beta, \quad \forall t \tag{3.3}$$

where $\alpha$ and $\beta$ are two positive constants.

**Remark 3.1.** We remark that Condition A1) implies that each state in $\{1, 2, \ldots, N\}$ can be visited by $\{\sigma_t\}$ with positive probability when $t$ is suitably large (this is the only property needed in the paper for the homogeneous Markov chain). Condition A2) is a sort of identifiability condition to be used later in the construction of stabilizing feedback laws (this condition may also be further weakened). Moreover, the lower bound to the noise covariance in Condition A3) is assumed for simplicity of derivations, and the noise-free case where $w_t = 0, \forall t$ can be treated analogously. □

For simplicity of presentation, we will continue to adopt the notation $\Lambda \triangleq \{1, 2, \dots, N\}$ throughout the sequel.

**Definition 3.2.** An input sequence $\{u_t\}$ is said to be an admissible feedback if $u_t \in \sigma\{x_0, x_1, \dots, x_t\}$ and $E\|u_t\|^2 < \infty$, $\forall t$, where $\sigma\{x_0, x_1, \dots, x_t\}$ is the $\sigma$-algebra generated by the observed state information $\{x_0, x_1, \dots, x_t\}$. Moreover, the system (3.1) is said to be stabilizable by (adaptive) feedback if there exists an admissible feedback law $\{u_t\}$ such that $\sup_{t\geq 0} E\|x_t\|^2 < \infty$. □

Now, we have the following result which gives a fairly complete characterization of the feedback stabilizability of the switched linear systems [18].

**Theorem 3.3.** *Let the above assumptions A1)–A3) hold for the switched linear system (3.1). Then the system is stabilizable by feedback if and only if the following coupled algebraic Riccati-like equations have a solution consisting of N positive definite matrices $\{M_i > 0, i \in \Lambda\}$:*

$$\sum_j A_j' p_{ij} M_j A_j - \Big(\sum_j A_j' p_{ij} M_j B_j\Big)\Big(\sum_j B_j' p_{ij} M_j B_j\Big)^+ \Big(\sum_j B_j' p_{ij} M_j A_j\Big) - M_i = -I, \quad (3.4)$$

*where $i \in \Lambda$ and $(\cdot)^+$ denotes the Moore–Penrose generalized inverse of the corresponding matrix.*

**Remark 3.4.** In contrast to most of the previous publications in the literature, we have neither restricted ourselves to the class of linear feedback laws, nor imposed any conditions on the rate of parameter changes. Hence, Theorem 3.3 provides us with a basis for exploring the full capability and limitations of the feedback mechanism. □

To understand more about the key equation (3.4), we denote

$$L_i^* \triangleq L_i^*(M) \triangleq \Big(\sum_{j=1}^N B_j' p_{ij} M_j B_j\Big)^+ \Big(\sum_{j=1}^N B_j' p_{ij} M_j A_j\Big), \quad i \in \Lambda, \qquad (3.5)$$

where and hereafter $M \triangleq [M_1, ..., M_N]'$. Then by properties of the generalized inverse, it is easy to see that (3.4) can be rewritten in the form

$$\sum_{j=1}^N (A_j - B_j L_i^*)' p_{ij} M_j (A_j - B_j L_i^*) - M_i = -I, \quad i \in \Lambda, \qquad (3.6)$$

which may be regarded as Lyapunov-like equations.

For the convenience of future discussion, we denote the first term on the left-hand side of (3.6) as $\psi_i(M)$, i.e.,

$$\psi_i(M) \triangleq \sum_{j=1}^{N}(A_j - B_j L_i^*)' p_{ij} M_j (A_j - B_j L_i^*), \quad i \in \Lambda. \tag{3.7}$$

Obviously, $\psi_i(\cdot)$ is a nonlinear mapping from $\prod_1^N R^{nn}$ to $R^{nn}$. Moreover, we denote

$$\psi(M) \triangleq (\psi_1(M), \psi_2(M), \ldots, \psi_N(M))' \tag{3.8}$$

which plays a key role in characterizing the properties of the Riccati-like equation (3.4) as will be seen shortly.

The following theorem gives several conditions equivalent to the existence of the solution of the Riccati-like equations (3.4) used in Theorem 3.3, which shows how the Riccati-like equations are related to Lyapunov-like equations and to some recursively defined equations. These may provide alternative ways for checking the stabilizability of the switched linear systems (see, [18]).

**Theorem 3.5.** *The following four facts are equivalent in characterizing the stabilizability of the switched linear system (3.1):*

*(i) There exists a set of $n \times n$ positive definite matrices $\{M_i > 0, i \in \Lambda\}$ which satisfies the following Riccati-like equations for $i \in \Lambda$ :*

$$\sum_j A_j' p_{ij} M_j A_j - \Big(\sum_j A_j' p_{ij} M_j B_j\Big)\Big(\sum_j B_j' p_{ij} M_j B_j\Big)^{+}\Big(\sum_j B_j' p_{ij} M_j A_j\Big) - M_i = -I. \tag{3.9}$$

*(ii) There exists a set of $m \times n$ matrices $\{L_i,\ i \in \Lambda\}$ such that the following coupled Lyapunov-like equations have a solution consisting of positive definite matrices $\{M_i > 0, i \in \Lambda\}$:*

$$\sum_{j=1}^{N}(A_j - B_j L_i)' p_{ij} M_j (A_j - B_j L_i) - M_i = -I, \quad i \in \Lambda. \tag{3.10}$$

*(iii) The sequence of matrices $\{V_t\}$ satisfies $\lim_{t \to \infty} V_t = 0$, where $\{V_t\}$ is defined recursively by*

$$V_{t+1} = \psi(V_t), \quad V_0 = I^*, \tag{3.11}$$

*and where $\psi(\cdot)$ is defined in (3.8), and $I^* \triangleq (I_{n \times n}, \ldots, I_{n \times n})$.*

*(iv) The sequence of matrices $\{M_t^*\}$ satisfies $\lim_{t\to\infty} M_t^* < \infty$, where $\{M_k^*\} \in R^{nN\times n}$ is defined recursively by*

$$M_k^* = \psi(M_{k-1}^* + I^*); \quad k \geq 1, \quad M_0^* = 0^*, \tag{3.12}$$

*with $\psi(\cdot)$ and $I^*$ being defined as in (iii) above, and $0^* \triangleq (0_{n\times n}, \dots, 0_{n\times n})$.*

**Remark 3.6.** The stabilizing adaptive feedback in Theorem 3.3 can be constructed as

$$u_t = -L(\hat{\sigma}_{t-1}) \cdot x_t \ , \tag{3.13}$$

where $\hat{\sigma}_{t-1}$ is the estimate of $\sigma_{t-1}$, defined by minimizing the "prediction error"

$$\hat{\sigma}_{t-1} = \operatorname*{argmin}_{1\leq i\leq N} \|x_t - (A_i x_{t-1} + B_i u_{t-1})\|, \tag{3.14}$$

and where $\{L(i), i = 1, 2, \dots, N\}$ is chosen to satisfy

$$\sum_{j=1}^{N} (A_j - B_j L_i)' p_{ij} M_j (A_j - B_j L_i) < M_i, \quad i \in \Lambda, \tag{3.15}$$

and that

$$\det\left[(A_i - A_j) - (B_i - B_j)L_k)\right] \neq 0, \quad \forall i \neq j, k \in \Lambda, \tag{3.16}$$

which is possible by Theorem 3.5 (ii) and the Assumption A2). □

**Remark 3.7.** It is interesting to compare the equation (3.6) with the following equation:

$$(A_i - B_i L_i)' [\sum_{j=1}^{N} p_{ij} M_j] (A_i - B_i L_i) - M_i = -I, \quad i \in \Lambda, \tag{3.17}$$

which was used in Ji and Chizeck [10] for the case where the Markov chain $\{\sigma_t\}$ is observable, to establish the stability of the closed-loop systems under linear feedbacks. Obviously, the existence of the solution of the equation (3.6) derived in the present (adaptive) case should imply the existence of the solution of (3.17). Indeed, it can be shown (see [18]) that if there exists a set of $m \times n$ matrices $\{L_i, i \in \Lambda\}$ such that (3.6) has a solution $\{M_i > 0, i \in \Lambda\}$, then so does (3.17). However, the converse assertion is not true in general(see [18]). This fact provides a rigorous comparison to the non-adaptive case, showing that the uncertainty in the system parameters $\{\sigma_t\}$ does indeed degrade the capability of the feedback mechanism. □

**Remark 3.8.** Theorem 3.3 shows that the capability of adaptive feedback depends on both the structure complexity measured by $\{A_j, B_j, 1 \leq j \leq N\}$ and the information uncertainty measured by $\{p_{ij}, 1 \leq i, j \leq N\}$, in a rather complicated way. To make it more clear in understanding how the capability of adaptive feedback depends on both the complexity and uncertainty of the system, we consider the following example. □

**Example 3.9.** Consider the simple scalar model

$$x_{t+1} = a(\sigma_t)x_t + u_t + w_{t+1} \tag{3.18}$$

where $\sigma_t \in \{1, 2\}$, $a(1) \triangleq a_1$ and $a(2) \triangleq a_2$, and the transition matrix of the Markov chain is

$$P = \begin{pmatrix} p_{11} & p_{12} \\ p_{12} & p_{22} \end{pmatrix}.$$

Obviously, each model in the model set $\{[a(i), 1], i = 1, 2\}$ is controllable. If the Markov chain $\{\sigma_t\}$ is observable, then the simple linear feedback $u_t = -a(\sigma_t)x_t$ can trivially stabilize the system. However, when $\{\sigma_t\}$ is not available, the system can not be stabilized by feedback for all $\{a(i), P\}$. In fact, it can be shown by Theorems 3.3 and 3.5 that the system is stabilizable if and only if

$$(a_1 - a_2)^2(1 - p_{12})p_{12} < 1$$

(see also [19]). Let us denote $I(p_{12})$ as

$$I(p_{12}) \triangleq 1 - (a_2 - a_1)^2(1 - p_{12})p_{12},$$

which may be regarded as a measure of the capability of adaptive feedback and may be further represented by

$$I(p_{12}) = 1 - CU, \tag{3.19}$$

where $C \triangleq (a_2 - a_1)^2$ and $U \triangleq (1 - p_{12})p_{12}$ can be interpreted as measures of the structure complexity (degree of dispersion) and the information uncertainty of the system, respectively. Obviously, the system is stabilizable $\Leftrightarrow I(p_{12}) > 0$. We now have the following interesting observations (cf. [19]):

(i) The capability of adaptive feedback defined by $I(p_{12})$ is a monotonic function of the uncertainty $U$ but is not monotonic in the rate of transition $p_{12}$. Furthermore, there does not exist a critical rate of transition $p_{12}^* \in (0, 1)$ such that the system is stabilizable or $I(p_{12}) > 0 \Leftrightarrow p_{12} \in [0, p_{12}^*)$. Moreover, the capability of adaptation $I(p_{12})$ achieves its maximum when the uncertainty $U$ reaches its minimum.

(ii) The uncertainty measure $U$ is closely related to the well-known Shannon information entropy, which is a measure of information uncertainty defined by

$$H = -\sum_{i=1}^{2} p_{1i} \log p_{1i}$$

in the current case. Note that $H$ can be rewritten as

$$H = -(1 - p_{12}) \log(1 - p_{12}) - p_{12} \log p_{12}.$$

It is not difficult to see that there exists a monotonically increasing function $m(\cdot)$ such that $U = m(H)$. This fact justifies why we refer to $U$ as the measure of information uncertainty here. Now, by (3.19) we have

$$I(p_{12}) = 1 - Cm(H),$$

which implies that the capability of adaptive feedback is also a monotonically decreasing function of the Shannon information entropy $H$.

□

**Remark 3.10.** In the present discrete-time case, one may naturally ask if we can establish some general results on stabilization of switched linear systems when the switchings are arbitrary, as those established in the last section for continuous-time systems. The answer is not simple, and it appears that the discrete-time case is more complicated to deal with, since the high-gain design methods which are so successful in the continuous-time case, fail in the discrete-time case. This is also the main reason why the theory of discrete-time adaptive nonlinear control is more difficult to establish than that in the continuous-time case (see also, [17] for related discussions). □

## References

[1] A. A. Agrachev and D. Liberzon, "Lie-algebraic stability criteria for switched systems," *SIAM J. Contr. Opt.*, vol. 40, no. 1, pp. 253–269, 2001.

[2] W. P. Blair, Jr. and D. D. Sworder, "Feedback control of linear discrete systems with jump parameters and quadratic criteria," *Int. J. Contr.*, vol. 21, no. 5, pp. 833–841, 1975.

[3] S. Boyd, L. E. Ghaoui, and E. Feron, *Linear Matrix Inequalities in System and Control Theory.* Philadelphia: SIAM Studies in Application of Mathematics, 1994.

[4] M. S. Branicky, "Multiple Lyapunov functions and other analysis tools for switched hybrid systems," *IEEE Trans. Automat. Contr.*, vol. 43, no. 4, pp. 475–482, 1998.

[5] P. E. Caines and H. F. Chen, "Optimal adaptive LQG control for systems with finite state process parameters," *IEEE Trans. Automat. Contr.*, vol. 30, pp. 185–189, 1985.

[6] P. E. Caines and J. F. Zhang, "On the adaptive control of jump parameter systems via nonlinear filtering," *SIAM J. Contr. Opt.*, vol. 33, no. 6, pp. 1758–1777, 1995.

[7] H. J. Chizeck, A. S. Willsky, and D. Castanon, "Discrete-time Markovian-jump linear quadratic optimal control," *Int. J. Contr.*, vol. 43, no. 1, pp. 213–231, 1986.

[8] W. P. Dayawansa and C. F. Martin, "A converse Lyapunov theorem for a class of dynamical systems which undergo switching," *IEEE Trans. Autmat. Contr.*, vol. 44, no. 4, pp. 751–760, 1999.

[9] L. Guo, Y. Wang, D. Cheng, and Y. Lin, "Stabilization of switched linear systems," *Proc. of the 21st Chinese Control Conference*, August 12–16, 2002, Hangzhou, China, pp. 429–434.

[10] Y. D. Ji and H. J. Chizeck, "Jump linear quadratic Gaussian control: Steady-state solution and testable conditions," *Control Theory and Advanced Technology*, vol. 6, no. 3, pp. 289–319, 1990.

[11] D. Liberzon, J. P. Hespanha, and A. S. Morse, "Stability of switched systems: A Lie-algebraic condition," *Sys. Contr. Lett.*, vol. 37, no. 3, pp. 117–122, 1999.

[12] D. Liberzon and A. S. Morse, "Basic problems on stability and design of switched systems," *IEEE Control Systems Magazine*, vol. 19, no. 5, pp. 59–70, 1999.

[13] A. S. Morse, "Supervisory control of families of linear set-point controllers—Part 1: Exact matchings," *IEEE Trans. Automat. Contr.*, vol. 41, pp. 1413–1431, October 1996.

[14] R. N. Shorten and K. S. Narendra, "Necessary and sufficient condition for the existence of a common quadratic Lyapunov function for M-stable linear second order systems," in *Proc. 2000ACC*, 2000, pp. 359–363.

[15] Z. Sun, S. S. Ge and T. H. Lee, "Controllability and reachability criteria for switched linear systems," *Automatica*, vol. 38, no. 5, pp. 775–786, 2002.

[16] D. D. Sworder, "Hybrid adaptive control," *Appl. Math. Comp.*, vol. 45, pp. 173–192, 1991.

[17] L. Xie and L. Guo, "How much uncertainty can be dealt with by feedback?" *IEEE Trans. Automat. Contr.*, vol. 45, no. 12, pp. 2203–2217, December 2000.

[18] F. Xue and L. Guo, "Necessary and Sufficient conditions for adaptive stabilizability of jump linear systems," *Communications in Information and Systems*, vol. 1, no. 2, pp. 205–224, April 2001.

[19] F. Xue, L. Guo, and M. Y. Huang, "Towards understanding the capability of adaptation for time-varying systems," *Automatica*, vol. 37, pp. 1551–1560, 2001.

Lei Guo
Institute of Systems Science
Chinese Academy of Sciences
Beijing 100080, P. R. China
Lguo@mail.iss.ac.cn

# Well-Posedness of a Class of Piecewise Affine Systems

**Jun-ichi Imura**

*Dedicated to Professor Hidenori Kimura on the occasion of his 60th birthday*

ABSTRACT The well-posedness problem (existence and uniqueness of solutions) of a class of piecewise affine (PWA) systems is addressed. First, two kinds of discrete transition rules, i.e., the mode-based rule and the switch-based rule, are introduced to define the concept of solutions of PWA systems. Next, after several fundamental tools for the well-posedness analysis are given, necessary and/or sufficient conditions for the system to be well-posed are presented.

**Keywords:** Hybrid systems, Piecewise affine systems, Well-posedness, Transition rule, Switches

## 1 Introduction

In hybrid systems with autonomous switching and jumps, as is well-known, there exist various phenomena such as no solutions, multiple solutions, sliding motions, Zeno trajectories, and multiple events by jump solutions. So the well-posedness (existence and uniqueness of solutions) problem is crucial for various developments of hybrid systems (see e.g., [18]). However, the existing standard theory (see e.g., [1, 15]) on well-posedness of such systems is not quite satisfactory. For example, the well-posedness problem of piecewise affine (PWA) systems, which is one of the simplest hybrid systems, has not completely been solved. In addition, there have been few works on how to define solutions, taking account of the discrete state as well as the continuous state from the viewpoints of hybrid systems. Thus several papers have recently addressed the well-posedness problem for various subclasses of hybrid systems. In [4, 3, 16, 17], the class of hybrid systems called complementarity systems, which is a generalization of mechanical systems with unilateral constraints, has been proposed, and several easily checkable, sufficient (and necessary) conditions for such a system to be well-posed have been derived. Furthermore, in the field of relay systems, the well-posedness problem has been solved in [12], and some issues on high-order sliding modes and fast switching have been studied in [2] and [11], respectively. For the more general classes such as differential automata and hybrid automata, several results on their execution have also been derived from the general points of view in [14, 10].

For the class of PWA systems, on the other hand, the author has addressed, partially with van der Schaft, the well-posedness problem under the definition of solutions in the sense of *extended* Carathéodory, which is different from that of the above existing works in the sense that the concept of sliding modes and jump solutions is not treated, and derived a necessary and sufficient condition for the *bimodal* system to be well-posed [5, 6, 7, 9], and a sufficient condition for the *multi-modal* system to be well-posed [8].

The well-posedness condition derived in the bimodal case is given in the form of an easily checkable, algebraic condition, which thus leads to a parametrization of all well-posed PWA systems in the bimodal case [5, 6]. These results have not been obtained so far even in the class of PWA systems. Furthermore, based on such a characterization, we have solved the feedback well-posedness problem of bimodal systems, i.e., the problem of determining which bimodal system can be transformed into a well-posed system by state feedback, and have discussed the stabilizability of a class of feedback well-posed systems [7, 9]. On the other hand, extensions of these results to the multi-modal case are not straightforward due to some kind of complexity that is explosively increased as the number of modes is increased. However, by introducing the discrete transition rule based on a feature on switching of the discrete state, we have recently derived in [8] a sufficient condition for the multi-modal PWA system to be well-posed in terms of well-posedness of its subsystems of lower complexity, i.e., "bimodal systems", which thus allows us to easily check well-posedness of the system.

This paper presents a series of the above results [5, 6, 7, 9, 8] on well-posedness of a class of PWA systems. First, two kinds of discrete transition rules are defined; the switch-based transition rule and the mode-based transition rule. Then the concept of solutions of the PWA systems is defined. After preparing several basic tools for the well-posedness analysis, we present some results on well-posedness.

# 2 Switch-driven piecewise affine systems

## 2.1 System description

Consider the system

$$\dot{\bar{x}} = (\bar{A} + \sum_{j=1}^{r} \delta_j \bar{G}_j \bar{F}_j)\bar{x}, \qquad \delta_j = \begin{cases} 0 & \text{if } \bar{S}_j \bar{x} \geq 0, \\ 1 & \text{if } \bar{S}_j \bar{x} \leq 0 \end{cases} \tag{2.1}$$

where

$$\bar{A} = \begin{bmatrix} A & a \\ 0 & 0 \end{bmatrix}, \ \bar{G}_j = \begin{bmatrix} G_j \\ 0 \end{bmatrix}, \ \bar{S}_j = \begin{bmatrix} S_j & s_j \end{bmatrix}, \ \bar{F}_j = \begin{bmatrix} F_j & f_j \end{bmatrix}, \ \bar{x} = \begin{bmatrix} x \\ 1 \end{bmatrix}$$

and $x \in \mathcal{R}^n$ is the state, $A \in \mathcal{R}^{n\times n}$, $G_j \in \mathcal{R}^{n\times l_j}$, $S_j \in \mathcal{R}^{1\times n}$, $F_j \in \mathcal{R}^{l_j\times n}$, $a \in \mathcal{R}^n$, $s_j \in \mathcal{R}$, and $f_j \in \mathcal{R}^{l_j}$. We call $\delta_j$ the *(binary) switch*, and $\delta = (\delta_1, \delta_2, \cdots, \delta_r) \in \{0,1\}^r$ also the *switch* or the *switch vector*. Let $\mathcal{J}$ be the index set of switches defined by $\mathcal{J} = \{1, 2, \dots, r\}$. The set $\mathcal{D}$ of all $\delta = (\delta_1,\ \delta_2,\ \cdots,\ \delta_r) \in \{0,1\}^r$ has $2^r$ elements, in other words, $2^r$ modes. So by numbering elements $\delta$ of $\mathcal{D}$ from 1 to $2^r$, $\delta$ is identified with mode $I \in \mathcal{I}$, where $\mathcal{I}$ is the set of the mode given by $\mathcal{I} \triangleq \{1, 2, \dots, 2^r\}$. Without loss of generality, mode $I$ is identified with $\delta$ in terms of the relation $I = \sum_{j=1}^{r} 2^{j-1}\delta_j + 1$. For example, we have for the case $r = 2$,

$$\dot{\bar{x}} = (\bar{A} + \delta_1 \bar{G}_1 \bar{F}_1 + \delta_2 \bar{G}_2 \bar{F}_2)\bar{x}, \quad \delta_j = \begin{cases} 0 & \text{if } \bar{S}_j \bar{x} \geq 0, \\ 1 & \text{if } \bar{S}_j \bar{x} \leq 0 \end{cases} \tag{2.2}$$

and $I = 1$ iff $\delta = (0,0)$, $I = 2$ iff $\delta = (1,0)$, $I = 3$ iff $\delta = (0,1)$, and $I = 4$ iff $\delta = (1,1)$. Thus $\delta \in \mathcal{D}$ and $I \in \mathcal{I}$ will hereafter be used depending on the occasion. We call $x$ the continuous state, $\delta$ or equivalently $I$ the discrete state, and $(\delta, x)$ or equivalently $(I, x)$ the hybrid state. An example shown in Figure 5.1 in Section 5 will provide a physical intuition for understanding such a system.

To define theoretically the solution behavior of the system (2.1), let us at first specify the transition rule that determines event (i.e., switching) times (times at which the discrete state will be switched) and the new discrete state selected at each event time.

**Definition 2.1 (Event time).** For the hybrid state $(\delta_0, x_0) \in \mathcal{D} \times \mathcal{R}^n$ at the current time $t_0 (\geq 0)$, consider the continuous state $x(t)$, $t \geq t_0$, of the system (2.1) with $\delta = \delta_0$ fixed, denoted by $x(t - t_0, \delta_0, x_0)$. Then we say that an event takes place at time $t = t_0 + \Delta t(\delta_0, x_0)$ if $\Delta t(\delta_0, x_0) < \infty$, where

$$\Delta t(\delta_0, x_0) = \inf\{\ \tau \geq 0 \mid \exists j \in \mathcal{J} \text{ s.t. } \phi(\tau, \delta_0, x_0) \notin \mathcal{C}_j^{\delta_{0,j}}\ \}. \tag{2.3}$$

Here $\delta_{0,j} \in \{0,1\}$ is the $j$th element of $\delta_0$ and $\mathcal{C}_j^k = \{\ x \in \mathcal{R}^n \mid (1-2k)\bar{S}_j \bar{x} \geq 0\ \}$. We call $t_0 + \Delta t(\delta_0, x_0)$ the *event time*, denoted by $T(t_0, \delta_0, x_0)$ or simply $T$. In addition, let $\mathcal{J}_{act}(T - t_0, \delta_0, x_0)$ be the set of indices of the switches defined as

$$\begin{aligned} \mathcal{J}_{act}(T - t_0, \delta_0, x_0) \ &= \ \{\ j \in \mathcal{J} \mid \phi(T - t_0, \delta_0, x_0) \in \mathcal{C}_j^{\delta_{0,j}}, \text{ and } \exists \varepsilon > 0 \text{ s.t.} \\ &\qquad \phi(t - t_0, \delta_0, x_0) \notin \mathcal{C}_j^{\delta_{0,j}}, \forall t \in (T, T + \varepsilon)\ \} \\ &\qquad \bigcup \{\ j \in \mathcal{J} \mid \phi(T - t_0, \delta_0, x_0) \notin \mathcal{C}_j^{\delta_{0,j}}\ \}. \end{aligned} \tag{2.4}$$

We call $\delta_j$, $j \in \mathcal{J}_{act}(T - t_0, \delta_0, x_0)$, the *active switches* at the event time $T$. □

The above definition is clearly interpreted based on the following two cases: (i) $x_0 \in \mathcal{C}_j^{\delta_{0,j}}$ for all $j \in \mathcal{J}$ and (ii) $x_0 \notin \mathcal{C}_j^{\delta_{0,j}}$ for some $j \in \mathcal{J}$.

In case (i), an event occurs at $t = T$ if there exist $T(\geq t_0)$, $j \in \mathcal{J}$ such that $x(t) \in \mathcal{C}_j^{\delta_j}$ for all $t \in [t_0, T]$ and, for some $\varepsilon > 0$, $x(t) \notin \mathcal{C}_j^{\delta_{0,j}}$ for all $t \in (T, T+\varepsilon)$. Case (ii) implies that the initial state $x_0$ initially violates at least one subregion specified by $\delta_0$, and thus an event occurs at $t = t_0$.

Only when an event takes place, every active switch $\delta_j$ is discretely changed from 0 to 1 or from 1 to 0; otherwise the switch $\delta$ remains constant. Thus let $T^-$ and $T^+$ denote the time just before and just after an event time $T$, respectively, so as to clearly express the discontinuous change of $\delta(t)$ at $T$. Note that $T = T^- = T^+$, $x(T) = x(T^-) = x(T^+)$, and $\delta_j(T^-) = \lim_{t \to T-0} \delta_j(t)$ and $\delta_j(T^+) = \lim_{t \to T+0} \delta_j(t)$. Then the following is considered as a transition rule of this system.

**Definition 2.2 (Switch-based transition rule).** Suppose an event takes place at time $T$ for $x(t - t_0, \delta_0, x_0)$ as in Definition 2.1. Then the switch vector $\delta(T^-) = \delta_0$ is changed at this event time $T$ into

$$\begin{aligned} \delta_j(T^+) &= 1 - \delta_j(T^-), \quad \forall j \in \mathcal{J}_{act}(T - t_0, \delta_0, x_0), \\ \delta_j(T^+) &= \delta_j(T^-), \quad \forall j \in \mathcal{J}/\mathcal{J}_{act}(T - t_0, \delta_0, x_0) \end{aligned}$$

where the symbol $\mathcal{J}/\mathcal{J}_{act}(T - t_0, \delta_0, x_0)$ expresses the difference set of $\mathcal{J}$ and $\mathcal{J}_{act}(T - t_0, \delta_0, x_0)$. □

This means that all active switches change from 0 to 1 (from 1 to 0) at the event time $T$, whereas the other switches keep the state just before the event. If, for the new state $\delta(T^+)$ that has been determined at the event time $T$, violation of at least one subregion with respect to $x(t - T^+, \delta(T^+), x(T^+))$ is also imminent at $t = T^+$ (i.e., for some $j \in \mathcal{J}$, $x(T^+) \in \mathcal{C}_j^{\delta_j(T^+)}$ and, for some $\varepsilon > 0$, $x(t) \notin \mathcal{C}_j^{\delta_j(T^+)}$ for all $t \in (T^+, T^+ + \varepsilon)$), then an event takes place again and the state of the switch changes at $t = T^+$. Thus the state of the switch will continue to change at the event time $T$, *not proceeding in time*, as long as violation of at least one of these subregions occurs, and as a result there may exist a finite or infinite transition sequence of the discrete state at some event time $T$, which is called *multiple events*. However notice that there exist no multiple solutions (the strict definition of solutions is given later) because in general the new mode is uniquely determined even in the case that more than two switches are changed simultaneously. We call the phenomenon that *infinite* transitions of the discrete state occur at a certain time instant, not proceeding in time, the *livelock*. This is undesirable because it is a kind of deadlock in some sense.

On the other hand, notice that the system (2.1) formally belongs to a class of the so-called PWA systems given by the form $\dot{x} = A_I x + a_I$ for all $x \in \mathcal{L}_I$, where each $\mathcal{L}_I$, called the location invariant in mode $I$, expresses the region assigned to mode $I$. If the system (2.1) is expressed in the form of PWA systems by focusing on the mode, we have, for example, for the case $r = 2$,

$$\begin{cases} \text{mode } 1: \dot{\bar{x}} &= \bar{A}\bar{x} & \text{if } \bar{S}_1\bar{x} \geq 0, \bar{S}_2\bar{x} \geq 0, \\ \text{mode } 2: \dot{\bar{x}} &= (\bar{A} + \bar{G}_1\bar{F}_1)\bar{x} & \text{if } \bar{S}_1\bar{x} \leq 0, \bar{S}_2\bar{x} \geq 0, \\ \text{mode } 3: \dot{\bar{x}} &= (\bar{A} + \bar{G}_2\bar{F}_2)\bar{x} & \text{if } \bar{S}_1\bar{x} \geq 0, \bar{S}_2\bar{x} \leq 0, \\ \text{mode } 4: \dot{\bar{x}} &= (\bar{A} + \bar{G}_1\bar{F}_1 + \bar{G}_2\bar{F}_2)\bar{x} & \text{if } \bar{S}_1\bar{x} \leq 0, \bar{S}_2\bar{x} \leq 0. \end{cases} \tag{2.5}$$

In general, the set $\mathcal{L}_I$ for the mode $I$ is given by, e.g., from the relation $I = \sum_{j=1}^{r} 2^{j-1}\delta_j + 1$, $\mathcal{L}_I = \bigcap_{j=1}^{r} \mathcal{C}_j^{\delta_j}$, where $\mathcal{C}_j^k$ is defined in Definition 2.1. Thus within the framework of the PWA systems, the discrete transition rule is usually based on the mode, not the switch as follows.

**Definition 2.3 (Mode-based transition rule).** Suppose an event takes place at time $T$ for $x(t - t_0, I_0, x_0)$ as in Definition 2.1. Then the mode $I(T^-) = I_0$ is changed at this event time $T$ into the new modes $I(T^+)$ given by all modes $I \in \mathcal{I}$ for which there exists an $\varepsilon > 0$ such that $x(t - T, I, x(T)) \in \mathcal{L}_I$ holds for all $t \in [T, T + \varepsilon)$. □

In the mode transition rule, at the event time $T$, every mode $I$ for which violation of the location invariant $\mathcal{L}_I$ from $x(T)$ is not imminent, is selected as a new mode $I(T^+)$. Therefore, there may exist more than two modes or no mode as a new mode $I(T^+)$, which leads to more than two solutions or no solution from $x(T^+)$. Note that, however, there exists neither finite nor infinite transition sequence of the discrete state at any event time $T$.

In this way, for the system with the same initial state $(I(0), x(0))$, both transition rules, i.e., the switch-based rule and the mode-based rule, provide the same first event time but in general different new discrete states at that event time. Furthermore, in some real systems with switches, the switching rules are given in the respective switches, in other words, the switching rule of each switch is individually realized by a digital device or by a mechanical device (see, e.g., a simple example in Section 5). In the behavior analysis of such a system, it may be more natural from the practical point of view to use the switch-based transition rule than the mode-based transition rule. On the other hand, the mode-based transition rule may be rather natural in the case of discontinuous physical systems. See [8] for more details.

In Section 3, after the strict definition of solutions in the next subsection, we will give a further discussion on the relation between these two transition rules from the viewpoints of well-posedness. We hereafter denote the system $\Sigma$ of (2.1) under the switch-based transition rule by $\Sigma_S$, on the other hand, the system $\Sigma$ of (2.1) under the mode-based transition rule by $\Sigma_M$. However, the symbol $\Sigma$ is used for both $\Sigma_S$ and $\Sigma_M$ in case we do not need to distinguish these systems.

## 2.2 Definition of solution concepts

Let us define the concept of solutions in the system $\Sigma$. A point $\hat{t}$ is said to be a right (left)-accumulation point of event times [3], if there exists a sequence $\{t_i\}$ of event times such that $t_i < (>) t_{i+1}$ and $\lim_{i\to\infty} t_i = \hat{t} < \infty$. The following kind of solution is defined.

**Definition 2.4.** We say that $(\delta(t), x(t))$ is a *hybrid solution (or simply, solution)* of $\Sigma_S(\Sigma_M)$ on $[t_0, t_1)$, if, for a given initial hybrid state $(\delta(t_0), x(t_0))$, $(\delta(t), x(t))$ satisfies on $[t_0, t_1)$ the following conditions.

(i) $x(t) = x(t_0) + \int_{t_0}^{t} f(x(\tau), \delta(\tau))d\tau$, where $f(x, \delta)$ expresses the right-hand side of (2.1) and $x(t_0) = x_0$.

(ii) Each event takes place according to Definition 2.1.

(iii) $\delta(t)$ is determined under the switch-(mode-)based transition rule at each event time.

(iv) There exist no left-accumulation points of event times. □

The condition of disallowing the existence of left-accumulation points (left-a.p.) of event times is necessary for the existence of a local solution in one mode. On the other hand, a solution with right-accumulation points (right-a.p.) of event times, which is called a Zeno trajectory (page 10, [18]), is allowed in the above definition of solutions. However, the following assumption is made throughout this paper.

**Assumption 2.5.** For the system $\Sigma$, no right-a.p. of event times exist. □

In [6], Assumption 2.5 is not required by considering a successively connected solution from the state $x(t_\infty)$, where $t_\infty$ is a right-a.p. of event times, since $x(t_\infty)$ always has a well-defined limit. On the other hand, in Definition 2.4, we consider the discrete state $\delta$ as well as the continuous state $x$ as a solution of the system. In this case, the discrete state $\delta(t_\infty)$ at a right-a.p. never has a well-defined limit under the switch-based transition rule; while it may have in some case a well-defined limit under the mode-based transition rule, because $I(t_\infty)$ is determined based on the possibility of violation of the location invariant from $x(t_\infty)$ in each mode. In either case, however, it will be difficult to characterize in a general way the non-existence condition of a right-a.p. $t_\infty$ or a well-defined limit of $I(t_\infty)$ if $t_\infty$ exists even in the bimodal case. Therefore, we here make this assumption (see [13, 4] for related discussion).

Now let us define well-posedness of the system $\Sigma$ as follows.

**Definition 2.6.** The system $\Sigma$ is said to be well-posed if for every initial state $(\delta(t_0), x(t_0)) \in \mathcal{D} \times \mathcal{R}^n$ there exists a right-unique solution of $\Sigma$ on $[t_0, \infty)$. □

It is remarked that we do not treat (higher-order) sliding modes in the sense of Filippov as is often discussed in relay systems. This is because the so-called equivalent control input, which forces the continuous state to stay in the switching surface, cannot be theoretically generated in the case of *binary* type of switches, and also unexpected phenomena such as chattering occur as a practical matter. So in the case that there exists a sliding mode from some $x_0$ when binary switches in (2.1) are replaced by relay switches such as $\delta_j = 0$ if $\bar{S}_j\bar{x} > 0$, $\delta_j \in [0,1]$ if $\bar{S}_j\bar{x} = 0$, and $\delta_j = 1$ if $\bar{S}_j\bar{x} < 0$, we judge under the above definition that no solution exists from $x_0$.

Let us consider a simple example: $\dot{x}(t) = 1 - 2\delta(t)$, $\delta = 0$ if $x \geq 0$ and $\delta = 1$ if $x \leq 0$ under the switch-based transition rule. Although the system appears to have two solutions of $x(t)$ from $x(0) = 0$, it turns out that a unique solution $(\delta(t), x(t))$ exists for every initial state $(\delta(0), x(0)) \in \{1,2\} \times \mathcal{R}$ (note that if $(\delta(0), x(0)) = (0, -1)$, an event takes place at $t = 0$ and $\delta(0^+) = 1$. Then $(\delta(t), x(t)) = (1, -t-1)$ holds for all $t \geq 0$.) Thus this system is well-posed. On the other hand, for the system $\dot{x}(t) = -1 + 2\delta(t)$, $\delta = 0$ if $x \geq 0$ and $\delta = 1$ if $x \leq 0$ under the switch-based transition rule, the system has solutions from $x(0) = 0$ in neither mode, which implies that the system is not well-posed. In this way, although the rule expressed by $\bar{S}_j\bar{x} \geq 0$ and $\bar{S}_j\bar{x} \leq 0$ at first sight seems to lead to nondeterminism, there does not necessarily exist a solution from the initial state $x(0)$ satisfying $\bar{S}_j\bar{x} = 0$ in each mode, so it is important to characterize, in each mode, the region of all the states $x$ from which there exists a solution, which will be called below the smooth-continuation set. Similarly for the case of the mode-based transition rule.

# 3 Smooth continuation and well-posedness

Let $\mathcal{L}$ be a set in $\mathcal{R}^n$. For an initial state $x(0) \in \mathcal{R}^n$, if there exists an $\varepsilon > 0$ such that $x(t) \in \mathcal{L}$ for all $t \in [0, \varepsilon)$, we say that smooth continuation is possible from $x(0)$ in $\mathcal{L}$ [16]. We also call a set of all $x(0)$ from which smooth continuation is possible in $\mathcal{L}_I$ the *smooth-continuation set* in $\mathcal{L}_I$ with respect to the system in mode $I$, denoted by $\mathcal{S}_I$. Thus by definition $\mathcal{S}_I$ expresses the region where a local unique solution exists in mode $I$.

The smooth-continuation set is characterized by using the lexicographic inequalities (see [6] for the details), as defined below: for $x \in \mathcal{R}^n$, if for some $i$, $x_j = 0$ $(j = 1, 2, \ldots, i-1)$, while $x_i > (<) 0$, we denote it by $x \succ (\prec) 0$. In other words, the set $\{\, x \in \mathcal{R}^n \mid x \succ (\prec) 0 \,\}$ expresses the set of all $x$ whose first nonzero element is positive (negative). Further, if $x = 0$ or $x \succ (\prec) 0$, we denote it by $x \succeq (\preceq) 0$. Let $y^{(k)}$ be the $k$th derivative of $y = \bar{S}\bar{x}$ along the trajectory of $\dot{\bar{x}} = \bar{A}\bar{x}$, i.e., $y^{(k)} = \bar{S}\bar{A}^k\bar{x}$. Then the smooth-continuation set $\mathcal{S}$ in $\{x \in \mathcal{R}^n \mid \bar{S}\bar{x} \geq 0,\ \bar{x} = [x^T\ 1]^T\,\}$ is expressed as

$$\mathcal{S} = \left\{ x \in \mathcal{R}^n \;\middle|\; [\, y \;\; y^{(1)} \;\; \cdots \;\; y^{(\gamma-1)} \,]^T \succeq 0 \right\} \tag{3.1}$$

where $\gamma$ is the observability index of $(\bar{S}, \bar{A})$. Note that the smooth-continuation set $\mathcal{S}$ expresses all $x$ satisfying $y > 0$, or $y = 0$ and $y^{(1)} > 0$, or $\cdots$, or $y = y^{(1)} = \cdots = y^{(\gamma-2)} = 0$ and $y^{(\gamma-1)} \geq 0$.

More detailed discussion leads to the following lemma. Let us define the following four sets for any given $k$: the sets of all $(\bar{S}, \bar{A})$ such that

$$\bar{S}\bar{A}^{k-1} = \sum_{i=1}^{k-1} \alpha_i \bar{S}\bar{A}^{i-1} + \beta[0_{1,n}\ 1] \tag{3.2}$$

with $\alpha_i \in \mathcal{R}$ and $\beta = 0$, $\beta > 0$, and $\beta < 0$, are denoted by $\mathcal{P}_0^k$, $\mathcal{P}_{2+}^k$, and $\mathcal{P}_{2-}^k$, respectively. Also $\mathcal{P}_1^k$ denotes the set of all $(\bar{S}, \bar{A})$ that do not belong to any of $\mathcal{P}_0^s$, $\mathcal{P}_{2+}^2$, $\mathcal{P}_{2-}^s$. Notice that if $(\bar{S}, \bar{A}) \in \mathcal{P}_1^k$, then $\bar{S}\bar{A}^{k-1} \notin \mathrm{span}\{\bar{S}, \bar{S}\bar{A}, \ldots, \bar{S}\bar{A}^{k-2}\}$ and $SA^{k-1} \notin \mathrm{span}\{S, SA, \ldots, SA^{k-2}\}$. Then we have the following lemma.

**Lemma 3.1 ([5, 9]).** *Let $\gamma$ be the observability index of the pair $(\bar{S}, \bar{A})$. The smooth-continuation set $\mathcal{S}$ in $\bar{S}\bar{x} \geq 0$ with respect to $\dot{\bar{x}} = \bar{A}\bar{x}$ is given by*

$$\mathcal{S} = \begin{cases} \{x \in \mathcal{R}^n \mid \bar{T}(\gamma)\bar{x} \succeq 0\} & \text{if } (\bar{S}, \bar{A}) \in \mathcal{P}_1^\gamma, \\ \{x \in \mathcal{R}^n \mid \bar{T}(\gamma-1)\bar{x} \succeq 0\} & \text{if } (\bar{S}, \bar{A}) \in \mathcal{P}_{2+}^\gamma, \\ \{x \in \mathcal{R}^n \mid \bar{T}(\gamma-1)\bar{x} \succ 0\} & \text{if } (\bar{S}, \bar{A}) \in \mathcal{P}_{2-}^\gamma, \end{cases} \qquad \bar{T}(i) \triangleq \begin{bmatrix} \bar{S} \\ \bar{S}\bar{A} \\ \vdots \\ \bar{S}\bar{A}^{i-1} \end{bmatrix}.$$

Using such smooth-continuation sets, the well-posedness property in each of $\Sigma_S$ and $\Sigma_M$ is characterized as follows.

**Lemma 3.2 ([8]).** *Suppose Assumption 2.5 holds. Then the following statements hold.*

*(i) The system $\Sigma_S$ is well-posed if and only if $\bigcup_{I=1}^{2^r} \mathcal{S}_I = \mathcal{R}^n$ and there exists no livelock in the discrete transition.*

*(ii) The system $\Sigma_M$ is well-posed if and only if $\bigcup_{I=1}^{2^r} \mathcal{S}_I = \mathcal{R}^n$, and for every $x \in \mathcal{R}^n$, there exist no $I, J, K \in \mathcal{I}$ such that $x \in cl\mathcal{S}_I - \mathcal{S}_I$ and $x \in \mathcal{S}_J \bigcap \mathcal{S}_K$, where $I, J, K$ are different, and $cl\mathcal{S}_I$ is the closure of $\mathcal{S}_I$.*

The condition $\bigcup_{I=1}^{2^r} \mathcal{S}_I = \mathcal{R}^n$ means that from every $x_0 \in \mathcal{R}^n$ smooth continuation is possible in at least one mode. The latter of condition (ii) guarantees that the mode selected as a new discrete state at each event time is unique.

Let $\mathcal{F}(\Sigma_S)$ be a set of all well-posed systems of $\Sigma_S$, and $\mathcal{F}^c(\Sigma_S)$ be the complementary set of $\mathcal{F}(\Sigma_S)$. Similarly for $\mathcal{F}(\Sigma_M)$ and $\mathcal{F}^c(\Sigma_M)$. Then the following relation between $\Sigma_S$ and $\Sigma_M$ is obtained.

**Lemma 3.3 ([8]).**

*(i) If $\Gamma \in \mathcal{F}(\Sigma_S) \bigcap \mathcal{F}(\Sigma_M)$, $\Gamma_S$ and $\Gamma_M$ have the same continuous states $x(t)$ of hybrid solutions.*

*(ii)* $\mathcal{F}(\Sigma_M)\bigcap\mathcal{F}^c(\Sigma_S)\neq\emptyset$ *and* $\mathcal{F}(\Sigma_S)\bigcap\mathcal{F}^c(\Sigma_M)\neq\emptyset$.

*(iii) In the bimodal case,* $\Sigma_S$ *and* $\Sigma_M$ *have the same hybrid solutions, and* $\Sigma_S$ *(or equivalently* $\Sigma_M$*) is well-posed if and only if* $\mathcal{S}_1\bigcup\mathcal{S}_2=\mathcal{R}^n$.

From (i), if the system is well-posed under both transition rules, the continuous states of hybrid solutions are the same in both rules. From (ii), we see that there exists a system that is (is not) well-posed under the mode-based transition rule and is not (is) well-posed under the switch-based transition rule. The claim (i) straightforwardly follows from Lemma 3.2. The claim (ii) is proven by showing examples in Example 3.4 below. The claim (iii) follows from the fact that if smooth continuation is impossible in one mode, the candidate of the next mode is uniquely determined at any event time. Note that nonexistence of livelock follows from $\mathcal{S}_1\bigcup\mathcal{S}_2=\mathcal{R}^n$.

**Example 3.4.** Consider the system given by

$$\Sigma_1:\ \dot{x}=\left(\begin{bmatrix}0&1\\-1&0\end{bmatrix}+\delta_1\begin{bmatrix}0&-2\\2&0\end{bmatrix}+\delta_2\begin{bmatrix}0&2\\0&0\end{bmatrix}\right)x+\begin{bmatrix}0\\-1\end{bmatrix}+\delta_1\begin{bmatrix}1\\0\end{bmatrix}$$

$$+\delta_2\begin{bmatrix}-1\\1\end{bmatrix},\quad \delta_1=\begin{cases}0 & \text{if } x_1\geq 0,\\ 1 & \text{if } x_1\leq 0,\end{cases}\quad \delta_2=\begin{cases}0 & \text{if } x_2\geq 0,\\ 1 & \text{if } x_2\leq 0.\end{cases}$$

Then from Lemma 3.1 the smooth-continuation sets are given by

$$\begin{aligned}
\mathcal{S}_1 &= \left\{x\in\mathcal{R}^2\,\middle|\,\begin{bmatrix}x_1\\x_2\end{bmatrix}\succ 0,\ \begin{bmatrix}x_2\\-x_1-1\end{bmatrix}\succeq 0\right\},\\
\mathcal{S}_2 &= \left\{x\in\mathcal{R}^2\,\middle|\,\begin{bmatrix}x_1\\-x_2+1\end{bmatrix}\prec 0,\ \begin{bmatrix}x_2\\x_1-1\end{bmatrix}\succeq 0\right\},\\
\mathcal{S}_3 &= \left\{x\in\mathcal{R}^2\,\middle|\,\begin{bmatrix}x_1\\3x_2-1\end{bmatrix}\succeq 0,\ \begin{bmatrix}x_2\\-x_1\end{bmatrix}\prec 0\right\},\\
\mathcal{S}_4 &= \left\{x\in\mathcal{R}^2\,\middle|\,\begin{bmatrix}x_1\\x_2\end{bmatrix}\preceq 0,\ \begin{bmatrix}x_2\\x_1\end{bmatrix}\preceq 0\right\}
\end{aligned}$$

where the correspondence of mode $I$ to the switch vector $\delta$ is the same as to that of (2.2). This leads to $\bigcup_{I=1}^4\mathcal{S}_I=\mathcal{R}^2$. As seen in Figure 3.1(a), where all four marks of white or black circles express the origin, and the arrows between these marks express the directions of the discrete transitions under the switch-based transition rule, condition (ii)(b) in Lemma 3.2 is satisfied for $\Sigma_{M1}$, while in $\Sigma_{S1}$ the livelock between mode 2 and mode 3 takes place just after a solution $x(t)$ from $x(0)$ in the interior of the region of mode 2 arrives at the origin. Hence $\Sigma_{M1}$ is well-posed, but $\Sigma_{S1}$ is not well-posed.

Consider the other example given by

$$\Sigma_2:\ \dot{x}=\left(\begin{bmatrix}0&1\\0&0\end{bmatrix}+\delta_1\begin{bmatrix}0&0\\1&0\end{bmatrix}+\delta_2\begin{bmatrix}0&0\\1&0\end{bmatrix}\right)x+\delta_1\begin{bmatrix}0\\-1\end{bmatrix}+\delta_2\begin{bmatrix}-1\\-1\end{bmatrix},$$

$$\delta_1=\begin{cases}0 & \text{if } x_1\geq 0,\\ 1 & \text{if } x_1\leq 0,\end{cases}\quad \delta_2=\begin{cases}0 & \text{if } x_2\geq 0,\\ 1 & \text{if } x_2\leq 0.\end{cases}$$

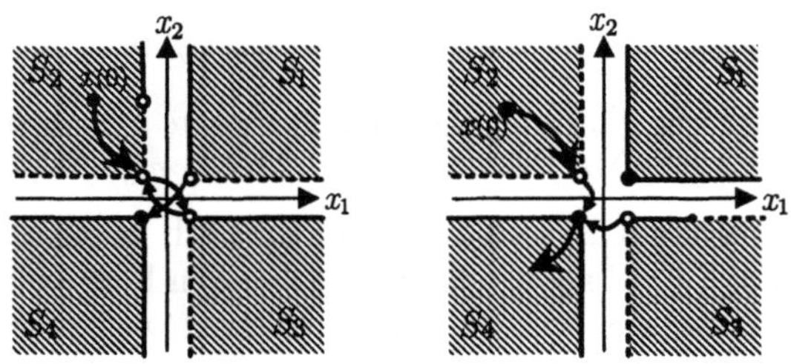


(a) $\Sigma_1$ (b) $\Sigma_2$

FIGURE 3.1. Example 3.4

Likewise, the smooth-continuation sets are given by

$$\begin{aligned}
\mathcal{S}_1 &= \left\{ x \in \mathcal{R}^2 \middle| \begin{bmatrix} x_1 \\ x_2 \end{bmatrix} \succeq 0,\ x_2 \geq 0 \right\}, \\
\mathcal{S}_2 &= \left\{ x \in \mathcal{R}^2 \middle| \begin{bmatrix} x_1 \\ x_2 \end{bmatrix} \preceq 0,\ \begin{bmatrix} x_2 \\ x_1 - 1 \end{bmatrix} \succeq 0 \right\}, \\
\mathcal{S}_3 &= \left\{ x \in \mathcal{R}^2 \middle| \begin{bmatrix} x_1 \\ x_2 - 1 \end{bmatrix} \succ 0,\ \begin{bmatrix} x_2 \\ x_1 - 1 \end{bmatrix} \preceq 0 \right\}, \\
\mathcal{S}_4 &= \left\{ x \in \mathcal{R}^2 \middle| \begin{bmatrix} x_1 \\ x_2 - 1 \end{bmatrix} \preceq 0,\ \begin{bmatrix} x_2 \\ 2x_1 - 2 \end{bmatrix} \preceq 0 \right\}
\end{aligned}$$

which implies that $\bigcup_{I=1}^{4} \mathcal{S}_I = \mathcal{R}^2$, and that $\Sigma_{M2}$ is not well-posed, but $\Sigma_{S2}$ is well-posed. In fact, for a solution $x(t)$ from $x(0)$ in the interior of the region of mode 2 that arrives at the origin, modes 1 and 4 can be taken under the mode-based rule as new modes. Hence $\Sigma_{M2}$ is not well-posed. $\Sigma_{S2}$ is well-posed since no livelock exists as seen in Figure 3.1(b). □

# 4 Easily checkable well-posedness conditions

## 4.1 Bimodal case

In this subsection, we consider the bimodal case, i.e., $r = 2$ in (2.1). So for brevity of notation, we consider the system given by

$$\Sigma_B \left\{ \begin{array}{lcll} \text{mode } 1 : \dot{\bar{x}} & = & \bar{A}\bar{x}, & \text{if } \bar{S}\bar{x} \geq 0, \\ \text{mode } 2 : \dot{\bar{x}} & = & (\bar{A} + \bar{G}\bar{F})\bar{x}, & \text{if } \bar{S}\bar{x} \leq 0 \end{array} \right. \tag{4.1}$$

where $\bar{G} = \begin{bmatrix} G^T & 0 \end{bmatrix}^T \in \mathcal{R}^{(n+1)\times \bar{l}}$, $\bar{F} = \begin{bmatrix} F & f \end{bmatrix} \in \mathcal{R}^{\bar{l}\times(n+1)}$, and $G \in \mathcal{R}^{n\times \bar{l}}$, $F \in \mathcal{R}^{\bar{l}\times n}$, and $f \in \mathcal{R}^{\bar{l}}$. Let $\bar{G}_k$ $(k = 1, 2, \ldots, \bar{l})$ be the $j$th column vector of $\bar{G}$, let $\bar{F}_k$ $(k = 1, 2, \ldots, \bar{l})$ be the $k$th row vector of $\bar{F}$. Let $q_k$ $(k = 1, 2, \ldots, \bar{l})$ be the relative degree of $(\bar{C}, \bar{A}, \bar{G}_k)$ (i.e., $q_k$ is defined by the condition of $\bar{S}\bar{G}_k = \bar{S}\bar{A}\bar{G}_k = \cdots = \bar{S}\bar{A}^{q_k-2}\bar{G}_k = 0$, and $\bar{S}\bar{A}^{q_k-1}\bar{G}_k \neq 0$). Let also $\mu_{q_k}$ be the leading Markov parameters of $(\bar{S}, \bar{A}, \bar{G}_k)$, namely, $\mu_{q_k} = \bar{C}\bar{A}^{q_k-1}\bar{G}_k$. Then using Lemmas 3.1 and 3.3(iii), we obtain the following result.

**Theorem 4.1 ([5, 9]).** *For some $l \leq \bar{l}$, assume $q_1 < q_2 < \cdots < q_l (\leq n)$, and $q_{l+1} = q_{l+2} = \cdots = q_{\bar{l}} = \infty$, without loss of generality. Then the system $\Sigma_B$ of (4.1) is well-posed if and only if there exists a $p \in \{1, 2, \ldots, l\}$ such that $\bar{F}_k$ $(k = 1, 2, \ldots, p)$ has the form*

$$\mu_{q_k}\bar{F}_k = \sum_{j=1}^{q_k+1} \alpha_j^k \bar{S}\bar{A}^{j-1} + \beta_k [0_{1,n} \;\; 1], \quad \alpha_j^k \in \mathcal{R}(j = 1, 2, \ldots, q_k) \qquad (4.2)$$

*with*

*(a) $1 + \alpha_{q_k+1}^k > 0$ and $\beta_k = 0$ for $k = 1, 2, \ldots, p-1$,*

*(b) $(1 + \alpha_{q_p+1}^p = 0, \beta_p = 0)$ or $(1 + \alpha_{q_p+1}^p \geq 0, \beta_p < 0)$ for $k = p$*

*where if $p = l$, condition (b) is replaced by*

*(b1) $\bar{F}_l$ is arbitrary if $(\bar{S}, \bar{A}) \in \mathcal{P}_0^{q_l+1}$ or $\mathcal{P}_{2+}^{q_l+1}$,*

*(b2) $1 + \alpha_{q_l+1}^l \geq 0$ and $\beta_l \leq 0$ if $(\bar{S}, \bar{A}) \in \mathcal{P}_1^{q_l+1}$,*

*(b3) $1 + \alpha_{q_l+1}^l \geq 0$ and $\beta_l = 0$ if $(\bar{S}, \bar{A}) \in \mathcal{P}_{2-}^{q_l+1}$.*

From Lemma 3.3(iii), we do not have to distinguish the switch-based rule and the mode-based rule in the bimodal case. Note also that the conditions obtained in the above theorem are algebraically checkable. Further, we can prove that the well-posedness property does not depend on the decomposition into $\bar{G}$ and $\bar{F}$.

## 4.2 Multi-modal case

In this subsection, a sufficient condition for well-posedness of the multi-modal system $\Sigma_S$ under the switch-based transition rule is derived in terms of well-posedness of bimodal systems.

In the following theorem, we use the three symbols $\delta_j$, $\hat{\delta}_j$, $\tilde{\delta}_j$, $j \in \mathcal{J}$; $\delta_j$ expresses the $j$th switch that changes from 0 to 1 or from 1 to 0 according to the condition on $\bar{S}_j\bar{x}$ of (2.1), $\hat{\delta}_j$ expresses the $j$th switch whose value is fixed with either 0 or 1, and $\tilde{\delta}_j$ expresses the $j$th switch defined by

$$\tilde{\delta}_j = \begin{cases} \tilde{\delta}_{j,h}^+ & \text{if } \bar{S}_h\bar{x} \geq 0, \\ \tilde{\delta}_{j,h}^- & \text{if } \bar{S}_h\bar{x} \leq 0, \end{cases}$$

where some $h \in \mathcal{J}/\{j\}$ (the difference set $\mathcal{J} - \{j\}$), and $\tilde{\delta}_{j,h}^+, \tilde{\delta}_{j,h}^- \in \{0, 1\}$.

**Theorem 4.2 ([8]).** *The system $\Sigma_S$ of (2.1) is well-posed under the switch-based transition rule if there exists an $h \in \mathcal{J}$ satisfying the following conditions.*

*(i) For every $\hat{\delta}_h \in \{0,1\}$, the following $2^{r-1}$-modal system is well-posed under the switch-based transition rule*

$$\dot{\bar{x}} = (\bar{A} + \hat{\delta}_h \bar{G}_h \bar{F}_h + \sum_{j \in \mathcal{J}/\{h\}} \delta_j \bar{G}_j \bar{F}_j)\bar{x}. \tag{4.3}$$

*(ii) For every $j \in \mathcal{J}/\{h\}$ and every $(\tilde{\delta}_j^+, \tilde{\delta}_j^-) \in \{0,1\} \times \{0,1\}$, the following bimodal system is well-posed:*

$$\dot{\bar{x}} = (\bar{A} + \delta_h \bar{G}_h \bar{F}_h + \sum_{j \in \mathcal{J}/\{h\}} \tilde{\delta}_j \bar{G}_j \bar{F}_j)\bar{x}. \tag{4.4}$$

In order to give an interpretation of Theorem 4.2, consider here at first the simplest case, i.e., $r = 2$.

The condition in Theorem 4.2 in this case is rewritten as follows: there exists a $(j_1, j_2) \in \{(1,2),(2,1)\}$ such that every bimodal system below is well-posed.

(i') (a) $\dot{\bar{x}} = (\bar{A} + \delta_{j_1} \bar{G}_{j_1} \bar{F}_{j_1})\bar{x}$. (b) $\dot{\bar{x}} = (\bar{A} + \bar{G}_{j_2} \bar{F}_{j_2} + \delta_{j_1} \bar{G}_{j_1} \bar{F}_{j_1})\bar{x}$.

(ii') (c) $\dot{\bar{x}} = (\bar{A} + \delta_{j_2} \bar{G}_{j_2} \bar{F}_{j_2})\bar{x}$. (d) $\dot{\bar{x}} = (\bar{A} + \delta_{j_2} \bar{G}_{j_2} \bar{F}_{j_2} + (1-\delta_{j_2}) \bar{G}_{j_1} \bar{F}_{j_1})\bar{x}$.
(e) $\dot{\bar{x}} = (\bar{A} + \delta_{j_2} \bar{G}_{j_2} \bar{F}_{j_2} + \delta_{j_2} \bar{G}_{j_1} \bar{F}_{j_1})\bar{x}$. (f) $\dot{\bar{x}} = (\bar{A} + \bar{G}_{j_1} \bar{F}_{j_1} + \delta_{j_2} \bar{G}_{j_2} \bar{F}_{j_2})\bar{x}$.

Here for $h = j_2$, (i') (a),(b) correspond to condition (i) in Theorem 4.2 with $\hat{\delta}_{j_2} = 0$ and $\hat{\delta}_{j_2} = 1$, respectively, and (ii')(c)–(f) to condition (ii) with $(\tilde{\delta}_{j_1,j_2}^+, \tilde{\delta}_{j_1,j_2}^-) = (0,0)$, $(1,0)$, $(0,1)$, and $(1,1)$, respectively.

For the general case, condition (i) in Theorem 4.2 corresponds to that, for some switch $\delta_h$, even if the value of that switch is fixed, the system with the other unfixed switches still is well-posed. Condition (ii) implies that even if, synchronized with the switching action of $\delta_h$, all the other switches change in an arbitrary way, the corresponding bimodal system is still well-posed. Notice that condition (i) on well-posedness of multi-modal systems can be verified by using recursively the result of Theorem 4.2.

From the other side, the result in Theorem 4.2 is interpreted as follows: in the situation that for a given multi-modal system that is well-posed, some binary-switch is newly added to it to produce the more complicated system, the result in Theorem 4.2 presents what condition is sufficient for the newly integrated switch to ensure well-posedness of the total system created in such a way.

Using Theorems 4.1 and 4.2, it can be verified that the system $\Sigma_{S2}$ in Example 3.4 is well-posed. Since $\Sigma_2$ is not well-posed under the mode-transition rule as seen in Example 3.4, it turns out that the condition in

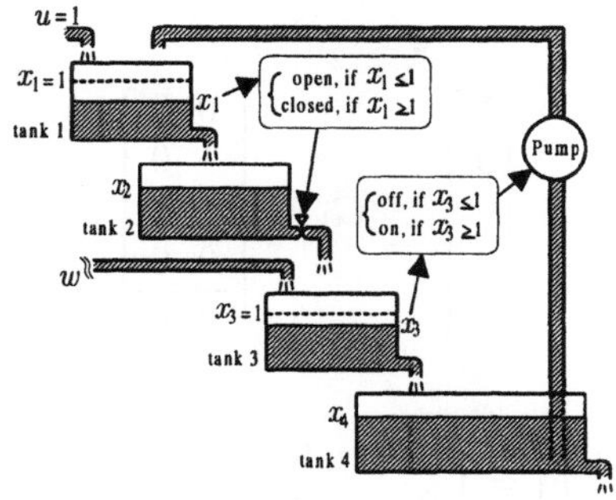


FIGURE 5.1. 4-tank system

Theorem 4.2 is not sufficient for well-posedness of the system $\Sigma_M$ under the mode-based transition. So in this sense, the sufficient condition in Theorem 4.2 is not restrictive. However, it is conservative at least in the following two senses. Concerning condition (i), it is in fact enough that each bimodal system is well-posed in some limited subregion of $\mathcal{R}^n$. However an algebraic characterization of such a condition requires more complicated calculations, although it is possible. On the other hand, condition (ii) guarantees that there exists no livelock for all discrete transitions, although some discrete transitions may not in fact occur. In this sense, condition (ii) is conservative. By taking these points into consideration, a more practical well-posedness condition will be derived.

# 5 Example: 4-tank system with binary logic

Consider an illustrative example of the 4-tank system in Figure 5.1, where $x_i$ is the water level of tank $i$, $i = 1, 2, 3, 4$, and the constant input $u = 1$ is the volume of water discharged into tank 1. The valve at tank 2 is *open* or *closed*, and the switch of the pump is *on* or *off* according to each rule shown in Figure 5.1. This switching logic implies that if the water of tank 1 is less than 1, the valve of tank 2 is open to increase the water of tank 3 and, as a result, to switch on the pump. Conversely if the water of tank 1 is greater than 1, the valve of tank 2 is closed in order to switch off the pump. We suppose that the switching action on the valve and the pump is executed independently by computer programming embedded in the respective switching devices, and also that the state of each switch is memorized for determining the new discrete state in the discrete transition. So we use the switch-based transition rule to analyze the solution behavior of this system.

Equations of motion of this system are supposed to be given by

$$\dot{\bar{x}} = (\bar{A} + \delta_1 \bar{G}_1 \bar{F}_1 + \delta_2 \bar{G}_2 \bar{F}_2)\bar{x} \tag{5.1}$$

where

$$\bar{A} = \left[\begin{array}{cccc|c} -1 & 0 & 0 & 0 & 1 \\ 1 & -1 & 0 & 0 & 0 \\ 0 & 1 & -1 & 0 & 0 \\ 0 & 0 & 1 & -1 & 0 \\ \hline 0 & 0 & 0 & 0 & 0 \end{array}\right], \quad \bar{G}_1 = \left[\begin{array}{cc} 0 & 0 \\ 1 & 0 \\ 0 & 1 \\ 0 & 0 \\ \hline 0 & 0 \end{array}\right], \quad \bar{G}_2 = \left[\begin{array}{cc} 1 & 0 \\ 0 & 0 \\ 0 & 0 \\ 0 & 1 \\ \hline 0 & 0 \end{array}\right],$$

$$\delta_i = \begin{cases} 0 & \text{if } \bar{S}_i\bar{x} \geq 0 \\ 1 & \text{if } \bar{S}_i\bar{x} \leq 0 \end{cases}, \quad \bar{F}_1 = \left[\begin{array}{cccc|c} 0 & 1 & 0 & 0 & 0 \\ 0 & -1 & 0 & 0 & 0 \end{array}\right], \quad \bar{F}_2 = \left[\begin{array}{cccc|c} 0 & 0 & 0 & 0 & 1 \\ 0 & 0 & 0 & 0 & -1 \end{array}\right],$$

$$\bar{S}_1 = \left[\begin{array}{ccccc} -1 & 0 & 0 & 0 & 1 \end{array}\right], \quad \bar{S}_2 = \left[\begin{array}{ccccc} 0 & 0 & -1 & 0 & 1 \end{array}\right].$$

For simplicity, all coefficients are normalized to 1. We also suppose that the pump supplies 1 unit of water per 1 second to tank 1 when the switch of the pump is on. The open/closed switch of the valve at tank 2 is expressed by $\delta_1 = 0/1$, and the on/off switch of the pump is expressed by $\delta_2 = 1/0$, respectively.

Now using Theorem 4.2 (i.e., conditions (i'),(ii') just after Theorem 4.2), let us check well-posedness of this system. We will show that conditions (i'),(ii') hold for $(j_1, j_2) = (2, 1)$. So first, in the case (i')(a), the relative degrees $q_{2i}$, $i = 1, 2$, of $(\bar{S}_2, \bar{A}, \bar{G}_{2i})$ are 3 and $\infty$, respectively, and $\eta_{q_{21}} = \bar{S}_2\bar{A}^2\bar{G}_{21} = -1$. Then let us apply Theorem 4.1. From $l = 1$ and $\bar{l} = 2$, we only have to consider the case $p = l = 1$. Then we have $(\bar{S}_2, \bar{A}) \in \mathcal{P}^4_{2-}$ from $\bar{S}_2\bar{A} = [0\ -1\ 1\ 0\ 0]$, $\bar{S}_2\bar{A}^2 = [-1\ 2\ -1\ 0\ 0]$, and $\bar{S}_2\bar{A}^3 = [3\ -3\ 1\ 0\ -1]$. In addition, we obtain $\eta_{q_{21}}\bar{F}_{21} = -\bar{S}_2 - 3\bar{S}_2\bar{A} - 3\bar{S}_2\bar{A}^2 - \bar{S}\bar{A}^3$. Thus since condition (b3) in Theorem 4.1 holds and no condition on $\bar{F}_{22}$ is required for well-posedness, the system is well-posed. On the other hand, in the case (i')(b), since the relative degrees of $(\bar{S}_2, \bar{A} + \bar{G}_1\bar{F}_1, \bar{G}_{2i})$, $i = 1, 2$, are all $\infty$, the corresponding system is well-posed.

In a similar way, it turns out every system in the cases (ii')(c)–(f) is well-posed. We have thus concluded that the system (5.1) is well-posed.

# Acknowledgement

The author thanks Professor A. J. van der Schaft, University of Twente, for the collaboration that started the research presented in this paper.

## References

[1] A. F. Filippov, *Differential equations with discontinuous righthand sides.* Dordrecht: Kluwer, 1988.

[2] L. Fridman and A. Levant, "Higher order sliding modes as a natural phenomenon in control theory," in *Robust Control via Variable Structure and Lyapunov Techniques*, F. Garofalo and L. Glielmo, Eds. Lecture Notes in Control and Information Science, vol. 217, 1996, pp. 107–133.

[3] W. P. M. H. Heemels, J. M. Schumacher, and S. Weiland, "Linear complementarity systems," *SIAM J. Applied Mathematics*, vol. 60, no. 4, pp. 1234–1269, 2000.

[4] W. P. M. H. Heemels, "Linear complementarity systems; a study in hybrid dynamics," Ph.D. thesis, Eindhoven Univ. of Technology, 1999.

[5] J. Imura and A. J. van der Schaft, "Well-posedness of a class of dynamically interconnected systems," in *Proc. of the 38th IEEE Conference on Decision and Control*, 1999, pp. 3031–3036.

[6] J. Imura and A. J. van der Schaft, "Characterization of well-posedness of piecewise linear systems," *IEEE Trans. on Automatic Control*, vol. 45, no. 9, pp. 1600–1619, 2000.

[7] J. Imura, "Feedback well-posedness and stabilizability of bimodal piecewise affine systems," in *the CD-ROM Proceedings at MTNS*, 2000.

[8] J. Imura, "Well-posedness analysis of switch-driven hybrid systems," in *Proc. of the 2001 American Control Conference*, pp. 862–867, 2001.

[9] J. Imura, "Classification and stabilizability analysis of bimodal piecewise affine systems," *Int. J. Robust and Nonlinear Control*, vol. 12, no. 10, pp. 897–926, 2002.

[10] K. H. Johansson, M. Egerstedt, J. Lygeros, and S. Sastry, "On the regularization of Zeno hybrid automata," *Systems & Control Letters*, vol. 38, no. 3, pp. 141–150, 1999.

[11] K. H. Johansson, A. Rantzer, and K. J. Åström, "Fast switches in relay feedback systems," *Automatica*, vol. 35, no. 4, pp. 539–552, 1999.

[12] Y. J. Lootsma, A. J. van der Schaft, and M. K. Çamlıbel, "Uniqueness of solutions of linear relay systems," *Automatica*, vol. 35, no. 3, pp. 467–478, 1999.

[13] A. Y. Pogromsky, W. P. M. H. Heemels, and H. Nijmeijer, "On well-posedness of relay systems," in *Proc. of the 5th IFAC Symposium on Nonlinear Control Systems (Nolcos'01)*, 2001, pp. 1537–1542.

[14] L. Tavernini, "Differential automata and their discrete simulations," *Nonlinear analysis, Theory, Methods, and Applications*, vol. 11, no. 6, pp. 665–683, 1987.

[15] V.I. Utkin, "Variable structure systems with sliding modes: a survey," *IEEE Trans. on Automatic Control*, vol. 22, no. 1, pp. 212–222, 1977.

[16] A. J. van der Schaft and J. M. Schumacher, "The complementary-slackness class of hybrid systems," *Mathematics of Control, Signals, and Systems*, vol. 9, pp. 266–301, 1996.

[17] A. J. van der Schaft and J. M. Schumacher, "Complementarity modeling of hybrid systems," *IEEE Trans. on Automatic Control*, vol. 43, no. 4, pp. 483–490, 1998.

[18] A. J. van der Schaft and J. M. Schumacher, *An Introduction to Hybrid Dynamical Systems*. Lecture Notes in Control and Information Sciences, vol. 251, Springer, 2000.

Jun-ichi Imura
Department of Mechanical and Environmental Informatics
Graduate School of Information Science and Engineering
Tokyo Institute of Technology
Oh-Okayama 2-12-1, Meguro-ku, Tokyo 152–8552, Japan
imura@cyb.mei.titech.ac.jp

# Robust Stabilization of Chaos via Delayed Feedback Control


## Shigeru Yamamoto and Toshimitsu Ushio


*With great admiration, this paper is dedicated to Professor Hidenori Kimura on the occasion of his 60th birthday*


ABSTRACT Delayed feedback control (DFC) is a useful method of stabilizing unstable fixed points of chaotic systems without their exact information. In this paper, we present a new recursive method for DFC, which enables us to easily design robust DFC. Since this recursive DFC is essentially dynamic feedback, it can overcome the so-called odd number limitation. Hence, it can robustly stabilize almost all unstable fixed points of chaotic systems.



**Keywords:** Chaos control, Robust stabilization, Delayed feedback control, Odd number limitation


## 1 Introduction

Delayed feedback control (DFC) is a strong method to stabilize unstable periodic orbits embedded in chaotic attractors via a small control input which is fed by the difference between the current state and the delayed state [18]. This method favorably requires no preliminary calculation of the unstable periodic orbit to be stabilized in contrast to the OGY method [15].

Although DFC is simple and convenient for controlling chaos, it has been reported that discrete-time DFC has the so-called *odd number limitation* [5], [23]. That is, DFC never stabilizes a target unstable fixed point of a chaotic discrete-time system, if the Jacobian matrix of the linearized system around the unstable fixed point has an odd number of real eigenvalues greater than unity. For continuous-time chaotic systems, it is known that there exists a similar limitation [12].

The odd number limitation comes from the fact that the original DFC scheme is constant state feedback. Since the original DFC has a constant feedback gain, we call it *static* DFC. Stabilization by static DFC is equivalent to that by static output feedback for an augmented system. Hence, we can intuitively realize that it is difficult to find stabilizing static DFC. Indeed by the limitation, we cannot stabilize a class of chaotic systems by static DFC.

To overcome the odd number limitation by static DFC, several methods have been proposed [3, 6, 10, 13, 14, 21, 24, 28]. In particular, avoiding the use of the delayed state and a constant feedback gain essentially resolves the limitation. In prediction-based feedback control [4, 25], the control input is determined by using the predicted future state instead of the delayed state. However, this method is no longer DFC. Other methods are based on the alternative to constant gains. For example, it has been reported that periodic gains are useful for a special class of chaotic systems [2, 19]. More effective methods are DFC with dynamical controllers [11, 26, 27, 28]. We call this *dynamic* DFC in contrast to static DFC.

In this paper, as a special class of dynamic DFC, we propose recursive delayed feedback control. Since it is essentially dynamic feedback, it does not have the odd number limitation. A necessary and sufficient condition for stabilization is relaxed by recursive DFC as conventional dynamic DFC. A primary advantage of recursive DFC is that the order of the dynamics is equal to the size of the control input. In addition, we can easily design recursive DFC to solve a stabilization problem by state feedback. Hence, we can apply it to a robust stabilization problem of chaotic systems.

# 2 Delayed feedback control

In this section, we review static delayed feedback control (DFC) proposed by Pyragas [18], its limitation and dynamic DFC [27].

We consider an $n$th-order nonlinear discrete-time system described by

$$\mathbf{x}(k+1) = f(\mathbf{x}(k), u(k)) \tag{2.1}$$

where $u(k) \in \Re^m$ is the input and $\mathbf{x}(k) \in \Re^n$ is the state. We assume that $f$ is differentiable. Let $x_f$ be a fixed point of the system (2.1) with $u(k) \equiv 0$, that is, $x_f = f(x_f, 0)$. The control purpose is to stabilize the fixed point $x_f$. Then, the linearized system around $x_f$ is given by

$$x(k+1) = Ax(k) + Bu(k) \tag{2.2}$$

where $A \in \Re^{n\times n}$, $B \in \Re^{n\times m}$,

$$A = \frac{\partial}{\partial x} f(x,u)|_{x=x_f,u=0}, \; B = \frac{\partial}{\partial u} f(x,u)|_{x=x_f,u=0}$$

and

$$x(k) = \mathbf{x}(k) - x_f$$

respectively.

Pyragas proposed a simple method that is feedback of

$$\mathbf{e}(k) = \mathbf{x}(k) - \mathbf{x}(k-1) \tag{2.3}$$

instead of $\mathbf{x}(k) - x_f$. It is a static feedback controller with a gain matrix $K$:

$$u(k) = K\mathbf{e}(k). \tag{2.4}$$

It is noted that $u(k) \to 0$ when $\mathbf{x}(k) \to x_f$. Hence, it ensures that the fixed point to be stabilized does not change by stabilization. To prevent the control input from destroying the chaotic dynamics without stabilizing $x_f$, it is applied to (2.1) only when $\|\mathbf{e}(k)\| < \varepsilon$ where $\varepsilon$ is a sufficiently small positive number.

The linearized system around the unstable fixed point $x_f$ of the closed-loop system with static DFC (2.4) can be described by

$$\begin{bmatrix} x(k+1) \\ x(k) \end{bmatrix} = \begin{bmatrix} A+BK & -BK \\ I_n & 0 \end{bmatrix} \begin{bmatrix} x(k) \\ x(k-1) \end{bmatrix}. \tag{2.5}$$

Therefore, the local stability of (2.1) with static DFC (2.4) can be reduced to that of the linear system (2.5). In addition, (2.5) is equal to the interconnection of

$$u(k) = Ky(k)$$

and

$$\begin{aligned} \mathsf{x}(k+1) &= \mathsf{A}\mathsf{x}(k) + \mathsf{B}u(k), \\ y(k) &= \mathsf{C}\mathsf{x}(k) \end{aligned} \tag{2.6}$$

where

$$\begin{aligned} \mathsf{x}(k) &= \begin{bmatrix} x(k) \\ x(k-1) \end{bmatrix} \in \Re^{2n}, \ \mathsf{A} = \begin{bmatrix} A & 0 \\ I_n & 0 \end{bmatrix} \in \Re^{2n\times 2n}, \\ \mathsf{B} &= \begin{bmatrix} B \\ 0 \end{bmatrix} \in \Re^{2n\times m}, \ \mathsf{C} = \begin{bmatrix} I_n & -I_n \end{bmatrix} \in \Re^{n\times 2n}. \end{aligned}$$

In general, it is difficult to derive a condition for the existence of a $K$ such that $\mathsf{A}+\mathsf{B}K\mathsf{C}$ is asymptotically stable. Particularly, this problem is closely related to an open fundamental problem in linear system theory called the problem of pole assignment by constant feedback in which we must find an appropriate $K$ such that $\mathsf{A}+\mathsf{B}K\mathsf{C}$ has a desired set of eigenvalues [8, 9]. In our problem, a special structure of $\mathsf{A}$, $\mathsf{B}$ and $\mathsf{C}$ yields a necessary condition for the existence of stabilizing (2.4) [23].

**Proposition 2.1 (Odd number limitation).** *If the linearized system (2.2) is stabilizable via (2.4), then*

$$\det(I_n - A) > 0. \tag{2.7}$$

*Equivalently, if*

$$\det(I_n - A) \leq 0, \tag{2.8}$$

*then there is no static DFC (2.4) stabilizing (2.2).*

The condition (2.8) implies that the number of real eigenvalues of $A$ greater than or equal to 1 is odd. By this proposition, it is impossible to stabilize any linearized system (2.2) with $\det(I_n - A) < 0$ via static DFC (2.4).

Dynamic DFC has been proposed to overcome the limitation. It can be realized by an $n_c$th-order linear dynamic controller

$$\begin{aligned} \hat{x}(k+1) &= \hat{A}\hat{x}(k) + \hat{B}\mathbf{e}(k), \\ u(k) &= \hat{C}\hat{x}(k) + \hat{D}\mathbf{e}(k) \end{aligned} \qquad (2.9)$$

where $\hat{A} \in \Re^{n_c \times n_c}$, $\hat{B} \in \Re^{n_c \times n}$, $\hat{C} \in \Re^{m \times n_c}$, $\hat{D} \in \Re^{m \times n}$, and $\hat{x} \in \Re^{n_c}$ is the state of dynamic DFC. Similar to static DFC, in the vicinity of the target fixed point $x_f$, the linearized closed-loop system with dynamic DFC (2.9) can be described by

$$x_c(k+1) = A_c x_c(k) \qquad (2.10)$$

where

$$A_c = \begin{bmatrix} \mathsf{A} + \mathsf{B}\hat{D} & -\mathsf{B}\hat{C} \\ \hat{B}\mathsf{C} & \hat{A} \end{bmatrix}, \ x_c(k) = \begin{bmatrix} \mathsf{x}(k) \\ \hat{x}(k) \end{bmatrix}. \qquad (2.11)$$

Thus, the local stability of (2.1) with (2.9) is reduced to that of (2.10). In the case where $n = n_c$, we have the following proposition.

**Proposition 2.2.** *Assume that $(A, B)$ is stabilizable. Then, the linearized system is stabilizable by dynamic DFC (2.9) if and only if $I_n - A$ is nonsingular.*

*Moreover, when the condition holds, one such dynamic controller is given by*

$$\begin{aligned} \hat{A} &= (I_n - A)^{-1}BK, \\ \hat{B} &= -(I_n - A)^{-1}BKA(I_n - A)^{-1}, \\ \hat{C} &= K, \\ \hat{D} &= -KA(I_n - A)^{-1} \end{aligned} \qquad (2.12)$$

*where $K$ is a matrix such that $A + BK$ is asymptotically stable.*

For the case where $n_c < n$, the nonsingularity of $I_n - A$ is also necessary for the existence of dynamic controllers. However, we have to solve a non-convex optimization problem to check the sufficiency and to obtain the controllers [20].

# 3 Recursive delayed feedback control

Recursive DFC is of the form

$$u(k) = K_e\mathbf{e}(k) + (K_u + I_m)u(k-1) \qquad (3.1)$$

where $K_e$ and $K_u$ are constant matrices. The control input is recursively determined. Motivational derivation of recursive DFC is as follows. In the vicinity of $x_f$, the chaotic system (2.1) can be linearized as

$$\begin{aligned} & x(k+1) - x(k) \\ & \quad = A\left(x(k) - x(k-1)\right) + B\left(u(k) - u(k-1)\right). \end{aligned} \tag{3.2}$$

Since $x(k) - x(k-1) = \mathbf{x}(k) - x_f - \mathbf{x}(k-1) + x_f = \mathbf{e}(k)$, we can rewrite (3.2) as

$$\mathbf{e}(k+1) = A\mathbf{e}(k) + B\left(u(k) - u(k-1)\right). \tag{3.3}$$

By letting $\delta u(k) = u(k) - u(k-1)$, we have

$$\begin{bmatrix} \mathbf{e}(k+1) \\ u(k) \end{bmatrix} = \bar{A} \begin{bmatrix} \mathbf{e}(k) \\ u(k-1) \end{bmatrix} + \bar{B}\delta u(k) \tag{3.4}$$

where

$$\bar{A} = \begin{bmatrix} A & 0 \\ 0 & I_m \end{bmatrix}, \ \bar{B} = \begin{bmatrix} B \\ I_m \end{bmatrix}.$$

For the system (3.4), state feedback control

$$\delta u(k) = \begin{bmatrix} K_e & K_u \end{bmatrix} \begin{bmatrix} \mathbf{e}(k) \\ u(k-1) \end{bmatrix} \tag{3.5}$$

yields (3.1). It is noted that the control input vanishes when the system (3.4) is stabilized. The linearized closed-loop system of the system (3.4) with (3.5) is

$$\begin{bmatrix} \mathbf{e}(k+1) \\ u(k) \end{bmatrix} = (\bar{A} + \bar{B}K) \begin{bmatrix} \mathbf{e}(k) \\ u(k-1) \end{bmatrix}, \tag{3.6}$$

where

$$\bar{x}(k) = \begin{bmatrix} \mathbf{e}(k) \\ u(k-1) \end{bmatrix}, \ K = \begin{bmatrix} K_e & K_u \end{bmatrix}.$$

Then, we have the following theorem.

**Theorem 3.1.** *Assume that $(A, B)$ is stabilizable. Then, there exists state feedback (3.5) stabilizing the linearized system (3.3) if and only if $I_n - A$ is nonsingular.*

**Proof.** Since $(A, B)$ is stabilizable, for $z \in \mathcal{C}$ such that $|z| \geq 1$, rank $\begin{bmatrix} zI_n - A & B \end{bmatrix} = n$. Hence, for $z \in \mathcal{C}$,

$$\operatorname{rank} \begin{bmatrix} zI_{n+m} - \bar{A} & \bar{B} \end{bmatrix} = \operatorname{rank} \begin{bmatrix} zI_n - A & 0 & B \\ 0 & (z-1)I_m & I_m \end{bmatrix}$$

$$= \begin{cases} \operatorname{rank} \begin{bmatrix} zI_n - A & B \end{bmatrix} + \operatorname{rank}(z-1)I_m = n + m & \text{if } |z| \geq 1 \text{ and } z \neq 1, \\ \operatorname{rank}\,(I_n - A) + \operatorname{rank} I_m = \operatorname{rank}\,(I_n - A) + m & \text{if } z = 1. \end{cases}$$

Therefore, $I_n - A$ is nonsingular if and only if $(\bar{A}, \bar{B})$ is stabilizable. □

Indeed, we can show that (3.6) is asymptotically stable if and only if the closed-loop system by recursive DFC (3.1) is asymptotically stable. Before showing the result, we should note that recursive DFC is dynamic control.

**Remark 3.2.** Recursive DFC can be realized by an $m$th dynamic controller. For example,

$$\begin{aligned} \hat{x}(k+1) &= \hat{A}\hat{x}(k) + \hat{B}\mathrm{e}(k), \\ u(k) &= \hat{C}\hat{x}(k) + \hat{D}\mathrm{e}(k) \end{aligned}$$

where $\hat{A} = \hat{C} = K_u + I_m$, $\hat{B} = \hat{D} = K_e$, $\hat{x}(k) = u(k-1)$. In dynamic DFC (2.9), we have to solve a non-convex problem to design a reduced-order controller [27]. However, this method easily enables us to design the $m$th order controller. Moreover, this control method also has an advantage in view of a pole assignment property. □

Since recursive DFC (3.1) belongs to a special class of dynamic DFC (2.9), substituting (3.1) to the closed-loop system (2.10) with the dynamic DFC, we have

$$x_C(k+1) = A_C x_C(k) \tag{3.7}$$

where

$$\begin{aligned} x_C(k) &= \begin{bmatrix} x(k) \\ x(k-1) \\ u(k-1) \end{bmatrix}, \\ A_C &= \begin{bmatrix} A + BK_e & -BK_e & B + BK_u \\ I_n & 0 & 0 \\ K_e & -K_e & I_m + K_u \end{bmatrix}. \end{aligned}$$

If $I_n - A$ is nonsingular, then so is

$$T = \begin{bmatrix} -I_n & A & B \\ I_n & -I_n & 0 \\ 0 & 0 & I_m \end{bmatrix}.$$

Hence, by the similarity transformation

$$\begin{bmatrix} 0 \\ \mathrm{e}(k) \\ u(k-1) \end{bmatrix} = T \begin{bmatrix} x(k) \\ x(k-1) \\ u(k-1) \end{bmatrix},$$

we can transform (3.7) into

$$\begin{bmatrix} 0 \\ \mathrm{e}(k+1) \\ u(k) \end{bmatrix} = T A_C T^{-1} \begin{bmatrix} 0 \\ \mathrm{e}(k) \\ u(k-1) \end{bmatrix} \tag{3.8}$$

where

$$TA_C T^{-1} = \begin{bmatrix} 0 & 0 & 0 \\ I_n & A+BK_e & BK_u \\ 0 & K_e & I_m+K_u \end{bmatrix}.$$

It is obvious that (3.8) is equal to (3.6). This yields the following theorem.

**Theorem 3.3.** *Assume that $(A, B)$ is stabilizable. Then, there exists recursive DFC (3.1) stabilizing (2.2) if and only if $I_n - A$ is nonsingular.*

**Remark 3.4.** A control method based on the Newton algorithm [26] has a relation to recursive DFC. Under the assumption that $B$ is of full column rank, it is given by

$$\begin{aligned} u(k) &= -(B^TB)^{-1}B^T\Big((A-I_n)^{-1}+I_n\Big) \\ &\quad \times\Big(A\left(\mathbf{x}(k)-\mathbf{x}(k-1)\right)-Bu(k-1)\Big). \end{aligned} \tag{3.9}$$

To compare (3.9) with recursive DFC (3.1), (3.9) can be realized by recursive DFC with

$$\begin{aligned} K_e &= -(B^TB)^{-1}B^TA(A-I_n)^{-1}A, \\ K_u &= (B^TB)^{-1}B^TA(A-I_n)^{-1}B-I_m. \end{aligned} \tag{3.10}$$

When we use (3.9) (or recursive DFC with (3.10)),

$$\begin{aligned} &\det\left(zI_{n+m}-\begin{bmatrix} A+BK_e & BK_u \\ K_e & I_m+K_u \end{bmatrix}\right) \\ &= z^m\det\Big(zI_n-\big(I_n-B(B^TB)^{-1}B^T\big)A\Big). \end{aligned}$$

This implies that (3.9) can stabilize unstable fixed points only when the matrix $(I_n - B(B^TB)^{-1}B^T)A$ is stable. However, it is obvious that (3.9) can always stabilize any fixed point of scalar chaotic systems and make the resulting closed-loop poles all zero. □

# 4 Robust control

In this section, we consider a robust control problem where exact information for the chaotic system (2.1) is not available. For example, $f$ and/or $x_f$ are uncertain. In this case, there is no other way to derive DFC based on the uncertain linearized system expressed as

$$x(k+1) = (A+\Delta A)x(k)+(B+\Delta B)u(k). \tag{4.1}$$

Due to the uncertainty in $f$ and/or $x_f$, (4.1) has uncertain terms $\Delta A$ and $\Delta B$. In this paper, we assume that they are norm-bounded and expressed by

$$\left[\, \Delta A \;\; \Delta B \,\right] = D\Delta \left[\, E \;\; F \,\right], \tag{4.2}$$

where the uncertain parameter $\Delta$ satisfies $\|\Delta\| \leq \gamma$ for a given positive scalar $\gamma$. It should be noted that $\Delta A = 0$ and $\Delta B = 0$ does not imply that $A = A_{true}$ and $B = B_{true}$ in the linearized system when we obtain the exact information of $f$ and $x_f$. $A$ and $B$ are simply the nominal of (4.1). Hence, we also assume that there exists $\Delta$ such that $A + \Delta A = A_{true}$ and $B + \Delta B = B_{true}$.

DFC can stabilize the true fixed point without its exact value. In addition, we can easily design recursive DFC by using the robust stabilization approach [7].

Similar to Section 3, we have the error system with the uncertain parameter written as

$$\bar{x}(k+1) = (\bar{A} + \Delta\bar{A})\bar{x}(k) + (\bar{B} + \Delta\bar{B})\delta u(k) \tag{4.3}$$

where

$$\bar{D} = \begin{bmatrix} D \\ 0 \end{bmatrix}, \; \left[\, \Delta\bar{A} \;\; \Delta\bar{B} \,\right] = \bar{D}\Delta \left[\, \bar{E} \;\; F \,\right], \tag{4.4}$$

$$\bar{E} = \left[\, E \;\; 0 \,\right]. \tag{4.5}$$

Then, we apply state feedback control (3.5) to the system (4.3). From (4.2), we have

$$\bar{x}(k+1) = (\bar{A} + \bar{B}K)\bar{x}(k) + \bar{D}\Delta(\bar{E} + FK)\bar{x}(k). \tag{4.6}$$

In order to discuss the stability analysis of (4.6), we give the following definition [1].

**Definition 4.1.** The uncertain system (4.3) is quadratically stabilizable, if there exist state feedback $\delta u = K\bar{x}$, a constant scalar $\alpha > 0$ and a positive definite matrix $P > 0$ such that

$$\bar{x}(k+1)^T P\bar{x}(k+1) - \bar{x}(k)^T P\bar{x}(k) \leq -\alpha\|\bar{\mathbf{x}}(k)\|^2 \tag{4.7}$$

for the resulting closed-loop system and all $\Delta\bar{A}$ in (4.4). □

**Theorem 4.2.** *The linearized system (4.3) is quadratically stabilizable for all $\Delta\bar{A}$ and $\Delta\bar{B}$ satisfying (4.2) if and only if there exist a positive definite matrix $Q > 0$ and a matrix $Y$ such that*

$$\mathrm{He}\left\{\begin{bmatrix} \frac{1}{2}Q & 0 & \bar{A}Q + \bar{B}Y & \bar{D} \\ 0 & \frac{1}{2}I_{n+m} & \bar{E}Q + FY & 0 \\ 0 & 0 & \frac{1}{2}Q & 0 \\ 0 & 0 & 0 & \frac{\gamma^{-2}}{2}I_m \end{bmatrix}\right\} > 0 \tag{4.8}$$

*where* He $\{M\} > 0$ *means* $M + M^T > 0$. *Moreover, one of the controllers is given by*

$$K = YQ^{-1}. \tag{4.9}$$

Thus, the robust stabilization problem can be reduced to a convex optimization problem

$$\min_{\gamma^{-2}\in\Re} \gamma^{-2} \text{ subject to (4.8) over } Q > 0 \text{ and } Y. \tag{4.10}$$

**Proof.** From Definition 4.1, if the uncertain system (4.1) is quadratically stabilizable, there exist $K$, $P > 0$, and $\alpha$ such that

$$\bar{x}(k+1)^T P\bar{x}(k+1) - \bar{x}(k)^T P\bar{x}(k) \leq -\alpha\|\bar{\mathbf{x}}(k)\|^2 \tag{4.11}$$

for $\bar{x}(k+1) = \{\bar{A} + \bar{B}K + \bar{D}\Delta(\bar{E} + FK)\}\,\bar{x}(k)$. This is equivalent to

$$\left(\bar{A} + \bar{B}K + \bar{D}\Delta(\bar{E} + FK)\right)^T P\left(\bar{A} + \bar{B}K + \bar{D}\Delta(\bar{E} + FK)\right) - P < 0. \tag{4.12}$$

Furthermore, by the Schur complement formula, it can be written as

$$\text{He}\left\{\begin{bmatrix} -\frac{1}{2}P^{-1} & \bar{A} + \bar{B}K + \bar{D}\Delta(\bar{E} + FK) \\ 0 & -\frac{1}{2}P \end{bmatrix}\right\} < 0. \tag{4.13}$$

Premultiplying and postmultiplying the above inequality by $\begin{bmatrix} I_{m+n} & 0 \\ 0 & P^{-1} \end{bmatrix}$, we obtain

$$\text{He}\left\{\begin{bmatrix} -\frac{1}{2}P^{-1} & (\bar{A} + \bar{B}K + \bar{D}\Delta(\bar{E} + FK))P^{-1} \\ 0 & -\frac{1}{2}P^{-1} \end{bmatrix}\right\} > 0.$$

Then, we have

$$\begin{bmatrix} -P^{-1} & (\bar{A} + \bar{B}K)P^{-1} \\ P^{-1}(\bar{A}^T + K^T\bar{B}^T) & -P^{-1} \end{bmatrix} + \text{He}\left\{\begin{bmatrix} \bar{D} \\ 0 \end{bmatrix}\Delta\begin{bmatrix} 0 & (\bar{E} + FK)P^{-1} \end{bmatrix}\right\} < 0 \tag{4.14}$$

for any $\|\Delta\| \leq \gamma$. Then, due to the results in [17] and [16], there exists a scalar $\beta > 0$ such that

$$\beta^{-1}\begin{bmatrix} -P^{-1} & (\bar{A} + \bar{B}K)P^{-1} \\ P^{-1}(\bar{A}^T + K^T\bar{B}^T) & -P^{-1} \end{bmatrix} + \gamma^2\begin{bmatrix} \bar{D} \\ 0 \end{bmatrix}\begin{bmatrix} \bar{D}^T & 0 \end{bmatrix} + \beta^{-2}\begin{bmatrix} 0 \\ P^{-1}(\bar{E}^T + K^T F^T) \end{bmatrix}\begin{bmatrix} 0 & (\bar{E} + FK)P^{-1} \end{bmatrix} < 0. \tag{4.15}$$

Defining $Q \triangleq \beta^{-1}P^{-1}$ and $Y \triangleq \beta^{-1}KP^{-1}$, we have

$$\begin{bmatrix} -Q & \bar{A}Q+\bar{B}Y \\ Q\bar{A}^T+Y^T\bar{B}^T & -Q \end{bmatrix} + \gamma^2 \begin{bmatrix} \bar{D} \\ 0 \end{bmatrix} \begin{bmatrix} \bar{D}^T & 0 \end{bmatrix} + \begin{bmatrix} 0 \\ Q\bar{E}^T+Y^TF^T \end{bmatrix} \begin{bmatrix} 0 & \bar{E}Q+FY \end{bmatrix} < 0. \tag{4.16}$$

By the Schur complement formula,

$$\mathrm{He}\left\{\begin{bmatrix} -\frac{1}{2}Q & \bar{A}Q+\bar{B}Y & \bar{D} & 0 \\ 0 & -\frac{1}{2}Q & 0 & Q\bar{E}^T+Y^TF^T \\ 0 & 0 & -\frac{\gamma^{-2}}{2}I_m & 0 \\ 0 & 0 & 0 & -\frac{1}{2}I_{m+n} \end{bmatrix}\right\} < 0. \tag{4.17}$$

Moreover, premultiplying and postmultiplying the above inequality by

$$\begin{bmatrix} I_{m+n} & 0 & 0 & 0 \\ 0 & 0 & -I_{m+n} & 0 \\ 0 & 0 & 0 & -I_m \\ 0 & I_{m+n} & 0 & 0 \end{bmatrix}$$

and its transpose, respectively, we obtain (4.8). □

# 5 Numerical simulations

In this section, we give numerical examples which illustrate the effectiveness of the proposed methods.

## 5.1 Stabilization of second-order chaotic system

We consider the following second-order discrete-time chaotic system [22].

$$\begin{cases} x_1(k+1) = 1.9x_1(k) - x_1^3(k) + x_2(k) + u(k), \\ x_2(k+1) = 0.5x_1(k). \end{cases} \tag{5.1}$$

This system has three fixed points given by

$$x_{f1} = \begin{bmatrix} 0 \\ 0 \end{bmatrix}, \; x_{f2} = \begin{bmatrix} \sqrt{1.4} \\ \sqrt{1.4}/2 \end{bmatrix}, \; x_{f3} = -x_{f2}.$$

The linearized system around the unstable fixed point $x_{f1}$ has a stabilizable pair

$$A = \begin{bmatrix} 1.9 & 1 \\ 0.5 & 0 \end{bmatrix} \text{ and } B = \begin{bmatrix} 1 \\ 0 \end{bmatrix}.$$

Since $\det(I - A) = -1.4 < 0$, there does not exist static DFC (2.4) stabilizing $x_{f1}$, but recursive DFC. Furthermore, since $(A, B)$ is controllable, it can arbitrarily assign the poles of the closed-loop system (3.6). Recursive DFC which makes eigenvalues of the closed-loop system (3.6) zero is obtained by

$$K_e = \begin{bmatrix} -3.6158 & -1.7149 \end{bmatrix} \text{ and } K_u = 0.7149.$$

By this recursive DFC, state trajectories of (5.1) starting from the initial state $[x_1(0),\ x_2(0)] = [0.6,\ -0.3]$ can be stabilized (Figure 5.1). Recursive DFC starts at $k = 188$ and the state converges to $x_{f1}$. The control action at $k = 106$ fails in the stabilization, since the state is not close to $x_{f1}$.

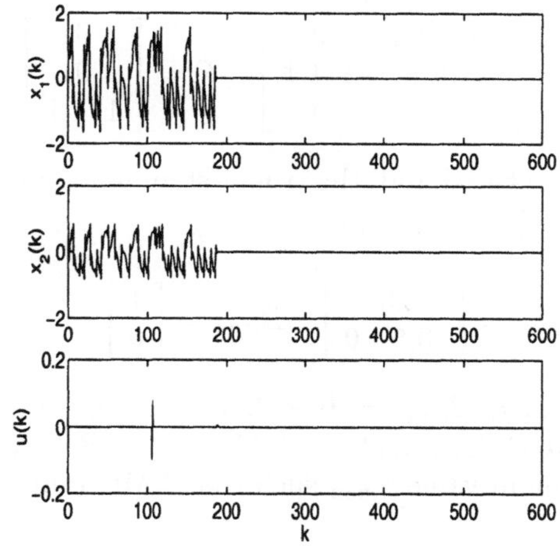


FIGURE 5.1. Stabilization of the fixed point $x_{f1}$ of (5.1)

## 5.2 Robust stabilization

We consider the robust control problem where we must stabilize the true chaotic system (5.1) based on uncertain information. We assume that we obtain the following uncertain model of (5.1).

$$\begin{cases} x_1(k+1) = (2+\delta_1)x_1(k) - x_1^3(k) \\ \qquad\qquad\quad + (0.9+\delta_2)x_2(k) + (1.1+\delta_3)u(k), \\ x_2(k+1) = 0.5x_1(k). \end{cases} \tag{5.2}$$

This nominal model has three fixed points given by

$$\mathsf{x}_{f1} = \begin{bmatrix} 0 \\ 0 \end{bmatrix}, \ \mathsf{x}_{f2} = \begin{bmatrix} \sqrt{1.45} \\ \sqrt{1.45}/2 \end{bmatrix}, \ \mathsf{x}_{f3} = -\mathsf{x}_{f2}$$

when all $\delta_i = 0,\ i = 1, 2, 3$. The corresponding system matrix $A$, $B$ to the UFP $\mathrm{x}_{f2}$ is given by

$$A = \begin{bmatrix} -2.35 & 0.9 \\ 0.5 & 0 \end{bmatrix},\ B = \begin{bmatrix} 1.1 \\ 0 \end{bmatrix}.$$

Then, the uncertainties can be described as

$$\Delta A = \begin{bmatrix} -2\delta_1 - 1.5\delta_2 & \delta_2 \\ 0 & 0 \end{bmatrix},\ \Delta B = \begin{bmatrix} \delta_3 \\ 0 \end{bmatrix}$$

where

$$\begin{aligned} \Delta &= \begin{bmatrix} \delta_1 & \delta_2 & \delta_3 \end{bmatrix},\ D = B, \\ E &= \begin{bmatrix} -2/1.1 & 0 \\ -1.5/1.1 & 1/1.1 \\ 0 & 0 \end{bmatrix},\ F = \begin{bmatrix} 0 \\ 0 \\ 1/1.1 \end{bmatrix} \end{aligned}$$

and $\|\Delta\| \leq \gamma$. On the other hand, the true system matrix corresponding to $x_{f2} = \begin{bmatrix} \sqrt{1.4} & \sqrt{1.4}/2 \end{bmatrix}^T$ is given by

$$A_{\mathrm{true}} = \begin{bmatrix} -2.3 & 1 \\ 0.5 & 0 \end{bmatrix},\ B_{\mathrm{true}} = \begin{bmatrix} 1 \\ 0 \end{bmatrix}.$$

Since $A_{\mathrm{true}} = A + \Delta A$, $B_{\mathrm{true}} = B + \Delta B$ when $\delta_1 = -0.1$, $\delta_2 = 0.1$, $\delta_3 = -0.1$, if $\gamma \geq \sqrt{0.03}$, $(A_{\mathrm{true}}, B_{\mathrm{true}}) \in \{(A + \Delta A, B + \Delta B) |\ \|\Delta\| \leq \gamma\}$.

Finding the maximum value $\gamma_{\max}$ such that LMI condition (4.8) is satisfied, we obtain

$$\begin{aligned} \gamma_{\max} &= 0.2005, \\ P &= \begin{bmatrix} 0.0828 & 0.052 & -0.065 \\ 0.052 & 0.2091 & 0.0397 \\ -0.065 & 0.0397 & 0.2720 \end{bmatrix}, \\ Y &= \begin{bmatrix} 0.1292 & -0.0502 & -0.1992 \end{bmatrix}. \end{aligned}$$

Moreover, we obtain the feedback gain matrix $K$ from (4.9)

$$K_e = \begin{bmatrix} 1.8079 & -0.6505 \end{bmatrix},\ K_u = -0.2049. \tag{5.3}$$

Since $\gamma_{max} > \sqrt{0.03}$, we can stabilize the chaotic system (5.1) by the controller that is obtained from the parameter of the uncertain model. Hence, we can stabilize the true chaotic system using the obtained gain $K$.

It is noted that this controller makes poles of the closed-loop system for the system (5.2) $\begin{bmatrix} 0.3959 & 0.3959 & 0.4611 \end{bmatrix}$. Figure 5.2 shows a behavior of the controlled system starting from the initial state $[x_1(0),\ x_2(0)] = [0.6,\ -0.3]$ where $\varepsilon = 0.1$. The controller starts at $k = 106$ and the state converges to the UFP $x_{f2}$ of the true chaotic system.

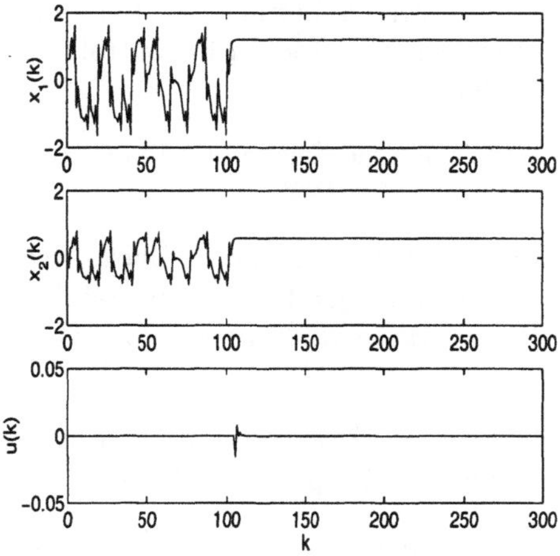


FIGURE 5.2. For the uncertain chaotic system (5.2), robust recursive DFC can stabilize the fixed point of the true system

# 6 Conclusion

In this paper, we have proposed to design robust stabilizing delayed feedback control for unstable fixed points of the discrete-time chaotic systems. We have also shown that the proposed method can overcome the limitation that the original static DFC has. Since the proposed controller is based on recursive delayed feedback control, the stabilizing feedback gains can be designed without solving any non-convex optimization problem.

## References

[1] B. R. Barmish, "Necessary and sufficient condition for quadratic stabilizability of an uncertain linear system," *J. Optimiz. Theory Appl.*, vol. 46, no. 4, pp. 399–408, 1985.

[2] S. Bielawski, D. Derozier, and P. Glorieux, "Experimental characterization of unstable periodic orbits by controlling chaos," *Physical Review A*, vol. 47, no. 2, pp. 2493–2495, 1993.

[3] G. Chen and X. Yu, "On time-delayed feedback control of chaotic systems," *IEEE Trans. Circuits and Systems I*, vol. 46, no. 6, pp. 767–772, 1999.

[4] T. Hino, S. Yamamoto, and T. Ushio, "Stabilization of unstable periodic orbits of chaotic discrete-time systems using prediction-based feedback control," in *Proc. AFSS2000*, 2000, pp. 347–352.

[5] W. Just, T. Bernard, M. Osthermer, E. Reibold, and H. Benner, "Mechanism of time-delayed feedback control," *Phys. Rev. Lett.*, vol. 78, no. 2, pp. 203–206, 1997.

[6] W. Just *et al.*, "Limits of time-delayed feedback control," *Physics Letters A*, vol. 254, pp. 158–164, 1999.

[7] P. P. Khargonekar, I. R. Petersen, and K. Zhou, "Robust stabilization of uncertain linear systems: Quadratic stabilizability and $H^\infty$ control theory," *IEEE Trans. Automat. Control*, vol. AC-35, no. 3, pp. 356–361, 1990.

[8] H. Kimura, "Pole-assignment by gain output feedback," *IEEE Trans. Automat. Control*, vol. AC-20, pp. 509–516, 1975.

[9] H. Kimura, "Pole-assignment by output feedback: A long-standing open question," in *Proc. 33rd IEEE Conf. Decision and Control*, 1994, pp. 2101–2105.

[10] K. Konishi, M. Ishii, and H. Kokame, "Stability of extended delayed feedback control for discrete-time chaotic systems," *IEEE Trans. Circuits and Systems I*, vol. 46, no. 10, pp. 1285–1288, 1999.

[11] K. Konishi and H. Kokame, "Observer-based delayed-feedback control for discrete-time chaotic systems," *Physics Letters A*, vol. 248, pp. 359–368, 1998.

[12] H. Nakajima, "On analytical properties of delayed feedback control of chaos," *Physics Letters A*, vol. 232, pp. 207–210, 1997.

[13] H. Nakajima, "A generalization of the extended delayed feedback control for chaotic systems," in *Proc. of COC2000*, vol. 2, 2000, pp. 209–212.

[14] H. Nakajima and Y. Ueda, "Half-period delayed feedback control for dynamical systems with symmetries," *Physical Review E*, vol. 58, no. 2, pp. 1757–1763, 1998.

[15] E. Ott, C. Grebogi, and J. A. Yorke, "Controlling chaos," *Phys. Rev. Lett.*, vol. 64, no. 11, pp. 1196–1199, 1990.

[16] I. R. Petersen, "A stabilization algorithm for a class of uncertain linear systems," *Systems and Control Letters*, vol. 8, pp. 351–357, 1987.

[17] I. R. Petersen and C. V. Hollot, "A Riccati equation approach to the stabilization of uncertain linear systems," *Automatica*, vol. 22, no. 4, pp. 397–411, 1986.

[18] K. Pyragas, "Continuous control of chaos by self-controlling feedback," *Physics Letters A*, vol. 170, pp. 421–428, 1992.

[19] H. G. Schuster and M. B. Stemmler, "Control of chaos by oscillating feedback," *Physical Review E*, vol. 56, no. 6, pp. 6410–6417, 1997.

[20] R. E. Skelton, T. Iwasaki, and K. Grigoriadis, *A Unified Algebraic Approach to Linear Control Design*. London: Taylor & Francis, 1998.

[21] J. E. S. Socolar, D. W. Sukow, and D. J. Gauthier, "Stabilizing unstable periodic orbits in fast dynamical systems," *Physical Review E*, vol. 50, pp. 3245–3248, 1994.

[22] T. Ushio, "Chaotic synchronization and controlling chaos based on contraction mappings," *Physics Letters A*, vol. 198, pp. 14–22, 1995.

[23] T. Ushio, "Limitation of delayed feedback control in non-linear discrete-time systems," *IEEE Trans. Circuits and Systems I*, vol. 43, no. 9, pp. 815–816, 1996.

[24] T. Ushio and S. Yamamoto, "Delayed feedback control with nonlinear estimation in chaotic discrete-time systems," *Physics Letters A*, vol. 247, pp. 112–118, 1998.

[25] T. Ushio and S. Yamamoto, "Prediction-based control of chaos," *Physics Letters A*, vol. 264, pp. 30–35, 1999.

[26] D. Xu and S. R. Bishop, "Self-locating control of chaotic systems using Newton algorithm," *Physics Letters A*, vol. 210, pp. 273–278, 1996.

[27] S. Yamamoto, T, Hino, and T. Ushio, "Dynamic delayed feedback controllers for chaotic discrete-time systems," *IEEE Trans. Circuits and Systems I*, vol. 48, no. 6, pp. 785–789, 2001.

[28] X. Yu, Y. Tian, and G. Chen, "Time delayed feedback control of chaos," in *Controlling Chaos and Bifurcations in Engineering Systems*, G. Chen Ed. CRC, Chapter 12, pp. 255–274, 1999.

Shigeru Yamamoto
Department of Systems and Human Science
Graduate School of Engineering Science
Osaka University
Machikaneyama 1–3 Toyonaka, Osaka 560–8531, Japan
yamamoto@sys.es.osaka-u.ac.jp

Toshimitsu Ushio
Department of Systems and Human Science
Graduate School of Engineering Science
Osaka University
Machikaneyama 1–3 Toyonaka, Osaka 560–8531, Japan
ushio@sys.es.osaka-u.ac.jp

# Time Varying Deadbeat Control of High Order Chained Systems with Applications to the Control of Underactuated Manipulators


## Tsutomu Mita and Taek Kun Nam


*Dedicated to Professor Hidenori Kimura on the occasion of his 60th birthday*


ABSTRACT A new digital control method for a time varying sampled data system called a time varying deadbeat control is proposed to control high order nonholonomic chained systems with drift terms. This control law has a potential to avoid an excessive overshoot to prevent the trajectories from getting close to the singular manifolds of the transformation needed to obtain the chained forms. The control will be applied to the control of a three link underactuated manipulator.



**Keywords:** Nonholonomic system, Underactuated manipulator, Multirate digital control, Time varying deadbeat control


## 1 Dedication

The first time Prof. Kimura talked to me (the first author) was about 25 years ago when I presented my new paper concerning the overshoot phenomenon in linear systems (the paper cited as [17] in this paper) in a domestic conference. He very much enjoyed my results and gave me several important questions and suggestions which moved and encouraged me much, because he was a very famous and great researcher even at that time. By this occasion, I knew the fun of research work. Several years after this conference, he asked me to join his research group and we held many workshops which brought up me as a researcher in the control theory field. I appreciate it much. He is very modest, kind and fair and we all love Prof. Kimura. I expect his new and great activities in many and wider fields even after this birthday. Thank you also Prof. Kimura for your kind friendship over 25 years.

# 2 Introduction

In mechanical systems subject to nonholonomic constraints, we can control a large number of generalized coordinates than the number of inputs, by taking advantage of the constraints. If the constraints are expressed in terms of the velocity, control systems are usually described by symmetrical affine systems (affine systems without drift terms) while they can be described by affine systems with drift terms when the constraints are given in terms of the acceleration [22, 14].

Since all symmetrical affine systems of order 3 or 4 can be transformed locally into the so-called chained system [21], many nonholonomic control laws are designed to control the chained systems. However, due to the theorem of Brockett [3], there is no time-invariant and smooth feedback control law to stabilize the symmetrical affine systems. There are many research results to overcome this limitation. Khennouf and de Wit [9] and Ikeda et al. [6] proposed control laws which use invariant manifolds; Astolfi [2] first made given systems discontinuous using the $\sigma$ process and stabilized them by continuous feedback control; Monaco and Normand-Cyrot [20] applied the multirate digital control; Tilbury and Chelouah [28] extended the multi-rate digital control to chained systems with multi-chains. Sordalen and Egeland [27], Pomet [25] and Samson [26] proposed time varying feedback control laws. However, in general, the time varying control takes a longer time before convergence and the responses are oscillatory.

On the other hand, there are several studies on underactuated manipulators (UAM) which represent mechanical systems subject to acceleration constraints. Oriolo and Nakamura [24] discussed the integrability of the acceleration constraint; Nakamura et al. [23] proposed a path planning of the UAM with two links laid on the horizontal plane. Luca et al. [15] also dealt with a two-link UAM. Arai et al. [1] developed a stabilizing control law of three-link UAM by paying attention to the impact point. While Imura et al. [7] developed a feedback control law for the same type of UAM using the high order chained form. Comparing to the symmetrical affine systems, there are few studies on the control of the systems having drift terms.

On the other hand, without distinguishing symmetrical affine systems from affine systems with drift terms, Khennouf and de Wit [10] investigated nonholonomic systems with single generator and multiple chains, and derived a feedback control law for an underwater vehicle. Laiou and Astolfi [12, 13] analyzed the so-called high order chained systems with drift terms based on the previous results [2] and derived general control laws including one which is applicable to the control of a different type of UAM.

In this paper, we will focus on the multirate digital control [20]. The multirate digital control may be easily seen to be applied to all the chained systems, including the multi-chain and high order multi-chain systems with drift terms as long as they have only single generators, since all the subchains become linear when the generator variable is set to be a non-zero constant. However, we have to use $N_i$ sampling intervals for the $N_i$th order sub-chain systems during one sampling interval of the generating system. This means that the sampling period of each sub-chain becomes shorter than the sampling period of the generating system. And this causes large overshoots for the input as well as some of the state variables of the sub-chain systems, and they often come close to or reach the singular manifolds of the transformation introduced to transform given plants to the chained systems.

To overcome this imbalance of the sampling period, we will first develop here a time varying deadbeat control (TVDC) which is equivalent to a deadbeat control for time varying discrete-time systems. Using this method, we can make the sampling interval of the generating system, as well as all sub-chain systems, uniform to improve the transient responses and to get around the singularity problem. We will then apply TVDC to the control of a three-link UAM which can be transformed into a high order chained system with the drift term.

# 3 Conventional multirate digital control

As an explanation of the background of this study, let us look at the control of the chained systems with single generators. The most popular chained system in this class with two inputs (i.e., single chain) is the following [22]:

$$
\begin{aligned}
\dot{z}_1 &= u_1, \\
\dot{z}_2 &= u_2, \\
\dot{z}_3 &= z_2 u_1, \\
&\vdots \\
\dot{z}_n &= z_{n-1} u_1.
\end{aligned}
\tag{3.1}
$$

This system is rewritten as

$$
\begin{aligned}
\Sigma_1 : \ \dot{z}_1 &= u_1, \\
\Sigma_2 : \ \dot{Z}_2 &= \begin{bmatrix} 0 & u_1 & 0 & \dots & 0 \\ 0 & 0 & u_1 & \dots & 0 \\ \vdots & \vdots & \vdots & \ddots & \vdots \\ 0 & 0 & 0 & \dots & u_1 \\ 0 & 0 & 0 & \dots & 0 \end{bmatrix} Z_2 + \begin{bmatrix} 0 \\ 0 \\ \vdots \\ 0 \\ 1 \end{bmatrix} u_2
\end{aligned}
\tag{3.2}
$$

where $Z_2 = (z_n, z_{n-1}, \dots, z_2)^T$. When $u_1$ is kept as a non-zero constant, the sub-system $\Sigma_2$ becomes a linear controllable canonical system. The stabilization problem of this chained system is simultaneously stabilization of the coupled systems $\Sigma_1$ and $\Sigma_2$. However, since $\Sigma_2$ becomes uncontrollable when $u_1 = 0$, the problem is not as easy as being solved by the linear robust control.

Therefore, there are many proposals to control this system as mentioned in the introduction. Hereafter, we will concentrate on the multirate digital deadbeat control proposed by Monaco and Normand-Cyrot [20]. In their method, as we can see in Figure 3.1, $\Sigma_1$ is controlled by the deadbeat control with the sampling interval $T$ while $\Sigma_2$ is controlled by the deadbeat control with the sampling interval $T/(n-1)$ to settle the whole system at $t = T$.

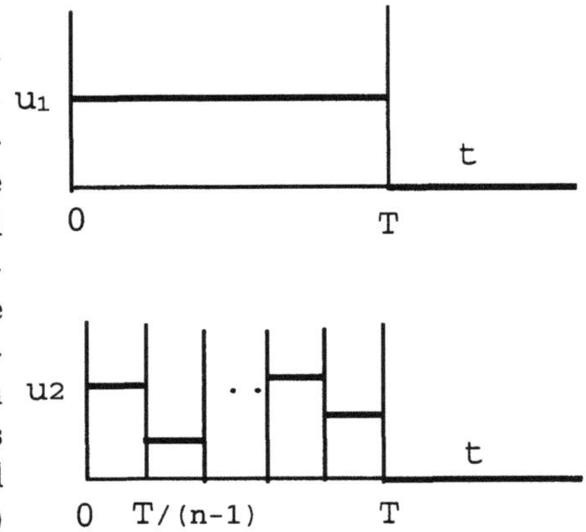


FIGURE 3.1. Inputs used in [20]

However, when $n$ is large, the sampling interval of $\Sigma_2$ becomes shorter compared to the sampling interval of $\Sigma_1$. In this paper, we will propose equalization of the sampling interval for the high order chained system with a single chain in the sequel.

# 4 Controlling high order chained systems with drift terms

## 4.1 Problems and time varying deadbeat control

In this paper, we will deal with the so called high order chained systems with drift terms represented by [12]

$$\begin{aligned} \Sigma_1 : \quad & z_1^{(k_1)} = u_1, \\ \Sigma_2 : \quad & \begin{bmatrix} z_2^{(k_2)} \\ z_3^{(k_3)} \\ z_4^{(k_4)} \\ \vdots \\ z_n^{(k_n)} \end{bmatrix} = \begin{bmatrix} 0 \\ z_2 u_1 \\ z_3 u_1 \\ \vdots \\ z_{n-1} u_1 \end{bmatrix} + \begin{bmatrix} 1 \\ 0 \\ 0 \\ \vdots \\ 0 \end{bmatrix} u_2. \end{aligned} \tag{4.1}$$

If we keep $u_1$ as a non-zero constant $\alpha$, this system can be described by

$$\begin{aligned} \Sigma_1 : \quad & z_1^{(k_1)} = u_1, \\ \Sigma_2 : \quad & z_n^{(k_n+\cdots+k_4+k_3+k_2)} = h u_2; \quad h = \alpha^{n-2} \end{aligned} \tag{4.2}$$

where $\Sigma_1$ and $\Sigma_2$ are the $k_1$th and $N$th order controllable systems, respectively, with

$$N = k_2 + k_3 + \cdots + k_{n-1} + k_n. \tag{4.3}$$

Therefore, as seen in Figure 4.1 (a), we can stabilize $\Sigma_1$ and $\Sigma_2$ by using $k_1$ and $N$ sampling intervals, respectively, if we apply the multirate digital control proposed by Monaco and Normand-Cyrot directly.

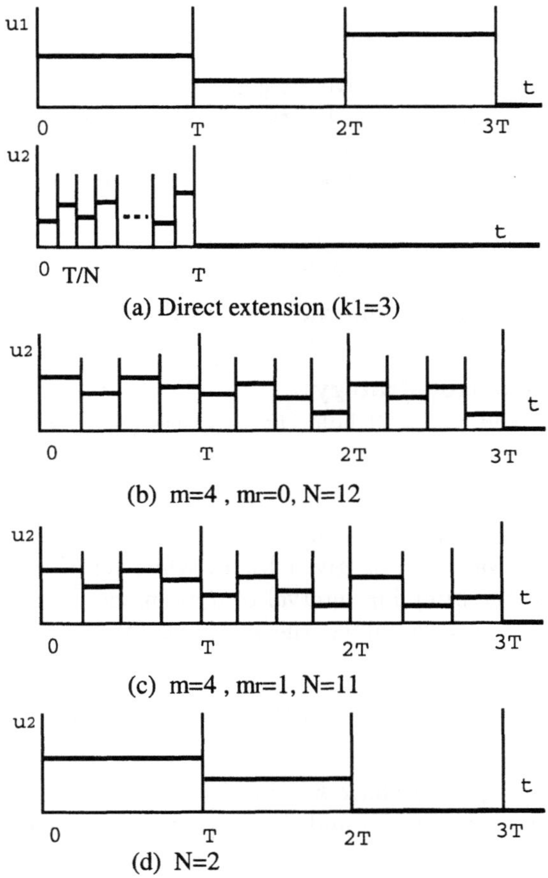


FIGURE 4.1. Conventional and improved sampling interval of $u_2$

However, by this sampling scheme, the sampling interval of $u_2$ becomes shorter compared to the sampling interval of $u_1$. As we know from [16], when the sampling interval becomes shorter, the responses and feedback input of the deadbeat controller inevitably have large overshoots which converge impulses when $T$ approaches to zero, depending on the initial states. The corresponding phenomenon for the continuous-time feedback control systems is investigated in [17, 4]. Furthermore, this overshoot often drives the responses and inputs to the singular points or manifolds which appear in the transformations to the high order chained form.

Therefore we need to avoid this problem by modifying the sampling strategy for $u_2$. To this end, we first suppose that $k_i \geq 1$ $(i = 1, \ldots, n)$.

And if

$$N \geq k_1 \tag{4.4}$$

we calculate so that

$$N = mk_1 - m_r \quad (1 \leq m_r \leq k_1 - 1) \tag{4.5}$$

holds, where $m$ and $m_r$ are positive integers.

Then, we will make the sampling interval of $u_2$ longer as seen in Figure 4.1 (b) and (c), where, in (b), the case $N = 12$, $m = 4$, $m_r = 0$, and, in (c), the case $N = 11$, $m = 4$, $m_r = 1$ are illustrated all for $k_1 = 3$, for an example.

When

$$N < k_1 \tag{4.6}$$

holds, we make the sampling interval of $u_2$ as the same as the sampling interval of $u_1$ as depicted in Figure 4.1 (d).

**Definition 4.1 (Control strategy).**

1. When (4.4) and $m_r = 0$ hold, as the case (b) in Figure 4.1, the sampling interval of $u_1$ is given by $T$, while all the sampling interval of $u_2$ is made $T/m$.
2. When (4.4) and $m_r \neq 0$ holds, as the case (c) in Figure 4.1, the sampling interval of $u_1$ is given by $T$, while for the first $m(k_1 - 1)$ sampling, the sampling interval of $u_2$ is made $T/m$, and for the remaining $m - m_r$ sampling, the sampling interval of $u_2$ is made $T/(m - m_r)$.
3. When (4.6) holds, as the case (d) in Figure 4.1, both the sampling intervals of $u_1$ and $u_2$ are made $T$. □

Note that $u_1$ and $u_2$ always have $k_1$ and $N$ samplings. Therefore we may render the deadbeat action to $\Sigma_1$ and $\Sigma_2$. However, we will face a technical problem as follows.

Since the parameters of $\Sigma_2$ (i.e., $u_1$) varies during each sampling interval of $u_1$, we cannot apply the conventional deadbeat control which makes the transition matrices nilpotent by assigning all poles at the origin. We can solve this problem by using the following theorem which gives a systematic design procedure of the deadbeat controller for time varying sampled data systems. We call this method a *time varying deadbeat control (TVDC).*

**Theorem 4.2 (TVDC).** *Suppose that*

$$\alpha_i(n \times n), \quad \beta_i(n \times 1) \tag{4.7}$$

*are constant coefficient matrices and vectors for a linear time invariant system and* $(\alpha_i, \beta_i)$ *is the controllable pair. Also let*

$$x(i+1) = A_i x(i) + b_i u(i) \tag{4.8}$$

*be the sampled data system derived from the linear system having the coefficient matrix and vector provided by (4.7) using the zero-order hold of sampling period $T_i$, where*

$$A_i = e^{\alpha_i T_i}, \quad b_i = \int_0^{T_i} e^{\alpha_i \tau} d\tau \beta_i. \tag{4.9}$$

*Then for almost all sampling periods $T_i$ $(i = 1, \dots, n)$, there exist feedback gain vectors $f_1, \dots, f_n$ satisfying*

$$(A_n - b_n f_n) \cdots (A_2 - b_2 f_2)(A_1 - b_1 f_1) = 0. \tag{4.10}$$

**Remark 4.3.** A remark is in order. Even if $A_2$ and $A_1$ are second order nilpotent matrices satisfying $A_1^2 = 0$ and $A_2^2 = 0$, $A_2 A_1 = 0$ rarely holds. □

To prove the theorem, we need the following simple lemma.

**Lemma 4.4.**

1. *$\gamma_i b_i \neq 0$ holds for arbitrary $\gamma_i \neq 0$ $(1 \times n)$ and almost all $T_i > 0$.*
2. *Suppose that $\gamma_i(A_i - b_i f_i) = 0$ holds and $p_1, \dots, p_{n-1}$ are arbitrary and linearly independent vectors which are also linearly independent of $\gamma_i$. Then, $p_k(A_i - b_i f_i)$ $(k = 1, \dots, (n-1))$ are linearly independent.*

**Proof of Lemma 4.4.** The first fact is proven in [18]. Also it is well known that $(A_i, b_i)$ is controllable when $(\alpha_i, \beta_i)$ is controllable for almost all $T_i$ [18]. This together with the assumption that $b_i$ is a vector lead to the fact that $(A_i - b_i f_i)$ is a non-degenerated matrix [5] and is of rank $n - 1$ even if it has multiple zero eigenvalues. Then it follows from this fact that the right-hand side of

$$\begin{bmatrix} \gamma_i \\ p_1 \\ \vdots \\ p_{n-1} \end{bmatrix} (A_i - b_i f_i) = \begin{bmatrix} 0 \\ p_1(A_i - b_i f_i) \\ \vdots \\ p_{n-1}(A_i - b_i f_i) \end{bmatrix} \tag{4.11}$$

is also of rank $n-1$. Therefore, $p_k(A_i - b_i f_i)$ $(k = 1, \dots, (n-1))$ are linearly independent. □

**Proof of Theorem 4.2.** The proof is constructive.

Equation (4.10) is equivalent to saying that

$$\begin{bmatrix} e_n \\ e_{n-1} \\ \vdots \\ e_1 \end{bmatrix} (A_n - b_n f_n)(A_{n-1} - b_{n-1} f_{n-1}) \cdots (A_2 - b_2 f_2)(A_1 - b_1 f_1) = 0 \tag{4.12}$$

holds for arbitrary $n$ dimensional linearly independent row vectors $e_1, \dots, e_n$. In the following, $f_n, f_{n-1}, \dots, f_1$ can be determined one by one uniquely to satisfy

$$\begin{aligned} & e_n(A_n - b_n f_n) = 0, \\ & e_{n-1}(A_n - b_n f_n)(A_{n-1} - b_{n-1} f_{n-1}) = 0, \\ & \vdots \\ & e_1(A_n - b_n f_n)(A_{n-1} - b_{n-1} f_{n-1}) \cdots (A_1 - b_1 f_1) = 0. \end{aligned} \tag{4.13}$$

At the beginning, let us define

$$\gamma_n := e_n. \tag{4.14}$$

Then $f_n$ can be found as

$$f_n = (\gamma_n b_n)^{-1} \gamma_n A_n \tag{4.15}$$

to satisfy

$$\gamma_n(A_n - b_n f_n) = 0. \tag{4.16}$$

This is always possible since $\gamma_n b_n \neq 0$ holds from the first item of Lemma 4.4.

Then it follows from the second item of Lemma 4.4 that

$$e_{n-1}(A_n - b_n f_n),\ e_{n-2}(A_n - b_n f_n), \dots, e_1(A_n - b_n f_n) \tag{4.17}$$

are linearly independent and they are non-zero at least. At this step, we define

$$\gamma_{n-1} = e_{n-1}(A_n - b_n f_n). \tag{4.18}$$

Since $\gamma_{n-1} \neq 0$, we have again $\gamma_{n-1} b_{n-1} \neq 0$, and we can determine

$$f_{n-1} = (\gamma_{n-1} b_{n-1})^{-1} \gamma_{n-1} A_{n-1} \tag{4.19}$$

to satisfy

$$\gamma_{n-1}(A_{n-1} - b_{n-1} f_{n-1}) = 0. \tag{4.20}$$

At this step, we also know from (4.17), (4.20) and Lemma 4.4 that

$$e_{n-2}(A_n - b_n f_n)(A_{n-1} - b_{n-1} f_{n-1}), \dots, e_1(A_n - b_n f_n)(A_{n-1} - b_{n-1} f_{n-1}) \tag{4.21}$$

are linearly independent and non-zero. Therefore, we define

$$\gamma_{n-2} = e_{n-2}(A_n - b_n f_n)(A_{n-1} - b_{n-1} f_{n-1}). \tag{4.22}$$

Continuing this procedure,

$$\gamma_i = e_i(A_n - b_n f_n)(A_{n-1} - b_{n-1} f_{n-1}) \cdots (A_{i+1} - b_{i+1} f_{i+1}) \qquad (4.23)$$

satisfies $\gamma_i b_i \neq 0$ and $f_i$ can be found as

$$f_i = (\gamma_i b_i)^{-1} \gamma_i A_i \qquad (4.24)$$

to satisfy

$$\gamma_i (A_i - b_i f_i) = 0. \qquad (4.25)$$

We can confirm that (4.12) holds by a backward substitution. In this procedure, $e_n, \ldots, e_1$ become free design parameters as long as they are linearly independent. □

Using this theorem, we can solve the original problem as follows.

**Theorem 4.5.** *All the states of $\Sigma_1$ and $\Sigma_2$ in (4.1) reach zero at $t = Tk_1$ when TVDC is applied, except for a set of singular initial states of $\Sigma_1$ which spans at most $k_1$ hyperplanes of dimension $k_1 - 1$ in $k_1$ dimensional space.*

**Proof.** Let us describe the state space representation of (4.1) as

$$\begin{aligned} \Sigma_1 : \quad & \dot{Z}_1 = \alpha_0 Z_1 + \beta_0 u_1, \\ \Sigma_2 : \quad & \dot{Z}_2 = \alpha(u_1) Z_2 + \beta u_2 \end{aligned} \qquad (4.26)$$

where $\alpha_0(k_1 \times k_1)$, $\beta_0(k_1 \times 1)$, $\alpha(N \times N)$ and $\beta(N \times 1)$. Since the sampling intervals of $u_1$ and $u_2$ are uniquely defined by Definition 4.1, the discrete time representation of each system is described by the forms

$$\begin{aligned} &\Sigma_1 : \ Z_1(l+1) = A_0 Z_1(l) + b_0 u_1(l) \quad (l = 0, \ldots, k_1 - 1), \\ &\Sigma_2 : \ Z_2(j+1) = A_{j+1}(u_1(l)) Z_2(j) + b_{j+1}(u_1(l)) u_2(j) \quad (j = 0, \ldots, N-1). \end{aligned} \qquad (4.27)$$

First, $\Sigma_1$ is controlled by the conventional deadbeat control to satisfy

$$(A_0 - b_0 f_0)^{k_1} = 0. \qquad (4.28)$$

Then, as far as $u_1 \neq 0$ for all sampling intervals, $\Sigma_2$ is controllable and it can be controlled to satisfy

$$(A_N - b_N f_N) \cdots (A_2 - b_2 f_2)(A_1 - b_1 f_1) = 0 \qquad (4.29)$$

using Theorem 4.2.

The initial states of $\Sigma_1$ which make $u_1 = 0$ in some intervals are those satisfying some of

$$f_0 Z_1(0) = 0, \quad f_0(A_0 - b_0 f_0) Z_1(0) = 0, \quad \ldots, \quad f_0(A_0 - b_0 f_0)^{k_1 - 1} Z_1(0) = 0 \qquad (4.30)$$

which will be proven to be at most $k_1$ hyperplanes of dimension $k_1 - 1$ in $k_1$ dimensional space. See Appendix.

When $Z_1(0)$ is on the hyperplanes defined by (4.30), the problem becomes singular. However, it is easily avoided by moving the initial state off the hyperplanes using an auxiliary control before TVDC, as in [9]. □

# 5 Example

We will examine the following system which will be used in the next section.

$$z_1^{(2)} = u_1, \quad z_2^{(2)} = u_2, \quad z_3^{(2)} = z_2 u_1. \tag{5.1}$$

This system can be rewritten as

$$\Sigma_1 : \quad \frac{d}{dt}\begin{bmatrix} z_1 \\ \dot{z}_1 \end{bmatrix} = \begin{bmatrix} 0 & 1 \\ 0 & 0 \end{bmatrix}\begin{bmatrix} z_1 \\ \dot{z}_1 \end{bmatrix} + \begin{bmatrix} 0 \\ 1 \end{bmatrix} u_1, \tag{5.2}$$

$$\Sigma_2 : \quad \frac{d}{dt}\begin{bmatrix} z_3 \\ \dot{z}_3 \\ z_2 \\ \dot{z}_2 \end{bmatrix} = \begin{bmatrix} 0 & 1 & 0 & 0 \\ 0 & 0 & u_1 & 0 \\ 0 & 0 & 0 & 1 \\ 0 & 0 & 0 & 0 \end{bmatrix}\begin{bmatrix} z_3 \\ \dot{z}_3 \\ z_2 \\ \dot{z}_2 \end{bmatrix} + \begin{bmatrix} 0 \\ 0 \\ 0 \\ 1 \end{bmatrix} u_2.$$

In this case, since $k_1 = 2$ and $N = 4$, we have $m = 2$ and $m_r = 0$ in (4.5). Therefore, the sampling intervals of $u_1$ and $u_2$ are given by $T$ and $T/2$ as seen in Figure 5.1 (b). The latter is twice as long as the conventional interval $T/4$ as we can see in Figure 5.1 (a).

Then the control strategy is given by

$$\begin{array}{ll} 0 \le t < T/2 : & u_1 = -f_0 Z_1(0), \quad u_2 = -f_1 Z_2(0), \\ T/2 \le t < T : & u_1 = -f_0 Z_1(0), \quad u_2 = -f_2 Z_2(T/2), \\ T \le t < 3T/2 : & u_1 = -f_0 Z_1(T), \quad u_2 = -f_3 Z_2(T), \\ 3T/2 \le t < 2T : & u_1 = -f_0 Z_1(T), \quad u_2 = -f_4 Z_2(3T/2), \\ 2T \le t : & u_1 = 0, \quad u_2 = 0 \end{array} \tag{5.3}$$

where $Z_1 = (z_1, \dot{z}_1)^T$, $Z_2 = (z_3, \dot{z}_3, z_2, \dot{z}_2)^T$ and $f_0 = (1/T^2, 3/2T)$ is the conventional deadbeat control gain for $u_1$.

In this case, the parameters of $\Sigma_2$ as the continuous time system are given by

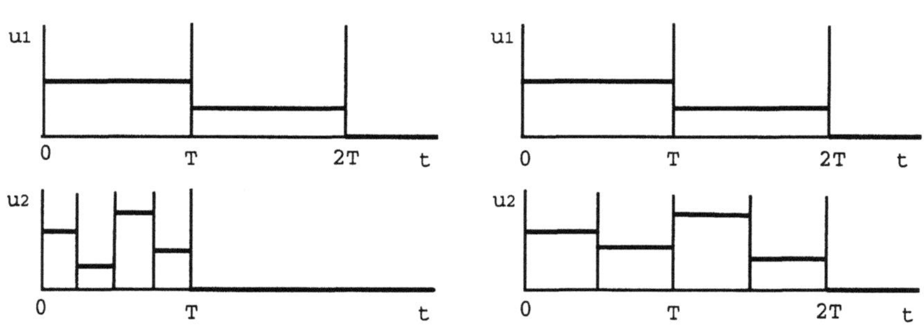


(a) Direct extension (b) Proposed method

FIGURE 5.1. Improvement of the sampling interval

$$\begin{array}{rrcl} 0 \le t < T/2: & \alpha_1 & = & \alpha(\delta_1), \quad \beta_1 = \beta, \\ T/2 \le t < T: & \alpha_2 & = & \alpha_1, \quad \beta_2 = \beta, \\ T \le t < 3T/2: & \alpha_3 & = & \alpha(\delta_2) \quad \beta_3 = \beta, \\ 3T/2 \le t < 2T: & \alpha_4 & = & \alpha_3 \quad \beta_4 = \beta \end{array} \tag{5.4}$$

where $\delta_1 = u_1(0)$, $\delta_2 = u_1(T)$ and

$$\alpha(\delta) = \begin{bmatrix} 0 & 1 & 0 & 0 \\ 0 & 0 & \delta & 0 \\ 0 & 0 & 0 & 1 \\ 0 & 0 & 0 & 0 \end{bmatrix}, \quad \beta = \begin{bmatrix} 0 \\ 0 \\ 0 \\ 1 \end{bmatrix}. \tag{5.5}$$

In the following, we choose $T = 10$ and

$$\begin{bmatrix} e_1 \\ e_2 \\ e_3 \\ e_4 \end{bmatrix} = \begin{bmatrix} 1 & 0 & 0 & 0 \\ 1 & 1 & 0 & 0 \\ 1 & 1 & 1 & 0 \\ 1 & 1 & 1 & 1 \end{bmatrix} \tag{5.6}$$

as an example. Then, (4.15)–(4.24) lead to

$$\begin{bmatrix} f_1 \\ f_2 \\ f_3 \\ f_4 \end{bmatrix} = \begin{bmatrix} -0.1371 & -1.9714 & 0.1186 & 0.4214 \\ 0.1811 & 2.8981 & -0.0491 & 0.1887 \\ 0.0334 & 0.3679 & 0.0535 & 0.3202 \\ 0.0557 & 0.3339 & 0.0654 & 0.3525 \end{bmatrix} \tag{5.7}$$

which satisfies $(A_4 - b_4 f_4)(A_3 - b_3 f_3)(A_2 - b_2 f_2)(A_1 - b_1 f_1) = 0$.

The conditions that $u_1(0) \neq 0$ and $u_1(T) \neq 0$ are violated by both or one of

$$2z_1(0) + 3T\dot{z}_1(0) = 0, \quad 2z_1(0) + T\dot{z}_1(0) = 0 \tag{5.8}$$

hold. The initial state satisfying these conditions composes two lines in $R^2$ at most. Therefore, we can avoid this singular condition by moving the initial condition using an auxiliary control at the starting time.

The responses are depicted in Figure 5.2 for the initial condition $z(0) = (1, 0, 1, 0, 1, 0)^T$, where the magnitude of $u_1$ is multiplied by the factor 20. This is the same as in Figure 5.3 shown next. In Figure 5.3, we also depict the responses when we use the conventional multirate digital control shown in Figure 5.1 (a).

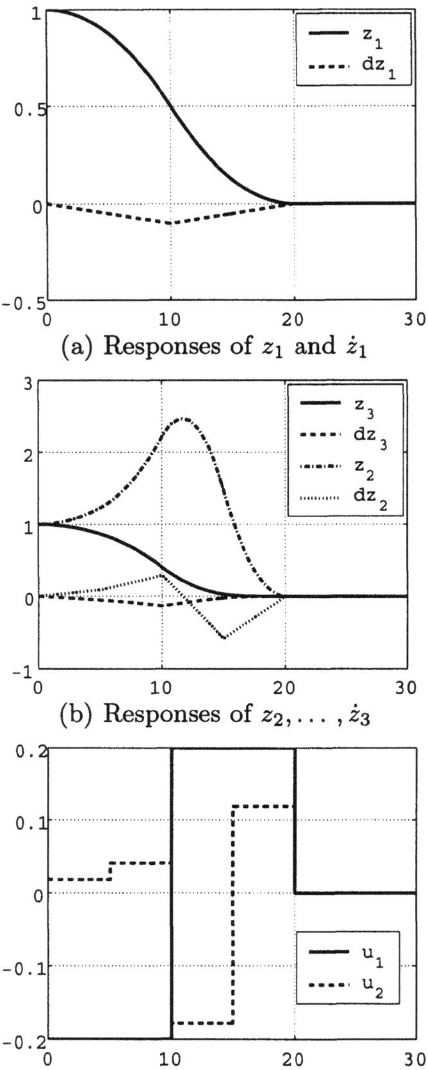


(a) Responses of $z_1$ and $\dot{z}_1$

(b) Responses of $z_2, \ldots, \dot{z}_3$

(c) Responses of $u_1$ and $u_2$

FIGURE 5.2. Obtained responses for the case (b) in Figure 5.1

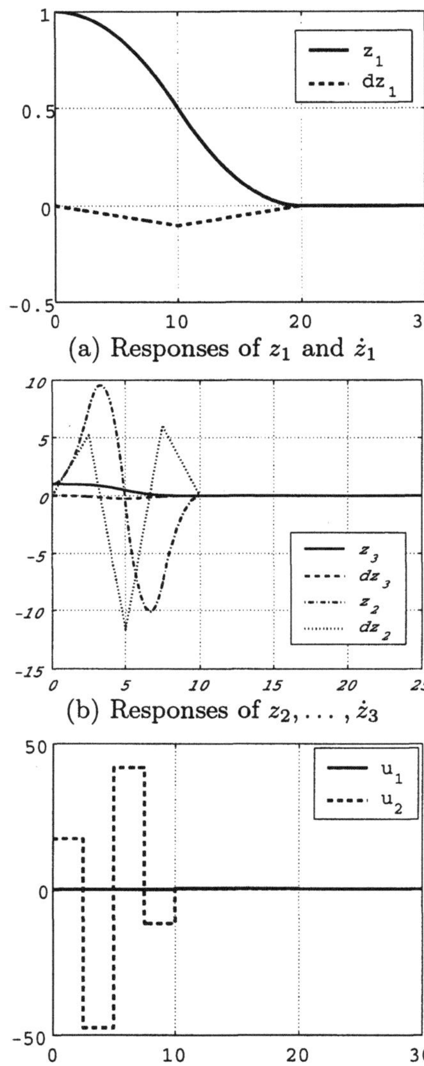


(a) Responses of $z_1$ and $\dot{z}_1$

(b) Responses of $z_2, \ldots, \dot{z}_3$

(c) Responses of $u_1$ and $u_2$

FIGURE 5.3. Obtained responses for the case (a) in Figure 5.1

Comparing Figure 5.2 with Figure 5.3, we can see that the peaks of the responses of $z_2$, $\dot{z}_2$ and $u_2$ in Figure 5.2 are much smaller than those in Figure 5.3. Therefore we can see that TVDC is effective to decrease the overshoots of the transient responses.

# 6 Control of an underactuated manipulator

We will deal with the control problem of a three link planar manipulator laid on the plane as shown in Figure 6.1 where the third (tip) link has no actuator to control.

Since we assume that the first and second links are controlled, we use the acceleration $(\ddot{x}, \ddot{y})$ at the position $(x, y)$ of the third joint as control variables [1].

Introducing the absolute angle $\theta$ of the third link, the equation of motion of the third link or the acceleration constraint of this system is described by

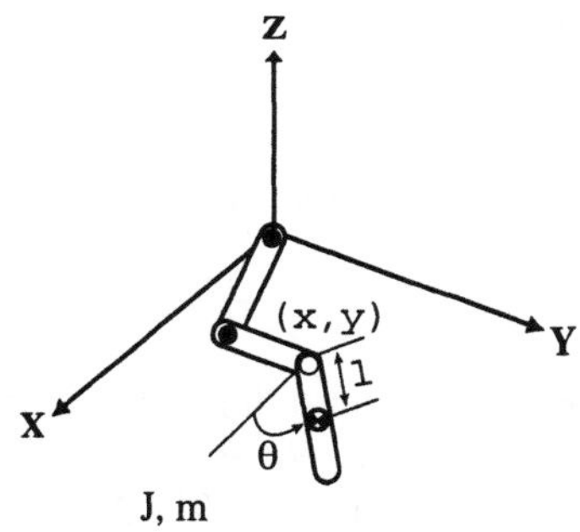

FIGURE 6.1. Underactuated manipulator

$$J\ddot{\theta} - ml\ddot{x}\sin\theta + ml\ddot{y}\cos\theta = 0 \tag{6.1}$$

where $m$, $l$ and $J$ are the weight, the distance between the joint and the center of mass (CM) and the moment of inertia around CM, all for the third link. This constraint is not integrable and becomes a second order nonholonomic constraint. Then (6.1) together with $\ddot{x} = u_1$ and $\ddot{y} = u_2$ lead to

$$\ddot{\theta} = \frac{ml}{J}(u_1\sin\theta - u_2\cos\theta) \tag{6.2}$$

which is known to be controllable [11].

We will introduce a series of transformations to obtain the higher order chained system. For this purpose we will use the impact point of the third link [11] expressed by

$$x_p = x + \lambda\cos\theta, \quad y_p = y + \lambda\sin\theta; \quad \lambda = \frac{J}{ml}. \tag{6.3}$$

This together with the input transformation

$$u = \begin{bmatrix} \cos\theta & \lambda\sin\theta \\ \sin\theta & -\lambda\cos\theta \end{bmatrix} v + \lambda\dot{\theta}^2 \begin{bmatrix} \cos\theta \\ \sin\theta \end{bmatrix} \tag{6.4}$$

yield

$$\ddot{x}_p = v_1\cos\theta, \quad \ddot{y}_p = v_1\sin\theta, \quad \ddot{\theta} = v_2. \tag{6.5}$$

We also introduce the reference $(x_d, y_d, \theta_d)$ for $(x, y, \theta)$ and define the error variables [7] as

$$\begin{aligned} z_1 &= \cos\theta_d(x_p - x_d - \lambda\cos\theta_d) + \sin\theta_d(y_p - y_d - \lambda\sin\theta_d), \\ z_2 &= \tan(\theta - \theta_d), \\ z_3 &= -\sin\theta_d(x_p - x_d - \lambda\cos\theta_d) + \cos\theta_d(y_p - y_d - \lambda\sin\theta_d). \end{aligned} \tag{6.6}$$

Then with the second transformation concerning the input

$$\mu = \begin{bmatrix} \cos(\theta-\theta_d) & 0 \\ 0 & \sec^2(\theta-\theta_d) \end{bmatrix} v + \begin{bmatrix} 0 \\ 2\tan(\theta-\theta_d)\sec^2(\theta-\theta_d) \end{bmatrix} \dot{\theta}^2 \tag{6.7}$$

we have the higher order chained system of the form

$$z_1^{(2)} = \mu_1, \quad z_2^{(2)} = \mu_2 \quad z_3^{(2)} = z_2\mu_1. \tag{6.8}$$

The singular manifold of this series of transformations is represented by the $\theta$ satisfying $\cos(\theta - \theta_d) = 0$, i.e., $\theta = \theta_d \pm \pi/2$.

Now we can apply TVDC developed in the previous section. Figure 6.2 shows the simulation result where we set $m = 2\,(\text{kg})$, $l = 0.5\,(\text{m})$, $J = 0.1622\,(\text{Nm})$ and $T = 10\,(\text{s})$. Also the initial and reference states for ($x$, $y$, $\theta$, $\dot{x}$, $\dot{y}$, $\dot{\theta}$) are given by (2, 0, 1, 0, 0, 0) and (0, 0, 0, 0, 0, 0), respectively. Figure 6.2 (d) is the animation of the motion of the third link. These figures prove that the proposed TVDC is useful since all the responses have no excessive overshoots, which implies that $\theta$ does not reach the singular point.

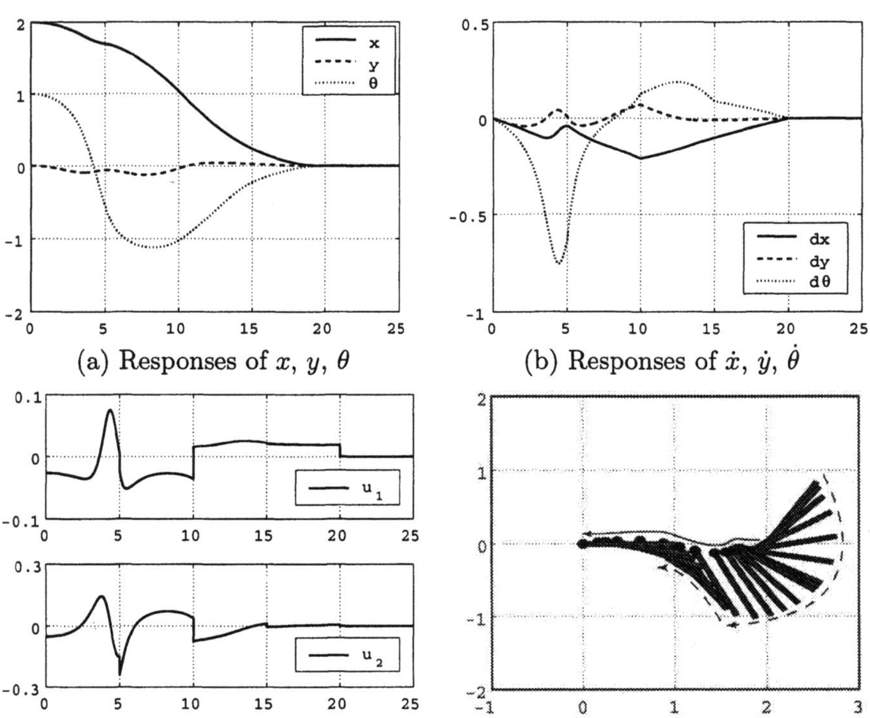


(a) Responses of $x$, $y$, $\theta$ (b) Responses of $\dot{x}$, $\dot{y}$, $\dot{\theta}$

(c) Responses of $u_1$ and $u_2$ (d) Motion of the third link

FIGURE 6.2. Responses of the underactuated manipulator

# 7 Conclusion

A control strategy of high order chained system with drift terms is proposed based on deadbeat control for the time varying discrete-time system. This control avoids excessive overshoots of the input as well as the state variables. Therefore it has a potential to get around the singularity problem. Remaining problems include the optimal choice of the free design parameters $e_n, \ldots, e_1$ and the robustness issue [8].

# Appendix

Since $\alpha_0(k_1 \times k_1)$ is given by

$$\alpha_0 = \begin{bmatrix} 0 & 1 & 0 & \ldots & 0 \\ 0 & 0 & 1 & \ldots & 0 \\ \vdots & \vdots & \vdots & \ddots & \vdots \\ 0 & 0 & 0 & \ldots & 1 \\ 0 & 0 & 0 & \ldots & 0 \end{bmatrix}, \quad \beta_0 = \begin{bmatrix} 0 \\ 0 \\ \vdots \\ 0 \\ 1 \end{bmatrix}, \tag{A.1}$$

all the eigenvalues of $A_0 = e^{\alpha_0 T}$ are located at 1, while all the eigenvalues of $A_0 - b_0 f_0$ are assigned to the origin, where $b_0 = \int_0^T e^{\alpha_0 \tau} d\tau \beta_0$. Therefore $(f_0, A_0)$ must be observable which also leads to $(f_0, A_0 - b_0 f_0)$ being observable. Therefore

$$\Gamma = \begin{bmatrix} f_0 \\ f_0(A_0 - b_0 f_0) \\ \vdots \\ f_0(A_0 - b_0 f_0)^{k_1 - 1} \end{bmatrix} \tag{A.2}$$

is of full rank and no row vector of $\Gamma$ is zero. This means that the vectors $Z_1(0)$ satisfying $f_0(A_0 - b_0 f_0)^j Z_1(0) = 0$ for some $j$'s constitute at most $k_1$ hyperplanes of dimension $k_1 - 1$.

## References

[1] H. Arai, K. Tanie, and N. Shiroma, "Feedback control of a 3-dof planar underactuated manipulator," in *Proc. of Int. Conference on Robotics and Automation*, 1997, pp. 703–709.

[2] A. Astolfi, "Discontinuous control of nonholonomic systems," *Systems Control Lett.*, vol. 27, pp. 37–45, 1996.

[3] R. W. Brockett, "Asymptotic stability and feedback stabilization," in *Differential Geometric Control Theory*. Birkhäuser, 1983, pp. 181–184.

[4] B. A. Francis and K. Glover, "Bounded peaking in the optimal linear regulator with cheap control," *IEEE Trans. Automat. Control*, vol. 23, no. 4, pp. 608–617, 1987.

[5] F. R. Gantmacher, *The Theory of Matrices*, vol. 1. Chelsea, 1960.

[6] T. Ikeda, T. K. Nam, T. Mita, and B. D. O. Anderson, "Variable constraint control of underactuated free flying robots: Mechanical design and convergence," in *Proc. of Conference on Decision and Control*, 1999, pp. 2539–2544.

[7] J. Imura, K. Kobayashi, and T. Yoshikawa, "Nonholonomic control of 3 link planar manipulator with a free joint," in *Proc. of Conference on Decision and Control*, 1996, pp. 1435–1436.

[8] N. Joliot and T. Mita, "Robust deadbeat control by an iterative control method," *Asian J. of Control*, to appear, 2002.

[9] H. Khennouf and C. C. de Wit, "On the construction of stabilizing discontinuous controllers for nonholonomic systems," in *Proc. of IFAC Nonlinear Control Systems Design Sympo.*, 1995, pp. 747–752.

[10] H. Khennouf and C. C. de Wit, "Quasi-continuous exponential stabilizers for nonholonomic systems," in *Proc. of 13th IFAC world Congress*, vol. 2b, 1996, pp. 49–54.

[11] K. Kobayashi, J. Imura, and T. Yoshikawa, "Controllability of planar manipulator with one underactuated joint," *J. of Robotics Society of Japan*, vol. 17, no. 6, pp. 53–57, 1999.

[12] M. C. Laiou and A. Astolfi, "Exponential stabilization of high-order chained systems," in *Proc. of IFAC Symposium on Nonlinear Control Systems Design*, 1998, pp. 649–654.

[13] M. C. Laiou and A. Astolfi, "Discontinuous control of high-order generalized chained systems," *Systems Control Lett.*, vol. 37, pp. 309–322, 1999.

[14] Z. Li and J. F. Canny, *Nonholonomic Motion Planning*. Kluwer Academic Publishers, 1993.

[15] A. D. Luca, R. Mattone, and G. Oriolo, "Stabilization of an underactuated planar 2R manipulator," *Int. J. of Nonlinear and Robust Control*, vol. 10, no. 4, pp. 181–198, 2000.

[16] T. Mita, "A relation between overshoot and sampling period in sampled data feedback control systems," *IEEE Trans. Automat. Control*, vol. 25, no. 3, pp. 603–604, 1980.

[17] T. Mita, "On zeros and responses of linear regulators and linear observers," *IEEE Trans. Automat. Control*, vol. 22, no. 3, pp. 423–428, 1997.

[18] T. Mita, Y. Chida, Y. Kaku, and H. Numasato, "Two-delay robust digital control and its applications: Avoiding the problem on unstable zeros," *IEEE Trans. Automat. Control*, vol. 35, no. 8, pp. 962–970, 1990.

[19] T. Mita and T. K. Nam, "Control of underactuated manipulator using variable period deadbeat control," in *Proc. Int. Conference on Robotics and Automation*, 2001, pp. 2735–2740.

[20] S. Monaco and D. Normand-Cyrot, "An introduction to motion planning using multirate digital control," in *Proc. of Conference on Decision and Control*, 1992, pp. 1780–1785.

[21] R. M. Murray, "Nilpotent bases for a class of non-integrable distributions with applications to trajectory generation for nonholonomic systems," Technical Memorandum, California Inst. of Tech., CIT-CDS 92-002, 1992, pp. 1-19.

[22] R. M. Murray, A. Li, and S. S. Sastry, *A Mathematical Introduction to Robotic Manipulation.* CRC Press, 1994.

[23] Y. Nakamura, T. Iwamoto, and K. Yoshimoto, "Control of nonholonomic mechanisms with drift," *J. Robotics Society of Japan*, vol. 13, no. 6, pp. 830–837, 1995.

[24] G. Oriolo and Y. Nakamura, "Control of mechanical systems with second-order nonholonomic constraints: Underactuated manipulators," in *Proc. of Conference on Decision and Control*, 1991, pp. 2398–2403.

[25] J.-B. Pomet, "Explicit design of time-varying stabilizing control laws for a class of controllable systems without drift," *Systems Control Lett.*, vol. 18, pp. 147–158, 1992.

[26] C. Samson, "Control of chained systems: Application to path following and time-varying point stabilization of mobile robots," *IEEE Trans. Automat. Control*, vol. 40, no. 1, pp. 628–633, 1995.

[27] O. J. Sordalen and O. Egeland, "Exponential stabilization of nonholonomic chained systems," *IEEE Trans. Automat. Control*, vol. 40, no. 1, pp. 35–49, 1995.

[28] D. Tilbury and A. Chelouah, "Steering a three-input nonholonomic system using multi-rate controls," in *Proc. of European Control Conference*, 1993, pp. 1428–1431.

Tsutmou Mita
Department of Mechanical and Control Systems
Tokyo Institute of Technology
2–12–1, Ookayama, Meguro-ku, Tokyo, 152-0033, Japan
mita@ctrl.titech.ac.jp

Teak Ken Nam
Korea Electrotechnology Research Institute,
Mechatronics group,
28–1, Seongju-dong, Changwon-city, Kyongnam, 641-120, Korea
tknam@keri.re.kr

# Control of Nonholonomic Systems with Bad Controllability Structure

**Mitsuji Sampei, Hisashi Date, and Shigeki Nakaura**

*Dedicated to Professor Hidenori Kimura on the occasion of his 60th birthday*

ABSTRACT There are many nonholonomic systems whose controllability structures are much worse than the chained form. Since it is quite hard to design feedback (type) controllers for those systems, many researchers worked on path planning rather than feedback stabilization. In this note, we will consider two of these systems, orientation control of a ball sandwiched and manipulated by two parallel plates, and propulsive control of a snake-like robot. We will analyze their controllability and design feedback type controllers.

**Keywords:** Nonholonomic system, Time-state control form, Ball-plate manipulation, Manipulability, Snake-like locomotion

## 1 Introduction

In general, kinematic models of systems under rolling constraints or the conservation law of angular momentum belong to the class of driftless nonlinear systems. Car trailers and space robots are well-known examples. According to Brockett's theorem, it is impossible to construct a stabilizing feedback controller using conventional design methods for such a class of systems [1]. In some cases, one can transform the system into the so-called chained form, a kind of canonical form, to design a feedback-base controller. Since the chained form has a structure with an exactly linearized subsystem, it has good controllability for designing a controller. There are many examples in the literature of control strategies that guarantee exponential or global convergence based on the chained form [9, 13, 12, 16, 10].

However, there still exist many nonholonomic systems whose controllability structure is much worse than systems that can be transformed or approximated to the chained form. Since it is quite hard to design feedback (type) controllers for those systems, many researchers worked on path planning rather than feedback stabilization. If we wish to design feedback based controllers for such systems, we should consider their controllability structure. In this note, we will deal with two of these systems, orientation control of a ball sandwiched and manipulated by two parallel plates, and propulsive control of a snake-like robot. We will analyze their controllability and design feedback type controllers for them.

# 2 Control of nonholonomic systems

We consider the following input-affine nonlinear system with $n$ states and $m$ inputs as a general model of driftless nonholonomic systems.

$$\frac{dx}{dt} = g_1(x)u_1 + g_2(x)u_2 + \cdots + g_m(x)u_m. \tag{2.1}$$

$x \in R^n$ is a state vector; $g_1, \cdots, g_m \in R^n$ are vector functions of $x$ which are linearly independent at the origin, i.e., $\mathrm{rank}[g_1(0), g_2(0), \cdots, g_m(0)] = m$; $u = [u_1, \cdots, u_m]^T$ is an input vector with $m < n$. According to Brockett's theorem, the origin of these systems cannot be stabilized by any kind of continuous static state feedbacks [1]. In order to control such systems by feedback, we need either a smooth time-varying controller, a discontinuous time-invariant controller, or a hybrid controller.

## 2.1 Time-state control form

In this subsection, we briefly summarize a feedback control strategy of the driftless nonholonomic system (2.1) based on the time-state control form.

There is a class of systems where there exist a coordinate transformation

$$\begin{bmatrix} \xi \\ \tau \end{bmatrix} = T(x),\ T(0) = 0, \quad \xi \in R^{n-1}, \quad \tau \in R, \tag{2.2}$$

and an input transformation

$$\mu_i = V_i(x, u_1, \cdots, u_m), \quad i = 1, 2, \cdots, m, \tag{2.3}$$

such that the system (2.1) can be split into the following two equations,

$$\frac{d\xi}{d\tau} = \bar{f}_0(\xi) + \bar{f}_1(\xi)\mu_1 + \cdots + \bar{f}_{m-1}(\xi)\mu_{m-1}, \tag{2.4}$$

$$\frac{d\tau}{dt} = \mu_m, \tag{2.5}$$

where $\xi \in R^{n-1}$ is a new state; $\tau \in R$ is a new time scale; the mapping $T$ is a (local) diffeomorphism; $V_i$s are scalar functions with respect to $x$ and $u$; $\bar{f}_i(\cdot)$s are vector functions with respect to only $\xi$ (not containing $\tau$). We call the former equation (2.4) *the state-control part* and the latter equation (2.5) *the time-control part.* This form of system expression is called *a time-state control form.* The necessary and sufficient conditions for the existence of the transformations to (2.2) and (2.3) were stated in [15, 5].

**Remark 2.1.** In the case of $m = 2$, the chained form is

$$\dot{\eta}_1 = \mu_1, \ \dot{\eta}_2 = \mu_2, \ \dot{\eta}_3 = \eta_2\mu_1, \ \ldots, \ \dot{\eta}_n = \eta_{n-1}\mu_1.$$

This system can also be represented in the time-state control form as

$$\begin{aligned} \frac{d}{d\eta_1}\begin{bmatrix} \eta_2 \\ \eta_3 \\ \vdots \\ \eta_n \end{bmatrix} &= \begin{bmatrix} \dot{\eta}_2/\dot{\eta}_1 \\ \dot{\eta}_3/\dot{\eta}_1 \\ \vdots \\ \dot{\eta}_n/\dot{\eta}_1 \end{bmatrix} = \begin{bmatrix} 0 \\ \eta_2 \\ \vdots \\ \eta_{n-1} \end{bmatrix} + \begin{bmatrix} 1 \\ 0 \\ \vdots \\ 0 \end{bmatrix} v_1, \\ \frac{d\eta_1}{dt} &= v_2, \end{aligned}$$

where $\mu_1 = v_2$ and $\mu_2 = v_1 v_2$. This implies that if the state-control part is exactly linearizable or approximated to a linear controllable system, then the original system is exactly or approximately transformed into the chained form, respectively. Note that exact linearizability is a very good property for constructing a feedback controller. On the other hand, it is a quite severe requirement. Thus the class of systems transformable to the chained form is limited. □

We can construct a feedback based controller using a time-state control form as follows. Assume that a first order approximation around the origin of the state-control part (2.4)

$$\begin{aligned} \frac{d\xi}{d\tau} &= \left.\frac{\partial \bar{f}_0}{\partial \xi}\right|_{\xi=0} \xi + \bar{f}_1(0)\mu_1 + \cdots + \bar{f}_{m-1}(0)\mu_{m-1} \\ &= A\xi + b_1\mu_1 + \cdots + b_{m-1}\mu_{m-1} \end{aligned} \tag{2.6}$$

is a controllable linear system. This means that the driftless system can be approximated to the chained form. Then, there exists a feedback $\mu_i = \alpha_i(\xi)$ $(i = 1, \cdots, m-1)$ which locally stabilizes the subsystem (2.4), i.e., $\xi$ converges to zero as the time scale $\tau$ monotonically increases.

On the other hand, if we define the backward time scale $\tau' = -\tau$ which monotonically increases while $\tau$ monotonically decreases, the state-control part (2.4) is rewritten as the backward state-control part

$$\frac{d\xi}{d\tau'} = \frac{d\xi}{d\tau}\frac{d\tau}{d\tau'} = -\bar{f}_0(\xi) - \bar{f}_1(\xi)\mu_1 - \cdots - \bar{f}_{m-1}(\xi)\mu_{m-1}. \tag{2.7}$$

Since the linear approximation of this equation is also controllable, there also exists a feedback $\mu_i = \beta_i(\xi)$ $(i = 1, \cdots, m-1)$ which locally stabilizes the subsystem (2.7), i.e., $\xi$ converges to zero as $\tau'$ monotonically increases (as $\tau$ monotonically decreases).

In the time-control part (2.5), we can freely increase or decrease the time scale $\tau$ by setting the appropriate sign of the input $\mu_m$.

Hence we can bring $\xi$ and $\tau$ sufficiently close to zero by switching the direction of the time scale $\tau$ and applying the corresponding feedbacks $\alpha_i$ and $\beta_i$. These feedbacks can be designed using conventional control theories such as LQ optimal controllers, or more advanced ones such as adaptive or robust controllers.

**Control Strategy 2.2.**

*Step 1.* Increase the new time scale $\tau$ with feed forward input $\mu_m > 0$ in the time-control part (2.5), and stabilize the state-control part (2.4) with the feedback $\mu_i = \alpha_i(\xi_i)$ $(i = 1, \cdots, m-1)$. At this step $\xi$ converges to zero, but $\tau$ does not.

*Step 2.* Decrease the time scale $\tau$, i.e., increase the backward time scale $\tau'$ with feed forward input $\mu_m < 0$, and stabilize the backward state-control part (2.7) with the feedback $\mu_i = \beta_i(\xi)$ $(i = 1, \cdots, m-1)$ until the time scale $\tau$ becomes zero.

*Step 3.* Repeat the above two steps until all the states, $\xi$ and $\tau$, are sufficiently close to zero. □

Note that if we need exponential convergence in the time-control part, we can take $\dot{\tau} = -\lambda\tau$ for some $\lambda > 0$ in the final stage. Convergence of the whole system was discussed in [6, 7] with Lyapunov functions.

Hence we can construct a feedback controller for the class of systems that can be approximated to the chained form. However, there still exist many systems that cannot be controlled by the above control strategy. In the following sections, we will show two examples that cannot be approximated to the chained form.

# 3 Example 1 — Ball-plate manipulation

Figure 3.1 shows a ball-plate manipulation problem. The objective of this problem is to control the position and the orientation of the ball by manipulating two plates. Its kinematic model can be expressed in a driftless input-affine system (2.1) with five states and two inputs. It turns out that this system cannot be approximated to the chained form due to the existence of an uncontrollable part in the linear approximation of the state-control part. To cope with the uncontrollability, we propose an extended control strategy with iterative coordinate change that varies the structure of controllability.

This system consists of a ball whose radius is $\rho$ and a pair of sufficiently wide horizontal plates. The lower plate is fixed on the ground and the upper

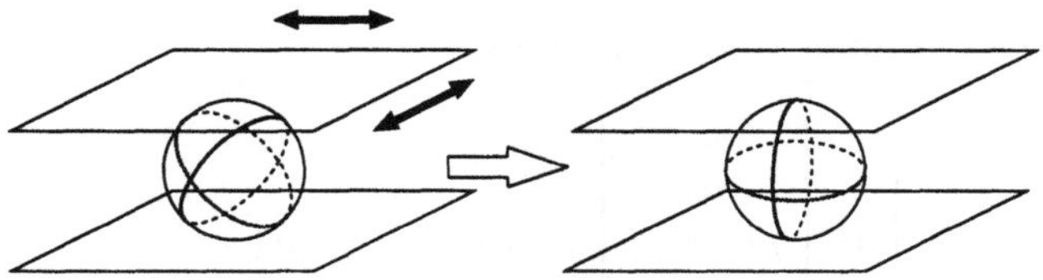

FIGURE 3.1. Ball-plate manipulation

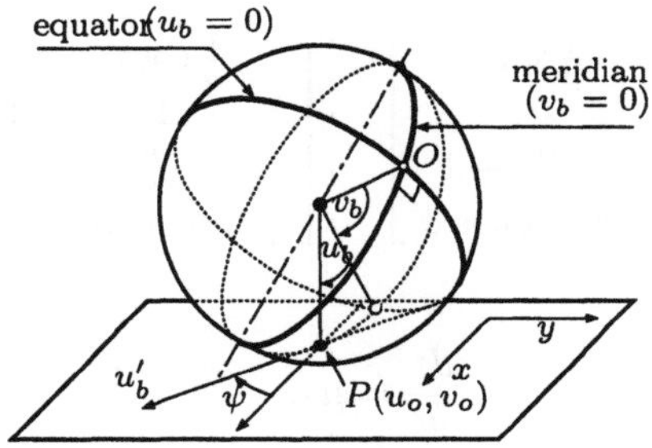


FIGURE 3.2. Orientation of the ball

plate moves in two translational directions which are assigned as control inputs. We take spherical coordinates on the ball which consist of latitude and longitude (Figure 3.2). $(u_b, v_b)$ denotes the contact point $P$ of the ball and the lower plate on the spherical coordinates. $P$ is also denoted by $(u_o, v_o)$ on the planar coordinates fixed on the lower plate. The skew angle between the spherical coordinates and the planar one is denoted $\psi$. Since the upper plate does not rotate, it has a holonomic relation to the position of the ball. Thus we can ignore the existence of the upper plate. Therefore these five variables complete the configuration of the system. The objective of the control is to bring $(u_b, v_b, u_o, v_o, \psi)$ to zero. If we assume that there is neither slip nor spin between the ball and the plates, a kinematic equation with rolling constraint is derived as follows [11].

$$\frac{d}{dt}\begin{bmatrix} u_b \\ v_b \\ u_o \\ v_o \\ \psi \end{bmatrix} = \begin{bmatrix} \frac{1}{\rho}\cos\psi \\ -\frac{1}{\rho}\frac{\sin\psi}{\cos u_b} \\ 1 \\ 0 \\ \frac{1}{\rho}\tan u_b \sin\psi \end{bmatrix} \dot{u}_o + \begin{bmatrix} -\frac{1}{\rho}\sin\psi \\ -\frac{1}{\rho}\frac{\cos\psi}{\cos u_b} \\ 0 \\ 1 \\ \frac{1}{\rho}\tan u_b \cos\psi \end{bmatrix} \dot{v}_o. \tag{3.1}$$

This system is a 5-dimensional 2-input driftless nonlinear system. We shall try to transform this system into time-state control form. Applying the coordinate and input transformations

$$\begin{bmatrix} \xi_1 \\ \xi_2 \\ \xi_3 \\ \xi_4 \\ \hline \tau \end{bmatrix} = \begin{bmatrix} u_b \\ u_o \\ \rho v_b + v_o \\ \psi \\ \hline v_o \end{bmatrix}, \quad \mu_1 = \frac{\dot{u}_o}{\dot{v}_o}, \quad \mu_2 = \dot{v}_o, \tag{3.2}$$

the system is transformed into the time-state control form

$$\frac{d\xi}{d\tau} = \begin{bmatrix} -\frac{1}{\rho}\sin\psi \\ 0 \\ -\frac{\cos\psi}{\cos u_b} + 1 \\ \frac{1}{\rho}\tan u_b \cos\psi \end{bmatrix} + \begin{bmatrix} \frac{1}{\rho}\cos\psi \\ 1 \\ -\frac{\sin\psi}{\cos u_b} \\ \frac{1}{\rho}\tan u_b \sin\psi \end{bmatrix} \mu_1, \tag{3.3}$$

$$\frac{d\tau}{dt} = \mu_2. \tag{3.4}$$

Here we see an obstruction. A linear approximation of the state control part,

$$\frac{d}{d\tau}\begin{bmatrix} u_b \\ u_o \\ \rho v_b + v_o \\ \psi \end{bmatrix} = \begin{bmatrix} 0 & 0 & 0 & \frac{1}{\rho} \\ 0 & 0 & 0 & 0 \\ 0 & 0 & 0 & 0 \\ \frac{1}{\rho} & 0 & 0 & 0 \end{bmatrix} \begin{bmatrix} u_b \\ u_o \\ \rho v_b + v_o \\ \psi \end{bmatrix} + \begin{bmatrix} \frac{1}{\rho} \\ 1 \\ 0 \\ 0 \end{bmatrix} \mu_1, \tag{3.5}$$

is not controllable. This prevents us from immediate application of the strategy in the previous section.

We cope with this difficulty in the following way. Since the derivative of the uncontrollable part $\rho v_b + v_o$ vanishes at the origin, this part does not move so much near the origin. On the other hand, the linear approximation of the rest part is controllable as

$$\frac{d}{d\tau}\begin{bmatrix} u_b \\ u_o \\ \psi \end{bmatrix} = \begin{bmatrix} 0 & 0 & \frac{1}{\rho} \\ 0 & 0 & 0 \\ \frac{1}{\rho} & 0 & 0 \end{bmatrix} \begin{bmatrix} u_b \\ u_o \\ \psi \end{bmatrix} + \begin{bmatrix} \frac{1}{\rho} \\ 1 \\ 0 \end{bmatrix} \mu_1. \tag{3.6}$$

Therefore we first ignore $\rho v_b + v_o$ and bring $u_b, u_o, \psi$ to zero using Control Strategy 2.2. As a consequence, a steady error will remain in the uncontrollable part. Moreover, only $v_b$ has the error since $v_o = 0$ holds at the end of this procedure.

Then, we will change the spherical coordinates as shown in Figure 3.3. $\tilde{\psi}$ is an angle between the $\tilde{u}_b$ axis and the $v_o$ axis. Note that the initial value of $\tilde{v}_b$ corresponds to the final value of $u_b$, which has been sufficiently small. The error remaining in $v_b$ becomes the initial error of $\tilde{u}_b$ in the

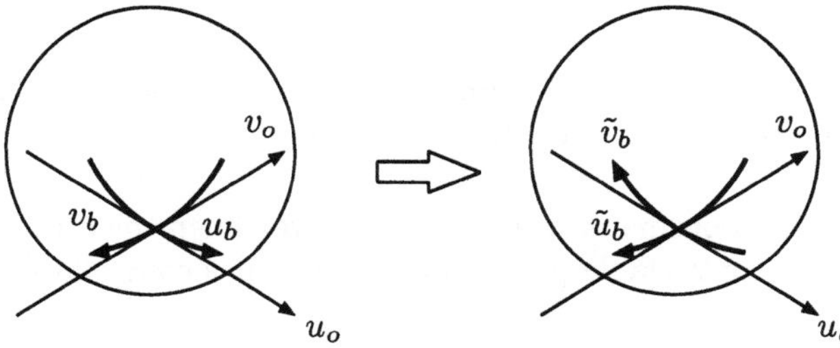


FIGURE 3.3. Coordinates change: the spherical coordinates rotate $\frac{\pi}{2}$ around the origin

new coordinates. It can be attenuated because $\tilde{u}_b$ is controllable as shown hereafter.

In the new coordinates, the original system (3.1) is rewritten as

$$\frac{d}{dt}\begin{bmatrix} \tilde{u}_b \\ \tilde{v}_b \\ u_o \\ v_o \\ \tilde{\psi} \end{bmatrix} = \begin{bmatrix} -\frac{1}{\rho}\sin\tilde{\psi} \\ -\frac{1}{\rho}\frac{\cos\tilde{\psi}}{\cos\tilde{u}_b} \\ 1 \\ 0 \\ \frac{1}{\rho}\tan\tilde{u}_b\cos\tilde{\psi} \end{bmatrix}\dot{u}_o + \begin{bmatrix} -\frac{1}{\rho}\cos\tilde{\psi} \\ -\frac{1}{\rho}\frac{\sin\tilde{\psi}}{\cos\tilde{u}_b} \\ 0 \\ 1 \\ \frac{1}{\rho}\tan\tilde{u}_b\sin\tilde{\psi} \end{bmatrix}\dot{v}_o. \tag{3.7}$$

Here we take $u_o$ as the new time scale (the former time scale was $v_o$). We apply the coordinates transformation as

$$\begin{bmatrix} \tilde{\xi}_1 \\ \tilde{\xi}_2 \\ \tilde{\xi}_3 \\ \tilde{\xi}_4 \\ \hline \tilde{\tau} \end{bmatrix} = \begin{bmatrix} \tilde{u}_b \\ v_o \\ \rho\tilde{v}_b + u_o \\ \tilde{\psi} \\ \hline u_o \end{bmatrix}, \quad \tilde{\mu}_1 = \frac{\dot{v}_o}{\dot{u}_o}, \quad \tilde{\mu}_2 = \dot{u}_o, \tag{3.8}$$

then the state-control part for the new coordinates is obtained as

$$\frac{d\tilde{\xi}}{d\tilde{\tau}} = \begin{bmatrix} -\frac{1}{\rho}\sin\tilde{\psi} \\ 0 \\ 1 - \frac{\cos\tilde{\psi}}{\rho\cos\tilde{u}_b} \\ \frac{1}{\rho}\tan\tilde{u}_b\cos\tilde{\psi} \end{bmatrix} + \begin{bmatrix} -\frac{1}{\rho}\cos\tilde{\psi} \\ 1 \\ -\frac{\sin\tilde{\psi}}{\rho\cos\tilde{u}_b} \\ \frac{1}{\rho}\tan\tilde{u}_b\sin\tilde{\psi} \end{bmatrix}\tilde{\mu}_1. \tag{3.9}$$

It can easily be seen that $\rho\tilde{v}_b + u_o$ is uncontrollable and $\tilde{u}_b, v_o, \tilde{\psi}$ are controllable. Since the initial value of $\rho\tilde{v}_b + u_o$ is sufficiently small and its linearization is uncontrollable at the origin, it will stay near the origin. We summarize the above argument as follows.

**Control Strategy 3.1.**

*Step 1* For the controllable part, apply Control Strategy 2.2. There will remain an error in the uncontrollable part.

*Step 2* Change the coordinates as described above. Then the error remaining in the uncontrollable part in Step 1 becomes controllable.

*Step 3* Repeat the above two steps until all variables are sufficiently close to zero. □

### 3.1 Simulation results

The control objective is to bring the state $(u_o, v_o, u_b, v_b, \psi)$ from $(0, 0, 0.7, 0, 0.7)$ to $(0, 0, 0, 0, 0)$. Feedback controllers for Control Strategy 2.2 are designed based on the LQ optimal method for the linear approximation. Figure 3.4 shows a trace of the ball on the ground. Starting from the origin, the ball first moves along the $v_o$ axis, which is assigned as a time scale, and then along the $u_o$ axis after the change of coordinates. Figure 3.5 shows the time response of the angles, $u_b, v_b, \psi$ in the original coordinates. Note that the change of coordinates occurs at 20 [s]. At this moment, there is an error on $v_b$ whereas $u_b, \psi$ are sufficiently close to zero. After 20 [s], $v_b$ converges to zero since it has become controllable through the coordinates change. Finally all states are brought sufficiently close to zero at 30 [s]. Figure 3.6 is a time response of the uncontrollable part $\rho v_b + v_o$ and $\rho \tilde{v}_b + u_0$. This value settles to a constant value as the controllable parameters converge to zero. The plot after 20 [s] is the value in the new coordinates $(\rho \tilde{v}_b + u_0)$.

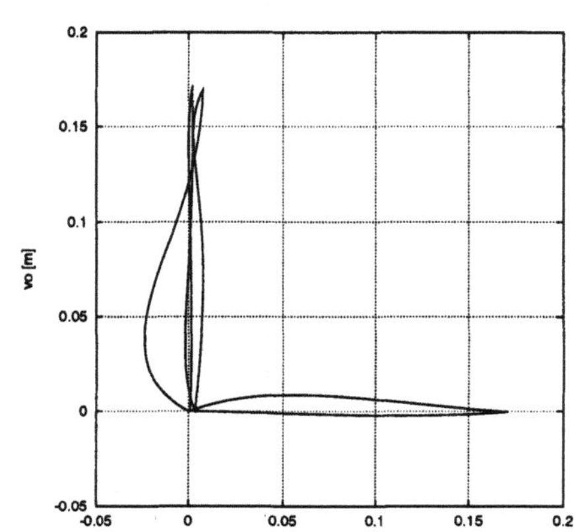


FIGURE 3.4. Trace of the ball $u_o, v_o$

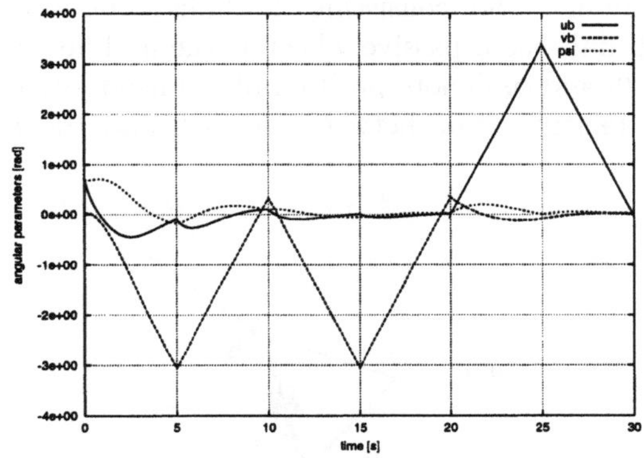


FIGURE 3.5. Time response of the angles $u_b, v_b, \psi$

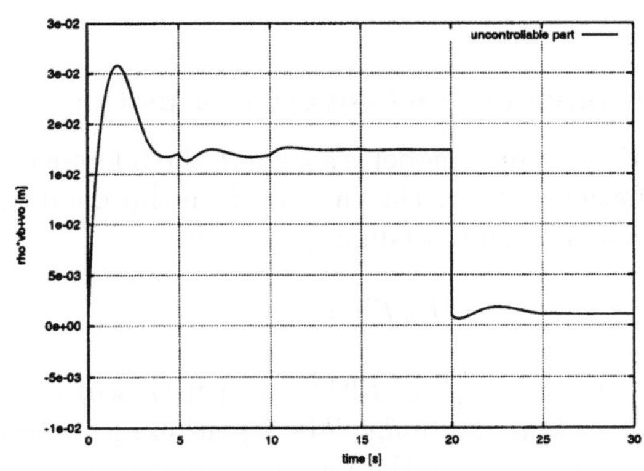


FIGURE 3.6. Time response of the uncontrollable variables $\rho v_b + v_o$ ($t < 20$) and $\rho \tilde{v}_b + u_O$ ($20 < t$)

# 4 Example 2 — Locomotion control of a snake-like robot

Figure 4.1 is a snake-like robot consisting of $n$ links serially connected with active joints. Each link has a passive wheel at the middle which does not sideslip. $(x_F, y_F)$, $\theta_i$ and $\phi_i$ denote position of the head, orientation of the $i$-th link and the relative angle between the $i$-th and the $i-1$-th link, respectively.

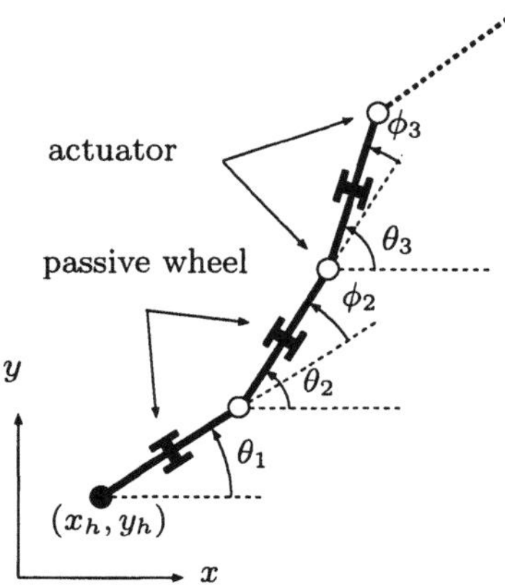


FIGURE 4.1. $n$-link model

## 4.1 Kinematic constraint and equation of motion

The derivation of the dynamic model is shown in the reference [14] in detail. Only essential equations will be shown here. Non-slip condition of wheels is expressed as a nonholonomic Pfaffian form [12]

$$[I_n, F]\dot{q} = 0, \tag{4.1}$$

where $q = [\theta_1, \cdots, \theta_n, x_F, y_F]^T \in R^{n+2}$ are generalized coordinates and $F \in R^{n \times 2}$ is a matrix function of $\theta_i$s. The equation of motion under such constraint can be expressed with the Lagrange multiplier $\lambda \in R^n$ as

$$M\ddot{q} + C\dot{\theta} + D\dot{q} + \begin{bmatrix} I_n \\ F^T \end{bmatrix} \lambda = \begin{bmatrix} E\tau \\ 0 \end{bmatrix}, \tag{4.2}$$

where $\tau \in R^{n-1}$, $M \in R^{(n+2)\times(n+2)}$, $C \in^{(n+2)\times n}$, $D \in R^{(n+2)\times(n=2)}$ are joint torque, inertia, Corioli's or centrifugal, and viscosity matrix, respectively. $E$ satisfies $\dot{\phi} = E^T\dot{\theta}$. Together with the equation of constraint (4.1), the whole motion of the robot can be represented as the motion of the head position $r = [x_F, y_F]^T$ with the reduction technique in [14] as

$$\tilde{M}\ddot{r} + (\tilde{C} + \tilde{D})\dot{r} = F^T E \tau, \tag{4.3}$$

where

$$\begin{aligned} \tilde{M} &= [F^T\ I_2] M \begin{bmatrix} F \\ I_2 \end{bmatrix} \in R^{2\times 2}, \quad \tilde{D} = [F^T\ I_2] D \begin{bmatrix} F \\ I_2 \end{bmatrix} \in R^{2\times 2}, \\ \tilde{C} &= [F^T\ I_2] M \begin{bmatrix} \dot{F} \\ 0 \end{bmatrix} + [F^T\ I_2] C F \in R^{2\times 2}. \end{aligned}$$

Note that we can generate arbitrary acceleration of the head $\ddot{r}$ as long as the matrix $F^T E \in R^{2\times(n-1)}$ is row full rank.

## 4.2 Similarity to and difference from a trailer system

If we take $\ddot{r}$ as an input to the snake robot, it can be easily found that its dynamics is the same as that of a multiple (odd hitched) trailer system, which is known to be approximated to the chained form. The fundamental difference between the snake robot and the trailer system consists in their input and the resultant singularity.

If the snake robot is in a straight or an arc posture, $rank(F^T E) = 1 < 2$ [14] and $\ddot{r}$ can not be arbitrarily controlled. Such postures are called singular posture. This implies that controllability drops at the origin (a straight shape). For this reason, the system cannot be transformed into the chained form and it is not easy to control its position and posture near the origin. Thus, hereafter, we will restrict our attention to the propulsion control, i.e., to control the robot to follow the desired path (mainly a straight line) while it moves forward. Since the robot tends to be straight while the head moves along a linear path, avoiding the singular posture is the main problem to construct a feedback controller.

## 4.3 Control strategy

If we take head position $r$ as the output of the system, then the zero dynamics is stable for forward movement as it is in a trailer system. Thus, we only control the head position $r$. As already shown, arbitrary head acceleration can be produced as long as the snake robot is not in the singular posture. Hence we consider the following acceleration $\ddot{r} = \alpha_{\text{next}}$ as a command input, which consists of two acceleration vectors $\alpha_t$ and $\alpha_m$ (Figure 4.2).

$$\alpha_{\text{next}} = w_t \alpha_t + w_m \alpha_m. \tag{4.4}$$

$\alpha_t$ is an acceleration vector for tracking a given path, as

$$\alpha_t = v_t - \dot{r}, \tag{4.5}$$

where $v_t$ is a desired velocity toward the desired trajectory. $\alpha_m$ is an acceleration improving the non-singularity. $\alpha_m$ is determined by a numerical search. In order to measure the non-singularity, we will introduce a notion of dynamic manipulability described in the next subsection.

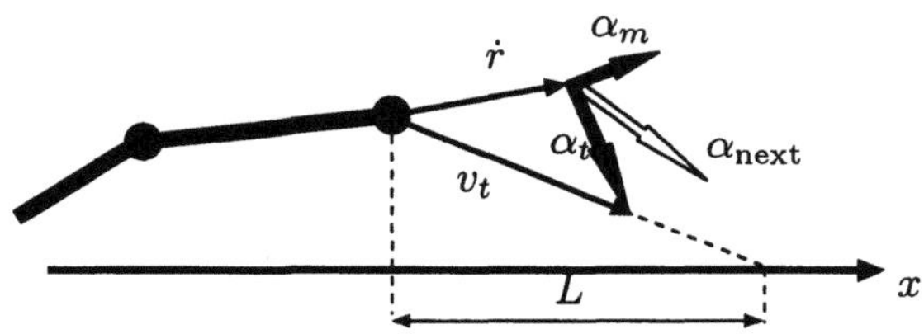


FIGURE 4.2. Two accelerations of the head

The input torques which generate $\alpha_{\text{next}}$ can be computed from (4.3) as the inverse dynamics of the head motion

$$\tau = (F^T E)^- \left[\tilde{M}\alpha_{\text{next}} + \tilde{C}\dot{r} + \tilde{D}\dot{r}\right], \tag{4.6}$$

where $(F^T E)^-$ is a generalized inverse of $F^T E$. If we choose the pseudo-inverse, the norm of $\tau$ is minimized whereas we can choose another type of inverse so that the norm of the constraint force is minimized [3].

In the following, we will describe the manipulability used for this strategy.

## 4.4 Dynamic manipulability

### *4.4.1 General formulation*

In this section, we will first show a brief summary of manipulability for a general $n$-d.o.f manipulator expressed by

$$M(\psi)\ddot{\psi} + h(\dot{\psi}, \psi) + g(\psi) = \tau, \tag{4.7}$$

where $\psi \in R^n$ and $\tau \in R^n$ denote joint angles and joint torques, respectively. $M \in R^{n\times n}$ is the moment of inertia which is always nonsingular, $h \in R^n$ and $g \in R^n$ are inertial force and gravity, respectively. The motion in the task coordinates $r \in R^m$ is expressed with a Jacobian $J(\psi) = \partial r/\partial\psi \in R^{m\times n}$ as

$$\dot{r} = J(\psi)\dot{\psi}, \tag{4.8}$$
$$\ddot{r} = J\ddot{\psi} + \dot{J}\dot{\psi}. \tag{4.9}$$

Eliminating $\ddot{\psi}$ from (4.7) and (4.9), we have

$$\ddot{r} = JM^{-1}[\tau - h - g] + \dot{J}\dot{\psi}. \tag{4.10}$$

Normalizing the input by

$$\tau = Nv \quad (|v_i| \leq 1), \tag{4.11}$$

$$N = \mathrm{diag}(\tau_{i\max} - |h_i - g_i|), \tag{4.12}$$

we obtain

$$\ddot{r} = JM^{-1}Nv + \dot{J}\dot{\theta}, \tag{4.13}$$

as a relation between the normalized input $v$ and the acceleration of the end effector $\ddot{r}$. This implies that the maximum acceleration is characterized by an ellipsoid $JM^{-1}Nv$ ($\|v\| \leq 1$) as long as $\dot{J}\dot{\psi}$ is relatively small. This ellipsoid is called a dynamic manipulability ellipsoid [17]. The larger and the more similar to a sphere the dynamic ellipsoid is, the higher the manipulability is. There have been proposed several measures of manipulability; the ratio of the length of the ellipsoid's minor axis to that of the major [8], the volume of the ellipsoid [17], and the length of the ellipsoid's minor axis [4].

We will first apply this manipulability to the snake-like robot. Assuming that the term $(\tilde{C} + \tilde{D})\dot{r}$ is sufficiently small in (4.3), the acceleration is yielded as

$$\ddot{r} = \tilde{M}F^T E\tau. \tag{4.14}$$

The manipulability ellipsoid is characterized by singular values $\sigma_1$ and $\sigma_2$ of the matrix $\tilde{M}F^T E \in R^{2\times n-1}$. The manipulability ellipsoid based on this matrix becomes high when the robot is in a shape that a large acceleration of the head can be produced by small input torques. However, a zigzag winding shape is associated with high manipulability under this definition [2]. In such a posture, it requires large constraint forces that may cause sideslip. Therefore, we will define another manipulability which takes the side constraint force into consideration.

### *4.4.2 Manipulability with consideration of the constraint force*

Ordinary dynamic manipulability becomes high when a large acceleration of the head can be obtained with small input torques. On the contrary, in this subsection, we consider another manipulability which becomes high when a large acceleration of the head can be obtained with small side forces on wheels. By straightforward calculation from the Lagrange multiplier $\lambda$ in (4.2), we obtain the constraint force $f$ as

$$f = Y(\theta, \dot{r}) + X(\theta)\tau. \tag{4.15}$$

$Y(\theta, \dot{r}) \in R^n$ is a force vector caused by the inertial force, and $X(\theta)\tau \in R^n$ is a term from the joint torque. Each element of $f \in R^n$ is a non-slip constraint force on the wheel perpendicular to each link. If we assume that $Y$ is negligible, then

$$f = X(\theta)\tau. \tag{4.16}$$

The constraint force can be normalized as

$$f = N\hat{f}, \tag{4.17}$$

$$N = \mathrm{diag}[f_{i\max}] \quad i = 1, \cdots, n, \tag{4.18}$$

where $f_{i\max}$ is the maximum friction of the $i$-th link. Hence the input torque $\hat{\tau}$ that will not cause slip is expressed as

$$\hat{\tau} = X^{+}N\hat{f}, \tag{4.19}$$

where $X^{+}$ is the pseudo-inverse of $X$. Replacing $\tau$ in (4.14) by this $\hat{\tau}$, we have the acceleration $\ddot{r}$ cased by the normalized constraint force $\hat{f}$.

$$\ddot{r} = \tilde{M}^{-1}F^{T}EX^{+}N\hat{f}. \tag{4.20}$$

If the normalized constraint force $\hat{f}$ is restricted in a unit sphere, the acceleration of the head $\ddot{r}$ draws an ellipsoid $\tilde{M}^{-1}F^{T}EX^{+}N\hat{f}$ ($\|\hat{f}\| \leq 1$). We regard this ellipsoid as the dynamic manipulability ellipsoid. The singular values $\sigma_1$ and $\sigma_2$ of the matrix $\tilde{M}^{-1}F^{T}EX^{+}N$ characterize the dynamic manipulability ellipsoid and the smaller one $\sigma_2$ is taken as the measure of manipulability here. Comparison between the ordinary manipulability based on (4.14) and the proposed one is stated in [2].

### 4.5 Simulation results

Figures 4.3 and 4.4 show the trace of the head and the body shape at a certain moment for a 10-link model. The design parameters $L, w_m, w_a$ are suitably chosen. Figure 4.3 is the result of tracking control based on the ordinary manipulability. Notice that the amplitude of winding motion is very small (The vertical scale is magnified five times as large as the horizontal scale). On the contrary, large winding motion can be observed in the results of the control based on the proposed manipulability (Figure 4.4: the vertical scale is the same as the horizontal one).

## 5 Conclusion

In this note we showed two examples of feedback control strategies for nonholonomic systems with bad controllability structure. Both examples have different structure of controllability. Therefore, we need to use a different control strategy for each system.

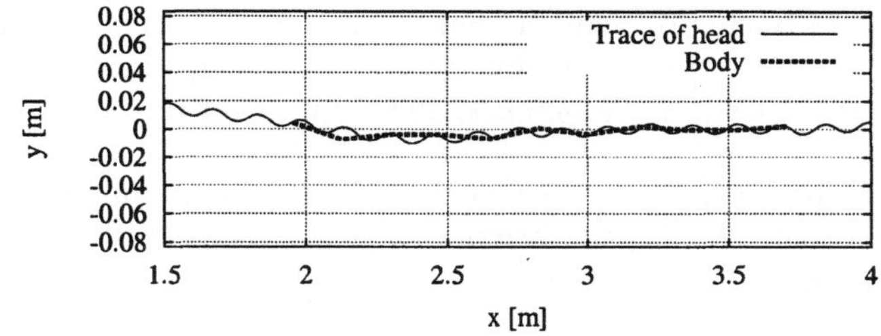


FIGURE 4.3. Trace of the head; with ordinary manipulability

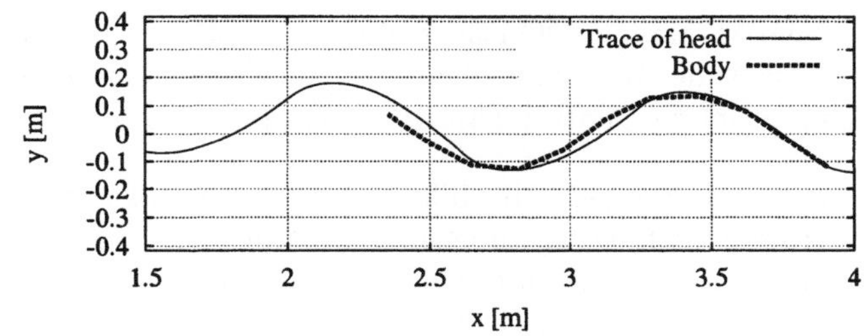


FIGURE 4.4. Trace of the head; with proposed manipulability

The first example, the ball-plate manipulation, is not transformable nor approximated to the chained form due to the uncontrollable part in the time-state control form. We proposed an iteration of coordinates change that permutes the controllable and uncontrollable parts.

The snake-like robot has a singular posture at the origin (a straight line shape), which is a stable equilibrium for locomotion along a linear path. We introduced and modified dynamic manipulability as a measure of non-singularity. A natural winding trajectory was automatically generated by the proposed feedback controller without giving any explicit gate.

## References

[1] R. W. Brockett, "Asymptotic stability and feedback stabilization," in *Differential Geometric Control Theory*. Progress in Mathematics, vol. 27, Springer-Verlag, 1983, pp. 181–191.

[2] H. Date, Y. Hoshi, and M. Sampei, "Locomotion control of a snake-like robot based on dynamic manipulability," in *Proceedings of the IEEE/RSJ Int. Conference on Intelligent Robots and Systems*, 2000, pp. 2236–2241.

[3] H. Date, Y. Hoshi, M. Sampei, and S. Nakaura, "Locomotion control of a snake robot with constraint force attenuation," in *Proceeding of the American Control Conference*, 2001, pp. 113–118.

[4] O. Khatib and J. Burdics, "Dynamic optimization in manipulator design: The operational space formulation," in *Proceedings of the ASME Winter Annual Meeting*, 1985.

[5] H. Kiyota and M. Sampei, "An attitude control of planar nonholonomic space robots," in *Proceedings of the 25th SICE Symposium on Control Theory*, 1996, pp. 51–56.

[6] H. Kiyota and M. Sampei, "On the stability of a class of nonholonomic systems using time-state control form," in *Proceedings of the 26th SICE Symposium on Control Theory*, 1997, pp. 409–412.

[7] H. Kiyota and M. Sampei, "Stabilization of a class of nonholonomic systems without drift using time-state control form" (in Japanese), *Trans. of the Inst. of Systems, Control and Information Engineers*, vol. 12, no. 11, pp. 647–654, 1999.

[8] K. Kosuge and K. Furuta, "Kinematic and dynamic analysis of robot arm," in *Proceedings of the IEEE International Conference on Robotics and Automation*, 1985, pp. 1039–1044.

[9] G. Lafferriere and H. Sussmann, "Motion planning for controllable systems without drift," in *Proceedings of the IEEE International Conference on Robotics and Automation*, 1991, pp. 1148–1153.

[10] R. T. M'Closkey and R. M. Murray, "Exponential stabilization of driftless nonlinear control systems using homogeneous feedback," *IEEE Trans. on Automatic Control*, vol. 42, no. 5, pp. 614–628, 1997.

[11] D. J. Montana, "The kinematics of contact and grasp," *The International Journal of Robotics Research*, vol. 7, no. 3, pp. 17–32, 1988.

[12] R. M. Murray, Z. Li, and S. S. Sastry, *A Mathematical Introduction to Robotic Manipulation.* CRC Press, 1994.

[13] R. M. Murray and S. S. Sastry. "Nonholonomic motion planning: Steering using sinusoids," *IEEE Trans. on Automatic Control*, vol. 38, no. 5, pp. 700–716, 1993.

[14] P. Prautsch and T. Mita, "Control and analysis of the gait of snake robots," in *Proceedings of the IEEE International Conference on Control Applications*, 1999, pp. 502–507.

[15] M. Sampei, H. Kiyota, M. Koga, and M. Suzuki, "Necessary and sufficient conditions for transformation of nonholonomic system into time-state control form," in *Proceedings of the IEEE International Conference on Decision and Control*, 1996, pp. 4745–4746.

[16] O. J. Sordalen and O. Egeland, "Exponential stabilization of nonholonomic chained systems," *IEEE Trans. on Automatic Control*, vol. 40, no. 1, pp. 35–49, 1995.

[17] T. Yoshikawa, "Analysis and control of manipulators with redundancy," in *Preprints of the 1st International Symposium of Robotics Research*, 1983.

Mitsuji Sampei
Department of Mechanical and Control Engineering
Graduate School of Science and Engineering
Tokyo Institute of Technology
2–12–1 Oh-okayama, Meguro-ku, Tokyo 152–8552, Japan
sampei@ctrl.titech.ac.jp

Hisashi Date
Department of Mechanical and Control Engineering
Graduate School of Science and Engineering
Tokyo Institute of Technology
2–12–1 Oh-okayama, Meguro-ku, Tokyo 152–8552, Japan
date@sc.ctrl.titech.ac.jp

Shigeki Nakaura
Department of Mechanical and Control Engineering
Graduate School of Science and Engineering
Tokyo Institute of Technology
2–12–1 Oh-okayama, Meguro-ku, Tokyo 152–8552, Japan
nakaura@ctrl.titech.ac.jp

# Part V

# Control Applications

# Computer Control of Vehicles: Overcoming Limitations of Human Drivers

Masayoshi Tomizuka

*Dedicated to Professor Hidenori Kimura on the occasion of his 60th birthday*

ABSTRACT This paper describes lateral control of automobiles for automated highway systems and automated roll-rate control of front and rear-wheel steered bicycles. Computer control of these vehicles provides significantly better performance than manual steering, overcoming limitations of the human driver such as the reaction time, limited actuation bandwidth and limited capability in comprehending the object dynamics with large phase shifts and/or delays.

**Keywords:** Modeling and control, Automated driving, Automated highway systems, Bicycles

## 1 Introduction

This paper presents automated driving of ground vehicles. More specifically, automated lateral control of automobiles for automated highway systems (AHS) and automated roll-rate control of front and rear-wheel steered bicycles. Automobiles and bicycles have both been in existence for many years, and they have been designed for manual driving. Driving of automobiles and bicycles represents excellent examples of manual feedback and predictive control, and the human driver has been proven to be an excellent controller. Computer control of these vehicles, however, may provide significantly better performance than manual steering, overcoming limitations of the human driver such as the reaction time, limited actuation bandwidth and limited capability in comprehending the object dynamics with large phase shifts and/or delays.

The California Partners for Advanced Transit and Highway (PATH) Programs was established in 1986 to promote the application of advanced technology to help meet California's growing need for increased highway capacity to relieve congestion. The major sponsor of this program has been the California Department of Transportation (Caltrans). The idea of Automated Highway Systems (AHS) was identified to be an attractive option for solving the congestion problem while improving the operation of highways in many regards including safety, fuel economy and pollution; AHS research has been a major activity of the PATH. In the United States, the National Automated Highway System Consortium (NAHSC) was formed in 1994, and the consortium demonstrated various AHS scenarios in the NAHSC Demonstration in San Diego in August 1997. The main demonstration in this event was an eight car platooning scenario, which included the two main PATH concepts: one was platooning for longitudinal control and the other was lateral control based on magnetic markers (nails) embedded along the center of highway lanes. A platoon is one or more vehicles traveling together as a group with relatively small spacing. Inter-platoon spacing is set large. For example, with an average platoon size of 15, intra-platoon distance of 2 m, inter- platoon distance of 60 m, vehicle length of 5 m, and speed of 72 km/h, the maximum flow or capacity through an automated lane is more than over 6,000 vehicles/lane/hour. This capacity is about three times as large as the maximum capacity under manual driving. The short intra-platoon distance is maintained by the longitudinal controller on-board each vehicle. The lateral controller provides the steering command based on the lateral error obtained by on-board magnetometers, the outputs of which depend on the lateral error relative to the magnetic markers. The lateral tracking error under automated driving is consistently small. Thus, the lane width of automated highway lanes may be set smaller than typical highway lanes designed for manual driving for better utilization of land space. For past and recent research activities on AHS at PATH, see Varaiya [20], Hedrick, et al. [5] and the homepage of PATH at http://www.path.berkeley.edu/.

The bicycle has been in existence for more than a century now. In recent years, there has been a significant interest in the development of highly efficient human powered vehicles. In the United States, several collegiate teams participate in an annual human powered vehicle contest. There have been other initiatives noted around the world towards developing efficient human powered bicycles. A key feature of such a bicycle is its low height to reduce the air drag force, and its target speeds are well above 100 km/hour (Suryanarayanan, et al.) [17]. In these initiatives, it has been noted that, because of limitations of human capability, roll control of low frame high-performance bikes was getting difficult. Even a rear-wheel steering design is under consideration for such bicycles to attain the optimal shape for minimizing the air drag force. It should be noted that it is practically impossible for humans to ride on rear-wheel steered bicycles [9]. It is expected that computer control of high speed bicycles for roll stability will significantly enlarge the operational envelope of speeds.

The remainder of this paper is organized as follows. The next section is concerned with the lateral control of vehicles for AHS; vehicle models for lateral control, fundamental control issues arising from the vehicle dynamics, and lateral control algorithms for forward and backward driving are presented. Section 3 is concerned with the roll rate control of high-speed bicycles. Concluding remarks are given in Section 4.

# 2 Vehicle lateral control for automated highway systems

A major objective of the AHS is to increase highway capacity and safety. To achieve this objective, PATH adopted the idea of platooning and magnetic markers. The overall control structure of automated highway systems may be defined as a hierarchical control system. The higher level controllers coordinate motions of neighboring vehicles and set the optimal operating conditions such as the size and speed of platoons. The higher level controllers assume that the low level controller (regulation level controller) maintains the desired intra-platoon spacing (longitudinal control) and small lateral error (lateral control).

The design of longitudinal control and lateral control provides a number of technical challenges. Longitudinal control laws must assure that a short intra-platoon distance can be robustly maintained. For this purpose, a sliding mode control has been found to be appropriate [5]. An important issue in longitudinal control is string stability. Namely, a platoon must be string stable in the sense that disturbance effect should be attenuated from one vehicle to the next as it propagates upstream. The propagation takes place because the primary input to the longitudinal controller is the measured distance between the vehicle and the preceding vehicle in front of it. String stability is now well understood [18].

The vehicle lateral control system for AHS has two major functions: 1) lane following and 2) lane change maneuvering. Primary focus in this article is on the lane following problem for vehicles with front wheel steering. In this case, technical challenges include: 1) the controller must assure good riding comfort as well as lane following capability; 2) despite the terminology, "vehicle lateral control," the steering input affects both the lateral motion and yaw motion, which must be simultaneously controlled to properly address the first challenge; and; 3) vehicle dynamics depend on other variables such as vehicle speed and tire pressures, as well as parameters such as tire-road adhesion coefficients. The second challenge implies that the vehicle dynamics for lateral control represents an under-actuated system.

## 2.1 Vehicle models for lateral control

The lateral dynamics of vehicles has been studied since the late 1950s. Segel [14] developed a three-degree-of-freedom vehicle model to describe the vehicle directional responses, which include yaw, lateral and roll motions.

The vehicle forward speed was considered a major stability parameter and was assumed to be constant. Most of previous works on vehicle lateral control have relied on a simplified dynamic model that retains the lateral and yaw motions (Fenton, et al. [4] and Shladover et al. [15] among others). This model is usually referred to as the bicycle model. The major control input for lateral *control* is the front steering angle. Secondary inputs include the rear steering angle for four wheel steering vehicles, the throttle angle and the brake force. Since actual control forces to cause vehicle motion come from the ground, the tire model is an important element of any vehicle lateral dynamic model. In this paper, we will consider bicycle models for a lane following situation as depicted in Figure 2.1. In the figure, $y_{CG}$ and $\varepsilon_r$ denote the lateral deviation of the vehicle from the road center line at the vehicle center of gravity (CG) and the vehicle yaw angle relative to the road, respectively, $\delta$ is the front steering angle and $y_s$ is the lateral tracking error measured by a sensor located ahead of the vehicle CG by a distance, $d_s$. Newton–Euler equations for the bicycle model are given by [7]

$$\begin{aligned} m\ddot{y} &= -\,m\dot{\varepsilon}\dot{x} - \frac{C_{\alpha r}(\dot{y} - \dot{\varepsilon}l_2)}{\dot{x}} - \frac{C_{\alpha f}(\dot{y} + \dot{\varepsilon}l_1)}{\dot{x}} + C_{\alpha f}\delta, \\ I_z\ddot{\varepsilon} &= l_2\frac{C_{\alpha r}(\dot{y} - \dot{\varepsilon}l_2)}{\dot{x}} - l_1\frac{C_{\alpha f}(\dot{y} + \dot{\varepsilon}l_1)}{\dot{x}} + l_1C_{\alpha f}\delta \end{aligned} \tag{2.1}$$

where $\dot{x}$ and $\dot{y}$ are the components of the vehicle velocity along the longitudinal and the lateral principle axis of the vehicle body and $\dot{\varepsilon}$ is the yaw rate. $m$ and $I_z$ are the mass and the yaw moment of inertia, respectively. $l_1$ and $l_2$ are distances of front and rear axle from CG, and $C_{\alpha f}$ and $C_{\alpha r}$ are the front and rear tire cornering stiffness, respectively.

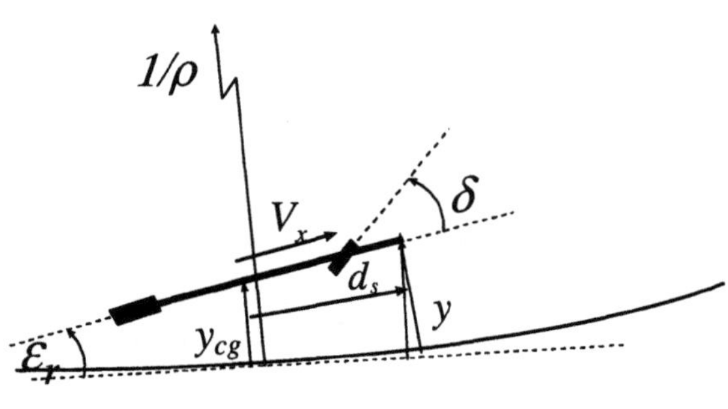


FIGURE 2.1. Bicycle model for land following

Transformation to the $y_s$ and $\varepsilon_r$ coordinates is given by

$$\dot{y}_s = \dot{y}\cos(\varepsilon_r) + \dot{x}\sin(\varepsilon_r) + d_s\dot{\varepsilon}_r.$$

Linearization of the resulting model yields

$$\begin{aligned}
\ddot{y}_s =& f_1 + b_1\delta, \\
\ddot{\varepsilon}_r =& -\frac{l_1 C_{\alpha f} - l_2 C_{\alpha r}}{I_z \dot{x}}\dot{y}_s + \frac{l_1 C_{\alpha f} - l_2 C_{\alpha r}}{I_z}\varepsilon_r + \frac{C_{\alpha f} l_1}{I_z}\delta \\
& - \frac{C_{\alpha f}(l_1^2 - l_1 d_s) + C_{\alpha r}(l_2^2 + l_2 d_s)}{I_z \dot{x}}\dot{\varepsilon}_r - \frac{C_{\alpha f} l_1^2 + C_{\alpha r} l_2^2}{I_z \dot{x}}\dot{\varepsilon}_d
\end{aligned} \tag{2.2}$$

where

$$\begin{aligned}
f_1 =& -\frac{\phi_1 + \phi_2}{\dot{x}}\dot{y}_s + (\phi_1 + \phi_2)\varepsilon_r \\
& + \frac{\phi_1(d_s - l_1) + \phi_2(d_s + l_2)}{\dot{x}}\dot{\varepsilon}_r + \frac{\phi_2 l_2 - \phi_1 l_1 - \dot{x}^2}{\dot{x}}\dot{\varepsilon}_d, \\
b_1 =& \phi_1, \\
\phi_1 =& C_{\alpha f}\left(\frac{1}{m} + \frac{l_1 d_s}{I_z}\right), \phi_2 = C_{\alpha r}\left(\frac{1}{m} - \frac{l_2 d_s}{I_z}\right).
\end{aligned}$$

It should be noted that this equation involves certain approximations, details of which are in [15]. Another way of writing these equations is in the standard state-space form

$$\dot{\xi} = A\xi + B\delta + W\rho \tag{2.3}$$

where $\xi = [y_s \quad \dot{y}_s \quad \varepsilon_r \quad \dot{\varepsilon}_r]^T$,

$$A = \begin{bmatrix} 0 & 1 & 0 & 0 \\ 0 & -\frac{a_{11}}{\dot{x}} & a_{11} & \frac{a_{12}}{\dot{x}} \\ 0 & 0 & 0 & 1 \\ 0 & -\frac{a_{41}}{\dot{x}} & a_{41} & \frac{a_{42}}{\dot{x}} \end{bmatrix}, \quad B = \begin{bmatrix} 0 \\ b_{21} \\ 0 \\ b_{41} \end{bmatrix}, \quad W = \begin{bmatrix} 0 \\ w_{21} \\ 0 \\ w_{41} \end{bmatrix},$$

$$\begin{aligned}
a_{11} &= (\phi_1 + \phi_2), \quad a_{12} = \phi_1(d_s - l_1) + \phi_2(d_s + l_2), \\
a_{41} &= l_1 C_{\alpha f} - l_2 C_{\alpha r}, \\
a_{42} &= l_1 C_{\alpha f}(d_s - l_1) + l_2 C_{\alpha r}(d_s + l_2), \\
b_{21} &= \phi_1, \quad b_{24} = \frac{l_1 C_{\alpha f}}{I_z}, \\
w_{21} &= -\frac{l_1^2 C_{\alpha f} + l_2^2 C_{\alpha r}}{I_z}, \\
w_{41} &= \phi_2 l_2 - \phi_1 l_1 - \dot{x}^2,
\end{aligned}$$

and $\rho = \dot{\varepsilon}_d/\dot{x}$ is the curvature of the road at the point on the road nearest to the center of gravity.

The transfer function between the steering angle $\delta$ and the acceleration at the sensor relative to the road, $\ddot{y}_s$, is given by

$$\ddot{y}_s = V_s(s)\delta(s) - \dot{x}^2\rho \tag{2.4}$$

where

$$V_s(s) = \left(C_{\alpha f}\dot{x}^2(ml_1d_s + I_z)s^2 + C_{\alpha f}C_{\alpha r}\dot{x}(l_1 + l_2)(d_s + l_2)s + C_{\alpha f}C_{\alpha r}(l_1 + l_2)\dot{x}^2\right)/D(s),$$

$$D(s) = I_z m\dot{x}^2 s^2 + \dot{x}\left(Iz(C_{\alpha f} + C_{\alpha r}) + m(C_{\alpha f}l_1^2 + C_{\alpha r}l_2^2)\right)s + m\dot{x}^2(C_{\alpha r}l_2 - C_{\alpha f}l_1) + C_{\alpha r}C_{\alpha f}(l_1 + l_2)^2.$$

For $d_s = 0$, we get $\ddot{y}_s = \ddot{y}_{CG}$. The block diagram of the vehicle lateral dynamics becomes as shown in Figure 2.2.

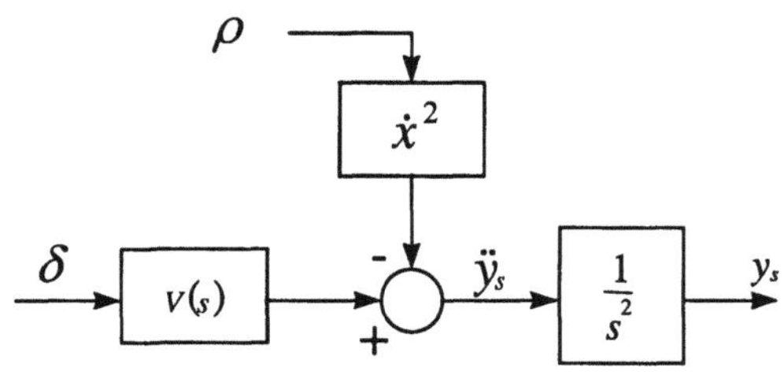


FIGURE 2.2. Block diagram of the vehicle dynamics

$V(s)$ is the transfer function from the wheel steering angle to the lateral acceleration at the sensor. Notice that $V(s)$ depends on both the vehicle speed and the distance between the vehicle's CG and the lateral position sensor. Figure 2.3 shows frequency responses from the wheel steering angle to the second derivative of the sensor signal for a typical passenger vehicle. Figure 2.4 shows the zero configuration of the corresponding transfer function. Note that the plant dynamics for feedback control is the combination of the frequency response in the figure and the double integrator dynamics: $-40$ db/decade gain and $-180$ degree phase characteristics. Figure 2.3 indicates that when the distance between the vehicle's CG and sensor is relatively short (2m is a typical number when the magnetometers are placed at the front bumper), the phase lag increases as the vehicle speed is increased. This implies that automation for high speed driving is more challenging than that for low speed driving. The figure also shows that at high speeds the input/output dynamics have smaller phase lags when the sensor is placed further ahead of the vehicle's CG.

Notice that the open loop system has a pair of weakly damped zeros, in particular at high speeds when $d_s$ is short. If a high gain controller is applied to this problem, two closed loop poles are attracted to those zeros. As a result, the closed loop system will have a weakly damped oscillatory mode, which the sensor does not see but the passengers feel. Such controllers are not acceptable. The same problem arises when the input-output linearization is applied to the nonlinear model in (2.1). In this case, the so called "zero dynamics" will contain such a weakly damped mode.

Similar to the linear case, zero dynamics is not seen by the sensor but is felt by passengers.

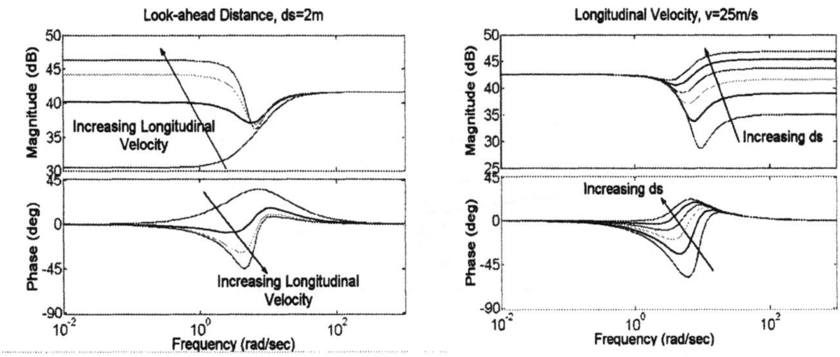


FIGURE 2.3. Frequency response — lateral dynamics

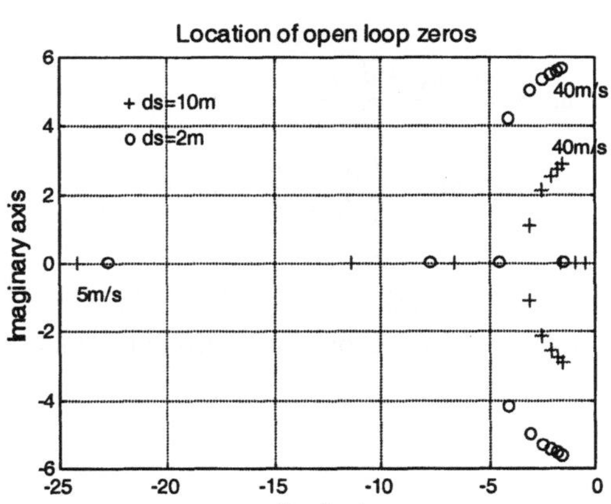


FIGURE 2.4. Open loop zeros (linearized model)

The road reference-sensing system based on magnetic markers is a look-down system. If the magnetometer is placed under the front bumper, the system allows only a small amount of "look ahead," and $d_s$ in Equation (2.4) is typically about 2 m. On the other hand, the lateral control system with a vision camera is a look-ahead system with an ample amount of look ahead. Recall that the open loop dynamics from the steering input to the sensor output defines dynamics easier for feedback control for large $d_s$'s, which explains why the vision based lateral controller has been reported to work well at high speeds [3]. The look-ahead distance by the magnetometer based road reference-sensing system may be increased by installing magnetometers under both the front and rear bumpers: i.e., both the lateral error at the front bumper, $y_s$, and one at the rear bumper, $y_r$, are measured (see Figure 2.5).

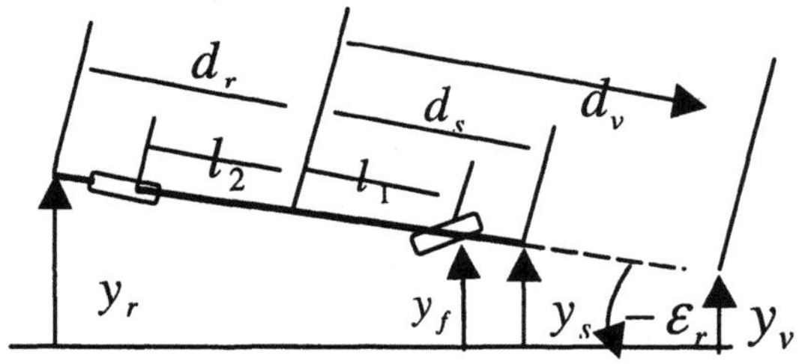


FIGURE 2.5. Construction of states from measurements

Under the assumption that the road is straight, these two measurements are converted to the lateral error at CG and relative yaw error by

$$
\begin{aligned}
y_{CG} &= \frac{d_r y_s + d_s y_r}{d_s + d_r}, \\
\varepsilon_r &= \tan^{-1}\left(\frac{y_s - y_r}{d_s + d_r}\right) \approx \frac{y_s - y_r}{d_s + d_r}
\end{aligned}
\tag{2.5}
$$

where the last approximation holds for small $\varepsilon_r$. These two quantities can be combined to synthesize the output of a virtual sensor located at any distance ahead of the vehicle.

## 2.2 Lateral controllers for AHS

In this section, we review several lateral controllers developed by PATH.

### *2.2.1 FSLQ controller*

The first version of the PATH lateral controller was a frequency shaped linear quadratic (FSLQ) optimal controller with preview [12]. The FSLQ approach was a natural choice because the lateral tracking error should be made small at low frequencies and the riding comfort critically depends on the lateral acceleration and yaw motion over a certain frequency range.

The frequency shaped performance index is thus,

$$J = \frac{1}{2\pi}\int_{-\infty}^{+\infty}\Big[a^*(j\omega)\frac{q_a^2}{1+\lambda_a^2\omega^2}a(j\omega) + y_{cg}^*(j\omega)\frac{q_y^2}{1+\lambda_y^2\omega^2}y_{cg}(j\omega)$$
$$+ \varepsilon_r^*(j\omega)\frac{q_\varepsilon^2}{1+\lambda_\varepsilon^2\omega^2}\varepsilon_r(j\omega) + y_s^*(j\omega)\frac{q_i^2}{(j\omega)^2}y_s(j\omega)$$
$$+ \delta(j\omega)\delta^*(j\omega)\Big]d\omega \tag{2.6}$$

where $a$ is the difference between the lateral acceleration $\ddot{y}_a$ and its desired value. $q$'s and $\lambda$'s are design parameters. This performance index combined with the state equation (2.3) defines an FSLQ problem. For the disturbance in (2.3), $\rho$, a finite preview length is assumed. The optimal control is obtained as

$$\delta(t) = -F_{fb}\xi_e(t) + \int_0^{t_p} F_1(\tau)\rho(\tau+t)d\tau + F_p\rho(t+t_p) \tag{2.7}$$

where $t_p$ is the preview time, the extended state vector $\xi_e$ contains the original state vector $\xi$, and the last two terms represent preview action [3]. The preview information was made available to the controller by alternating the polarity of the magnets. This controller was successfully tested by experiments (on a Pontiac 6000 STE sedan) at speeds up to about 60 km/hour [11].

### *2.2.2 Sliding mode controllers*

Even though FSLQ incorporates preview and achieves good tracking while preserving passenger comfort, gain scheduling for longitudinal velocity will be required to ensure good performance over the entire speed range. Furthermore, the robustness of the controller to parametric uncertainties is not quantified. In order to take into account the parametric uncertainties such as mass and cornering stiffness, Sliding Mode Control (SMC) technique was proposed for lateral control [13, 1, 8]. SMC controller is robust to parametric uncertainties, and the gain scheduling with respect to longitudinal velocity is not required. SMC is designed such that the system states reach a stable manifold called the "sliding surface" in finite time and stay on it thereafter. A typical sliding surface for lateral tracking (regulation of $y_s$ to zero), is shown in Figure 2.6.

The sliding surface is represented by the sliding surface variable $s(\dot{y}_s, y_s) = \dot{y}_s + \alpha_1 y_s = 0$. Thus, for $\alpha_1 > 0$ and $s = 0$, $y_s \to 0$ asymptotically.

The control law that achieves $s = 0$ in finite time is

$$\delta = \frac{1}{\hat{b}_1}\left(-\hat{f}_1 - \alpha_1\dot{y}_s - K\text{sign}(s)\right) \tag{2.8}$$

where $\hat{f}_1$ and $\hat{b}_1$ are the values of $f_1$ and $b_1$ in Equation (2.2) for nominal values of the plant parameters. $K$ is the gain of the robustness term sign$(s)$

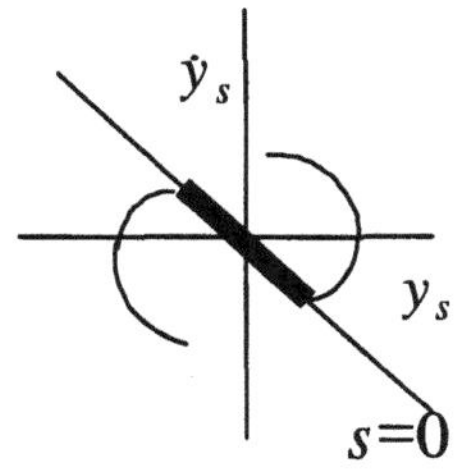


FIGURE 2.6. Sliding surface and chatter in sliding mode

and is dependent on the uncertainty bounds on $f_1$ and $b_1$. Robustness of the SMC is achieved because of the presence of the term $K\text{sign}(s)$. However, this term causes rapid switching or "chatter" in the control command. This undesired phenomena may be removed by replacing the sign function, which is discontinuous at $s = 0$, by a continuous (see Hingwe and Tomizuka [8] for use of this idea for vehicle lateral control). Experiments with this controller were done on a Pontiac 6000 STE sedan and compared well with the FSLQ algorithms at speeds up to 60 km/h.

*2.2.3 Lateral controllers for high speed driving*

As mentioned already, the magnetometer under the front bumper permits only a small look-ahead distance, i.e., $d_s$, is small. This becomes a limitation in particular at high speeds because the weakly damped zeros of the transfer function would attract the closed loop poles if high gain feedback control were attempted for disturbance rejection. The preview control action in the FSLQ solution is for incoming road curvatures only, and it is not for unmeasurable disturbances such as wind gust. Accordingly, PATH researchers decided to work on the two-sensor scheme introduced in Section 2.1. Once the two quantities in Equation (2.5) are computed from the outputs of two sensors, it is possible to synthesize the virtual measurement quantity,

$$y_v(t) = y_{CG}(t) + d_v \varepsilon_r(t) \tag{2.9}$$

where $d_v$ may take any value. Actually, $d_v$ can be dynamic in the sense that the transform domain expression of (2.9) is

$$y_v(s) = y_{CG}(s) + D_v(s)\varepsilon_r(s). \tag{2.10}$$

Combining Equations (2.5) and (2.10), the control law can be written as

$$\delta(s) = C_s(s)y_s(s) + C_r(s)y_r(s). \tag{2.11}$$

PATH researchers tuned $C_s(s)$ and $C_r(s)$, and the resulting controller performed extremely well even at high speeds [19].

### *2.2.4 Remark*

The three controllers described above show the evolutionary development of lateral control technology in PATH. At every stage, the designed controller was experimentally verified. The experimentation at each stage was constrained by test tracks available at that moment. For the FSLQ controller and early versions of SMCs, the test track was only about 300 m long, and the maximum speed was limited to about 60 km/h. As the 1997 San Diego Demonstration plan was finalized, the fundamental issues and limitations coming from the look-down system were addressed, and it was concluded that the most reliable and robust controllers may be developed by measuring lateral deviations at both the front and rear bumpers. This does not imply that high speed driving is not possible if the lateral deviation is measured only at the front bumper [16].

## 2.3 Automated backward driving

PATH has conducted research on various topics related to vehicle lateral control from lane change maneuvers, backward driving and control of heavy vehicles. A recent emphasis has been longitudinal and lateral control of heavy vehicles [6] such as tractor-semitrailer combinations, for the following reasons: 1) while the control methodologies for platooning are not discussed in this paper, it is noted here that the wind drag coefficients of vehicles, which are a key factor in deciding the fuel cost, drop dramatically for all vehicles in the platoon including the lead vehicle as the headway is reduced to about one half of the vehicle length, 2) relative equipment cost for the automation of the heavy vehicles is far less than that of the passenger vehicles, and 3) there are many key highway routes where the concentration of truck traffic is heavy. In terms of the theme of this paper, overcoming limitations of human drivers, an interesting topic is automated backward driving [10].

For relieving traffic congestion caused by an accident on an automated highway, the vehicles sufficiently upstream of the incident will be notified well in advance and will be diverted to alternate routes or to neighboring clear lanes if available. However, there may be a large number of vehicles trapped immediately behind the obstacle produced by the accident. In order for these vehicles to leave the blockade, they should either (i) change lane or (ii) drive backwards to the nearest exit point upstream. As we know from driving experience, backward driving is much more difficult than forward driving. Figure 2.7 shows a virtual mechanical linkage by Partwardhan for automated backward driving for front wheel steered vehicles [10]. If the reference path end of the long link D of this linkage moved along the path, the front wheel is steered so that the vehicle motion converges to the reference path. Notice that the front steering angle is expressed in terms of $y_f'$,

$y'_r$ and kinematic parameters of the linkage: i.e.,

$$\delta = \tan^{-1}\left(\frac{b\sin(\gamma+\varepsilon_r)}{L-\cos(\gamma+\varepsilon_r)}\right) \tag{2.12}$$

where

$$\sin(\gamma) = \frac{-y_b}{b}, \quad \tan(\varepsilon_r) = \frac{y'_f - y'_r}{L_T}, \quad y_b = y'_r + \frac{l_r}{L_T}(y'_f - y'_r).$$

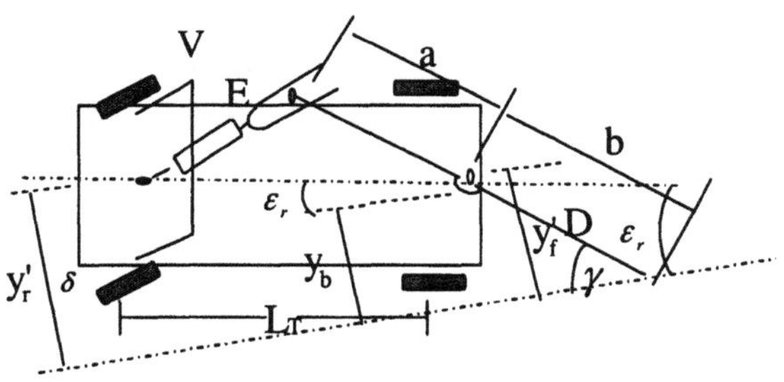


FIGURE 2.7. Backward driving

$L_T$ is the wheel base of the vehicle and $l_r$ is the distance between the pivot of the link D and the rear end of the vehicle. Equation (2.12) is a nonlinear equation that defines the controller. Assuming that the angles, $\gamma$, $\varepsilon_r$ and $\delta$ are small, it is linearized to

$$\delta = \frac{b}{a(L-b)}\left[-y'_r + \frac{a-l_r}{L_T}(y'_f - y'_r)\right]. \tag{2.13}$$

Backward driving based on this control law was demonstrated at the 1997 NAHSC Demonstration.

# 3 Roll-rate control of high-speed bicycles

Figure 3.1 shows one of Matt Weaver's high speed bicycles. Key features are its low height and the absence of a look-through wind shield. A low height design was adopted to minimize the total surface area of the bike, which, in turn, results in lesser drag.

## 3.1 Roll dynamics of bicycles

Roll motion of a bicycle can be modeled as that of an inverted pendulum, the pendulum inertia being comprised of the rider and the bicycle frame. The lateral acceleration acting at the "base of the pendulum" is generated

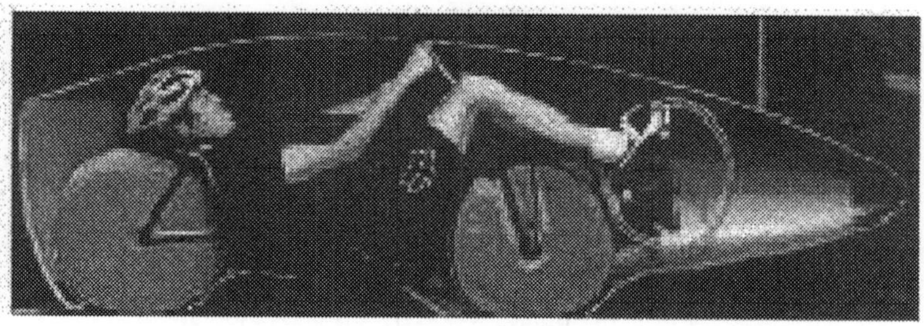

FIGURE 3.1. Matt Weaver's high speed bicycle

by lateral tire-road interaction forces, generated because of a steering action. Figure 3.2 summarizes this idea. Further details of modeling bicycles are given in Suryanarayanan, et al. [17]. The linearized model of the front steered bicycle is obtained as

$$\frac{s\theta(s)}{\delta(s)} = \frac{amlV_0}{bJ}\frac{s(s+\frac{V_0}{a})}{s^2-\frac{mgl}{J}} \tag{3.1}$$

where $\delta(s)$ is the steering angle, $s\theta(s)$ is the roll rate, $m$ and $J$ represents the mass and moment of inertia, respectively, $l$ is the distance between the mass center and the pendulum pivot, and other kinematic parameters are as defined in Figure 3.3.

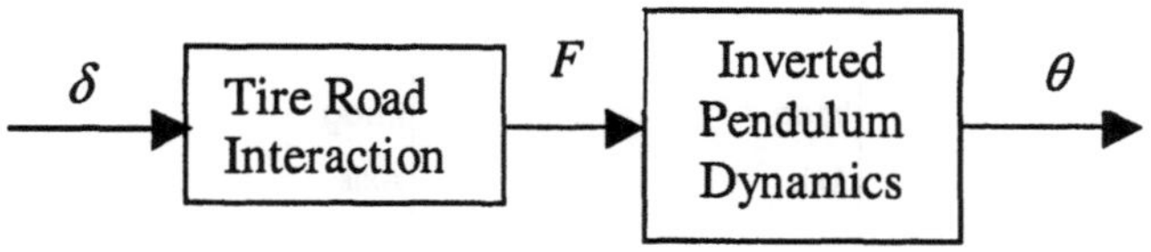


FIGURE 3.2. Effect of steering angle on roll dynamics

As stated in the Introduction, a rear-wheel steering design is under consideration for high-speed bicycles to attain the optimal shape for minimizing the air drag force. The transfer function from $\delta(s)$ to $s\theta(s)$ for a rear-wheel steered bike is

$$\frac{s\theta(s)}{\delta(s)} = \frac{amlV_0}{bJ}\frac{s(-s+\frac{V_0}{a})}{s^2-\frac{mgl}{J}}. \tag{3.2}$$

The transfer function model of the front-wheel steered bike (Equation (3.1)) has two poles and two zeros. Both the poles and one of the zeros (at $s=0$) are fixed for all forward speed $V$. The other zero at $s=-V_0/a$ moves away from the imaginary axis for larger $V$. Figure 3.4 (a) shows the pole-zero locations for typical parameter values (of high-speed bicycles).

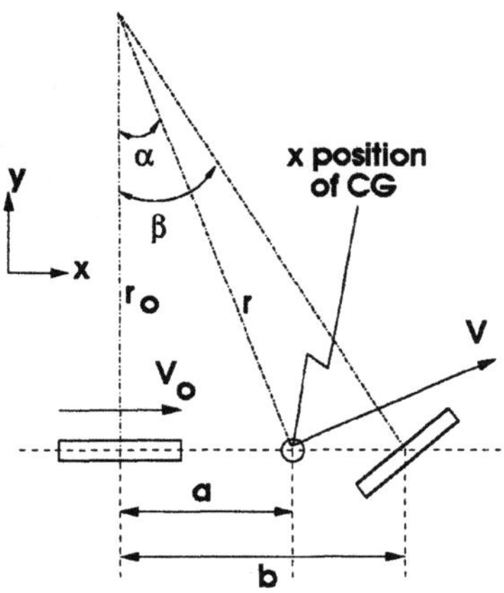


FIGURE 3.3. Top view of bicycle (rider removed)

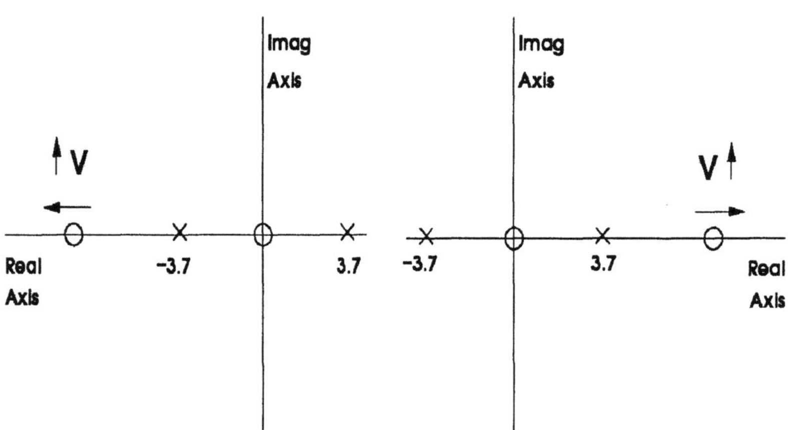


(a) Front wheel steered case (b) Rear wheel steered case

FIGURE 3.4. Pole-zero locations of bicycle dynamics

In a conventional bicycle, the trail (the offset between the tire contact patch and the steering axis) gives a torque that turns the fork into the lean. This torque is opposed by the caster-camber action and the elasticity of tire-road interaction force. A simple model for this behavior can be expressed as

$$\delta = -k\theta. \tag{3.3}$$

Thus, in a conventional bicycle, the front fork design results in a feedback mechanism. Incorporating this feedback equation into the linearized tilt dynamics, we obtain

$$\frac{d^2\theta}{dt^2} + \frac{amlkV_0}{bJ}\frac{d\theta}{dt} + \frac{ml}{J}\left(\frac{kV_0^2}{b} - g\right)\theta = 0. \tag{3.4}$$

This equation represents a stable system when $V_0 > V_c = \sqrt{bg/k}$. This explains why it is easier to ride a conventional front-wheel steered bicycle at higher speeds as compared to lower speeds. The equation also shows that the damping is weaker as $l$ gets smaller, which is the case of Matt Weaver's bicycle.

The transfer function model of the rear-wheel steered bicycle (Equation (3.2)) indicates that the system has two poles and two zeros. Both the poles and one zero (at $s= 0$) are fixed for all forward speeds $V$. The other zero at $s = V_0/a$ (real RHP zero) moves away from the imaginary axis for larger $V$ (Figure 3.4 (b)). Actually, rear-wheel steered dynamics are very interesting. We note the following points:

1. Since the RHP zero moves away from the imaginary axis with increase in forward speed ($V$) and the poles remain fixed for all $V$, it must be that for a certain velocity, it happens that the RHP zero cancels the unstable pole. This makes the system uncontrollable.

2. The zero-pole ratio ($\zeta$) defined as the distance of the zero from the origin divided by the distance of the pole from the origin is

$$\zeta = \frac{V_0}{a}\sqrt{\frac{J}{mgl}}. \tag{3.5}$$

Astrom [2] shows that the system is difficult to control if this ratio is in the range of 0.25 to 4 and alludes to why Klein's rear-wheeled bicycle [9] was unridable. For parameter values of Weaver's high speed bicycle, the range 0.25 to 4 corresponds to vehicle velocity (3 mph – 40 mph), i.e., control becomes extremely difficult in this range. However, for higher speed (> 45 mph), since the RHP zero moves significantly away from the imaginary axis, control becomes relatively easier.

3. Youla et al. [21] showed that an LTI plant can not be stabilized by a stable controller if there exist an odd number of real unstable poles between any two real RHP zeros. This is exactly the case we encounter in this problem. Therefore, any stabilizing controller for a rear-wheeled steered bicycle will be unstable.

## 3.2 Controller design

The aim of the control system is to regulate the roll rate. The bicycle is required to travel, more or less, in a straight line. For this reason, regulation of roll angle is not desired and the right sensor is a rate gyro. If roll angle is regulated, in the presence of a strong wind gust from one side, a large steering angle will be required which will, in turn, steer the bicycle off course. We will show in this section the design of controllers based on the root locus technique.

### *3.2.1 Front-wheel steered bicycle*

Figure 3.5 shows the root locus plot of the bicycle at 112 km/h under lead control. The zero and pole of the lead compensator are fixed at $s = -1$ and $s = -8$, and the controller gain is varied. An appropriate gain can be chosen for the controller to yield satisfactory dominant pole configuration for the closed loop system.

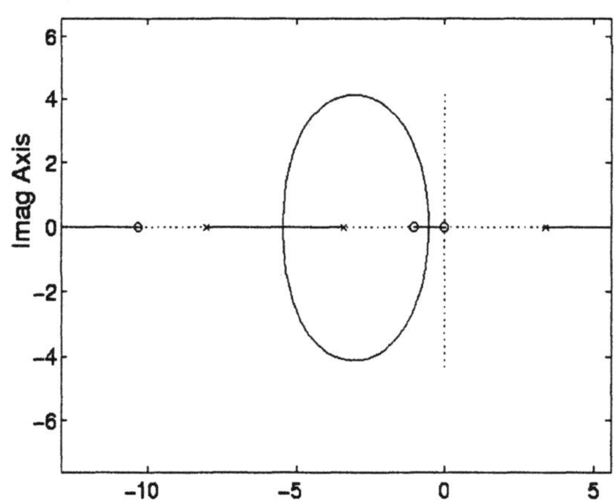


FIGURE 3.5. Root locus plot (front wheel steered bicycle at speed = 112 km/h)

### *3.2.2 Rear-wheel steered bicycle*

Figure 3.6 shows the root locus plot of the rear-wheel steered bicycle at 70 mph. The zero and pole of the controller are at $s = 0.8$ and $s = -3$. Notice that the controller is unstable, but the closed loop system is asymptotically stable for relatively large values of the control gain.

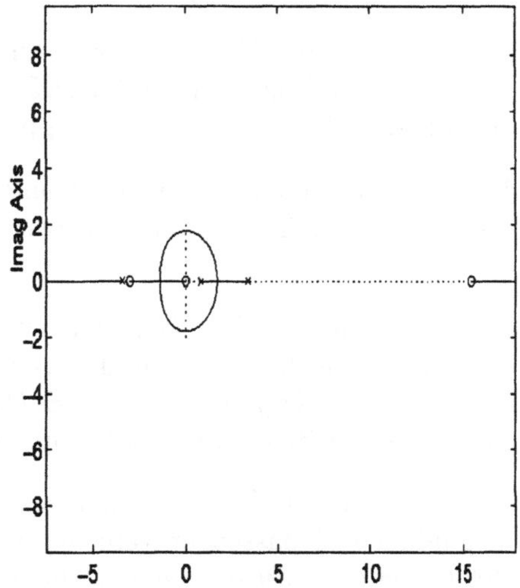


FIGURE 3.6. Root locus plot (rear wheel steered bicycle at speed = 112 km/h)

For simulation results of the two design examples, see [17].

# 4 Concluding remarks

A human is known to be an intelligent controller, hard to beat by automation. In case that human is placed in the feedback loop, the human as a feedback controller has several limitations. They include 1) reaction time (time delay), 2) comprehending dynamics with large phase shifts and/or delays, and 3) limited bandwidth of actuation. Performance limits of manual driving are due to these limitations. Computer control of vehicles and bicycles makes it possible to operate them at performances in a large operating range not attainable by human.

## References

[1] J. Ackermann, J. Guldner, W. Sienel, and R. Steinhauser, "Linear and nonlinear controller design for robust automatic steering," *IEEE Transactions on Control Systems Technology*, vol. 3, pp. 132–143.

[2] K. J. Astrom, "Limitations on control system performance," *European Journal of Control*, vol. 6, no. 1, 2000, pp. 2–20.

[3] E. D. Dickmans and A. Zapp, "Autonomous high speed road vehicle guidance by computer vision," in *Proceedings of IFAC $10^{th}$ Triennial World Congress*, Munich, FRG, 1987, pp. 221–226.

[4] R. E. Fenton, G. C. Melocik, and K. W. Olson, "On the steering of automated vehicles: theory and experiment," *IEEE Transactions on Automatic Control*, vol. AC-21, no. 3, June 1976, pp. 306–315.

[5] J. K. Hedrick, M. Tomizuka, and P. Varaiya, "Control issues in automated highway systems," *IEEE Control Systems*, pp. 21–32, Dec. 1994.

[6] P. Hingwe, M. Tai, J-Y. Wang, and M. Tomizuka, "Lateral control of tractor-semitrailer combination for automated highway systems—an experimental study," in *Proceedings of the ASME Dynamic Systems and Control Division—1999*, DSC-vol. 67, 1999, pp. 195–202.

[7] P. Hingwe and M. Tomizuka, "Robust and gain scheduled H4 controllers for lateral guidance of passenger vehicles in AHS," in *Proceedings of the ASME Dynamic Systems and Control Division*, DSC-vol. 61, November 1997, pp. 707–713.

[8] P. Hingwe and M. Tomizuka, "Experimental evaluation of a chatter free sliding mode controller for lateral control in AHS," in *Proceedings of the American Control Conference*, Albuquerque, NM, 1997, pp. 3365–3369.

[9] R. E. Klein, "Using bicycles to teach system dynamics," *IEEE Control Systems Magazine*, April 1989, pp. 4–8.

[10] S. Patwardhan, H-S. Tan, J. Guldner, and M. Tomizuka, "Lane following during backward driving for front wheel steered vehicles," in *Proceedings of the American Control Conference*, June 1997, pp. 3348–3353.

[11] H. Peng, T. Hessburg, M. Tomizuka, W-B. Zhang, Y. Lin, P. Devlin, and S. E. Shladover, "A theoretical and experimental study on vehicle lateral control," in *Proceedings of the American Control Conference*, 1992, pp. 1738–1742.

[12] H. Peng and M. Tomizuka, "Preview control for vehicle lateral guidance in highway automation," *ASME Journal of Dynamic Systems, Measurement and Control*, vol. 115, no. 4, pp. 679–686, December 1993.

[13] H. Pham, K. Hedrick, and M. Tomizuka, "Combined lateral and longitudinal control of vehicles," in *Proceedings of the American Control Conference*, 1994, Baltimore, MD, pp. 1205–1206.

[14] L. Segel, "Theoretical prediction and experimental substantiation of the response of the automobile to steering control," *Automobile Division*, The Institute of Mechanical Engineers, pp. 26–46, 1956.

[15] S. E. Shladover, D. N. Wormley, H. H. Richardson, and R. Fish, "Steering controller design for automated guideway transit vehicles," *ASME Journal of Dynamic Systems, Measurement and Control*, vol. 100, March 1978, pp. 1–8.

[16] S. Suryanarayanan and M. Tomizuka, "Lateral control of automated vehicles: on degraded mode control problems," in *Proceedings of 2001 ASME Int. Mechanical Engineering Congress and Exposition*, New York, Nov. 2001, IMECE2001/DSC-24518.

[17] S. Suryanarayanan, M. Tomizuka, and M. Weaver, "System dynamic and control of bicycles at high speeds," in *Proceedings of the American Control Conference*, Anchorage, AK, May 2002, pp. 845–850.

[18] D. Swaroop and J. K. Hedrick, "String stability of interconnected systems," *IEEE Transactions on Automatic Control*, vol. 41, no. 3, March 1996, pp. 349–357.

[19] H-S. Tan, Private Communication, 1997.

[20] P. Varaiya, "Smart cars on smart roads: problems of control," *IEEE Transactions on Automatic Control*, vol. 38, no. 2, Feb. 1993.

[21] D. C. Youla, J. J. Bongiorno, and C. N. Liu, "Single-loop feedback stabilization of linear multivariable plants," *Automatica*, vol. 10, no. 2, 1974, pp. 159–173.

Masayoshi Tomizuka
Department of Mechanical Engineering
5100B Etcheverry Hall
University of California
Berkeley, CA 94720-1740, USA
tomizuka@me.berkeley.edu

# Modeling, Control and Experiment of a Feedback Active Noise Control System for Free Sound Fields

**Shuichi Adachi**

*Dedicated to Professor Hidenori Kimura on the occasion of his 60th birthday*

ABSTRACT Feedback active noise control (FB-ANC) systems need no reference signal sensors and their implementation cost is therefore attractive due to the relatively simple configuration. This paper reports the design of an FB-ANC system for free sound fields by $\mathcal{H}_\infty$ control theory based on a lower-order infinite impulse response (IIR) model of the acoustic plant. The acoustic plant, which is an acoustic transmission system from a loudspeaker (i.e., the actuator) to an error microphone (i.e., the sensor), is first devised by using the first-principle modeling method in conjunction with a system identification result. The feedback controller is next designed to attenuate the noise level in the vicinity of the error sensor by reducing the sensitivity function over the frequency range of interest. $\mathcal{H}_\infty$ control theory is applied to achieve the control objective, because it can formulate the control specifications in terms of the frequency weighting functions in the frequency domain. Finally, the effectiveness of the proposed design procedure is demonstrated by experimental tests.

**Keywords:** Active noise control, Acoustics, First-principle modeling, System identification, $\mathcal{H}_\infty$ control, Automobile

## 1 Introduction

Many studies have been reported on active noise control (ANC) systems, especially feed-forward ANC (FF-ANC) systems [8], [11]. Figure 1.1 shows a single-input, single-output, and single-error (SISOSE) system, where $H$ and $P$ are respectively the transfer functions from the noise source to the error sensor (microphone), and from the secondary source (loudspeaker) to the error sensor. $C$ represents the transfer function of the feed-forward controller [12]. The control objective of the ANC system is to reduce the noise level in the vicinity of the error microphone over the frequency range of interest. To achieve this objective, the feed-forward controller calculates

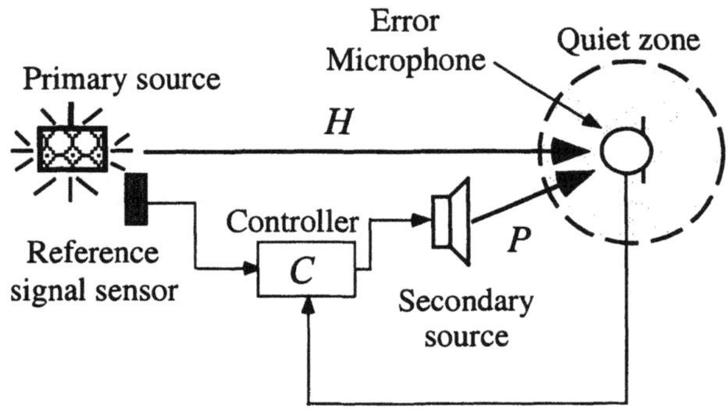


FIGURE 1.1. Configuration of an SISOSE FF-ANC system

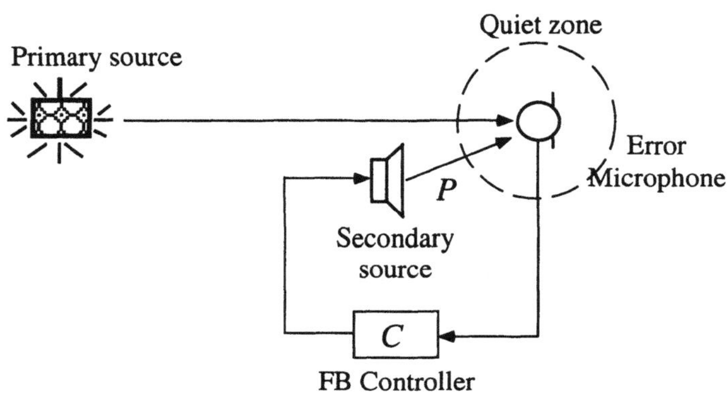


FIGURE 1.2. Configuration of an SISOSE FB-ANC system

the control signal based on the reference and error signals. The best-known FF-ANC system is based on the filtered-x least mean square (LMS) algorithm which has been proposed for adaptive filtering [16].

One of the problems restricting industrial applications of the FF-ANC system is its high cost, for example, numerous reference signal sensors, i.e., input signals, are required when an FF-ANC system is employed for reducing road noise in the cabin of an automobile [2], [15], because the input paths are very complex. It is also known that the optimum placement of the reference signal sensors is a very difficult task.

In order to solve these problems, this study focuses on the feedback (FB) ANC system shown in Figure 1.2. A typical FB-ANC system is based on the tight-coupled monopole (TCM) method [4], in which error sensors (i.e., microphones) are placed near the control noise sources (i.e., speakers). However, it seems that little research has been reported on the FB-ANC system from the point of view of feedback control engineering. In recent years, some studies based on the feedback control theory have been attempted [1], [7], although FB-ANC research is still less extensive than FF-ANC research.

The main advantage of the FB-ANC system is its relatively simple configuration because it needs no reference signal sensors (see Figure 1.2). On the other hand, the drawback of FB-ANC is that its noise reduction performance is usually inferior to that of FF-ANC. Therefore, clarifying the performance of an FB-ANC system by applying feedback control theory is considered to be an important prerequisite for further advancements in ANC technology.

The objective of this present study is to design a feedback controller ($C$ in Figure 1.2) by applying $\mathcal{H}_\infty$ control theory [3] based on a low-order infinite impulse response (IIR) model of the plant. The plant in the FB-ANC system, which is denoted by $P$ in Figure 1.2, is an acoustic transmission system from a loudspeaker (i.e., the actuator) to an error microphone (i.e., the sensor). The control objective of the FB-ANC system is to reduce the noise level in the vicinity of the error microphone over a low-frequency range below 500 Hz, since noise at more than 500 Hz can be attenuated by conventional passive noise control methods.

In Section 2, the acoustic plant, which contains a loudspeaker, sound field and microphone, is devised by using the first-principle modeling method in cooperation with a system identification result. The loudspeaker and the microphone are modeled by physical insight with equivalent electrical circuits. The sound field is also represented by the first-principle modeling method, assuming that the distance from the speaker to the microphone is sufficiently short relative to the wavelength of the sound to be controlled. Since the phase characteristics of the model are very important for designing the feedback controller, these are fitted to the system identification result that is obtained by the well-known LMS algorithm that assumes a high-order finite impulse response (FIR) model. In Section 3, a feedback controller is designed to attenuate the noise level in the vicinity of the error microphone by reducing the sensitivity function over the frequency range of interest. $\mathcal{H}_\infty$ control theory is applied to achieve the control objective, because it can formulate the control specifications in terms of the frequency weighting functions in the frequency domain. In Section 4, the effectiveness of the proposed design procedure is demonstrated by tests. Finally, in Section 5, the conclusions are presented. Throughout this paper, the terms "speaker" and "microphone" respectively refer to the control sound source and error sensor used in this study.

# 2 Modeling of the acoustic plant

## 2.1 Conventional modeling methods

The most commonly used modeling method in ANC studies is system identification [10]. In particular, the celebrated LMS algorithm for an FIR model is widely employed because it is simple and does not require squaring, averaging or differentiating operations [16]. Moreover, robustness of the LMS algorithm has also been revealed [6]. However, an FIR model with numerous parameters, for example 64 or 256 or more, is required to precisely describe the plant dynamics. Furthermore, it is not necessarily an FIR model, but a low-order IIR model that is used to design a feedback controller by $\mathcal{H}_\infty$ control theory.

The simplified hyperstable adaptive recursive filter (SHARF) algorithm [9] has been applied as an IIR modeling method for ANC systems to reduce the parameters in the model. The actual number of parameters, however, was determined by trial and error. As a means of determining the number of parameters with quantitative indicators, an IIR model identification method employing a subspace method has been proposed [13]. Using this method, the acoustic plant was expressed by an IIR model with twelve parameters, i.e., a sixth-order IIR filter.

Nevertheless, it is not clear to what extent the number of parameters can be reduced while still maintaining the required precision; determining this not only simplifies the design of the FB controller, but is also an important factor in achieving low-cost ANC. Moreover, a discrete-time model was obtained by the system identification methods just mentioned, although a continuous-time model is usually more suitable for controller design. For this purpose, the present study employs first-principle modeling to obtain a reasonable-precision, low-order and continuous-time model to design the feedback control system.

## 2.2 Modeling procedure

The proposed modeling procedure for the acoustic plant is summarized as follows.

Step 1: First-principle modeling

Step 2: System identification experiment

Step 3: Modeling of implementation devices

Step 4: Parameter tuning

Step 5: Evaluation of model uncertainty

*Step 1: First-principle modeling*

The acoustic plant is composed of three parts: a loudspeaker, an acoustic field, and a microphone. First-principle models of these components are derived in the next steps.

*Step 1.1: First-principle model for the loudspeaker*

The loudspeaker, which is considered as a mechanical system, is converted to an equivalent electrical circuit. Transfer function $P_1(s)$ from applied voltage $e_0$ [V] to vibration velocity $v_0$ [m/s] of the diaphragm can be described by the following second-order system:

$$P_1(s) = \frac{As}{a_2 s^2 + a_1 s + a_0} \tag{2.1}$$

where

$$\begin{aligned} a_2 &= \frac{8}{3} a^3 \rho r + mr + L r_M, \\ a_1 &= r \cdot r_M + C_M L + A^2, \\ a_0 &= C_M r \end{aligned}$$

and $\rho$ [kg/m$^3$] is the air density, $a$ [m] is the cone radius, $m$ [kg] is the diaphragm and coil mass, $A$ is the electrodynamic converter's force factor, $r$ [W] is the coil resistance, $L$ [H] is the coil inductance, $r_M$ [Ns/m] is the mechanical resistance, and $C_M$ [N/m] is stiffness of the diaphragm support.

*Step 1.2: First-principle model for the sound field*

It is assumed in this study that the distance from the speaker to the microphone is sufficiently short relative to the wavelength of the sound to be controlled. Since the effect of direct sounds rather than reflected sounds is predominant even in an enclosed space such as an indoor chamber, the effect of the natural acoustic mode is considered to be minimal. Therefore, a free sound field is assumed for the sake of simplicity.

It is also assumed that the radiative surface of the loudspeaker can be approximated by a planar piston, because the vibration velocity of the diaphragm is equivalent to that of air particles on the diaphragm. Transfer function $P_2(s)$ from diaphragm air particle velocity $v_0$ [m/s] to the sound pressure at a point on the central axis located distance $l$ [m] from the planar piston can be written by

$$\bar{P}_2(s) = \rho c \left\{ \exp\left(-\frac{l}{c}s\right) - \exp\left(-\frac{\sqrt{l^2+a^2}}{c}s\right) \right\} \tag{2.2}$$

where $c$ [m/s] is the speed of sound. If the distance to the measurement point is extremely large relative to the speaker radius, or if the sound contains only low-frequency components, Equation (2.2) can be approximated by

$$P_2(s) = \rho \frac{a^2}{2l} s \cdot \exp\left(-\frac{l}{c}s\right). \tag{2.3}$$

Equation (2.3) shows that the sound radiation characteristics consist of derivative element $(\rho a^2/2l)s$ and time delay element $\exp\{-(l/c)s\}$. This is a reasonable result since the loudspeaker has a radiative property and the acoustic transmission system is a dead-time type. To examine the approximation accuracy of Equation (2.3), the sound pressure levels as a function of distance $l$ are plotted in Figure 2.1 for sound frequencies of 40 Hz and 500 Hz, respectively. The reason why these frequencies were selected is that the target frequency range of the noise reduction considered in this study is from 40 to 500 Hz. The radius of the cone was $a = 0.0475$ m which is the same as that of the actual speaker used in the experiments which will be described later. From Figure 2.1, it is clear that Equation (2.3) is fully capable of expressing the transfer function of a sound field when the distance is larger than 0.1 m which is shaded in the figures.

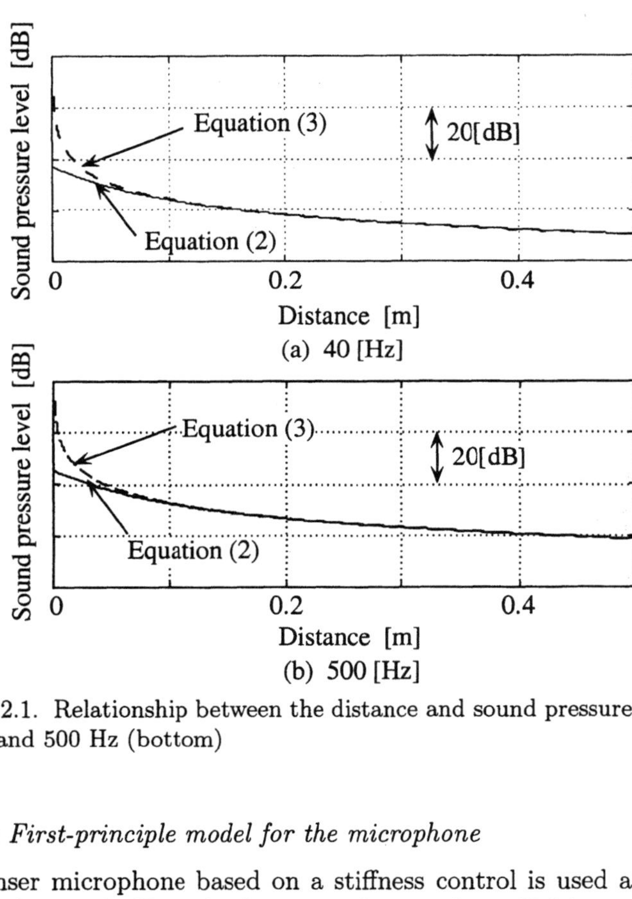


FIGURE 2.1. Relationship between the distance and sound pressure level for 40 Hz (top) and 500 Hz (bottom)

*Step 1.3: First-principle model for the microphone*

A condenser microphone based on a stiffness control is used as the error sensor in this study. Transfer function of microphone $P_3(s)$ can be written by the proportional gain:

$$P_3(s) = 1, \tag{2.4}$$

because its frequency response function has flat gain and no phase delay over the frequency range of interest.

*Step 1.4: First-principle model for the acoustic plant*

The first-principle model for the acoustic plant made up from the three elements, $P_1(s)$, $P_2(s)$ and $P_3(s)$, can be described by

$$\begin{aligned} P_a(s) &= P_1(s)P_2(s)P_3(s) \\ &= \frac{(\rho A a^2/2l)s^2}{a_2 s^2 + a_1 s + a_0} \cdot \exp\left(-\frac{l}{c}s\right) = P_c(s)\exp(-\tau s). \end{aligned} \tag{2.5}$$

Substituting the values for the physical parameters of the loudspeaker which is used for the experiments summarized in Table 2.1, $P_c(s)$ and $\tau$ for $l = 0.2$ m can be written as

$$\begin{aligned} P_c(s) &= \frac{1.02 \times 10^{-2}s^2}{9.03 \times 10^{-3}s^2 + 12.43s + 3512.6}, \\ \tau &= 5.88 \times 10^{-4}. \end{aligned}$$

Therefore, the acoustic plant (2.5) consists of second-order lumped-parameter system $P_c(s)$ and a distributed-parameter system with time delay element $\exp(-\tau s)$.

TABLE 2.1. Parameters of the loudspeaker (Panasonic WS-A10-K)

| Parameter | | Value |
|---|---|---|
| Mass | $(m)$ | $3.1 \times 10^{-3}$ kg |
| Mechanical resistance | $(r_m)$ | 2.6 Ns/m |
| Mechanical stiffness | $(C_m)$ | 1,351 N/m |
| Coil resistance | $(r)$ | 2.6 Ω |
| Coil inductance | $(L)$ | $5.4 \times 10^{-5}$ H |
| Radius of cone | $(a)$ | 0.0475 m |
| Force factor | $(A)$ | 2.98 Tm, N/A |
| Density of air | $(\rho)$ | 1.21 kg/m$^3$ |
| Speed of wave | $(c)$ | 340 m/s |

*Step 2: System identification experiment*

In order to refine the first-principle model (2.5), a system identification experiment on the acoustic plant was carried out. The experiment was set up as shown in Figure 2.2.

A pseudo-random signal whose power spectral density is flat below 1 kHz was applied as the identification input signal. The input and output data were measured at a sampling rate of 11,025 Hz, the number of data being 160,000. The sampling rate of the data was reduced by a factor of 8 to 1,378 Hz, this decimation processing being applied in order to focus the identification frequency range below 500 Hz. This correspondingly reduced

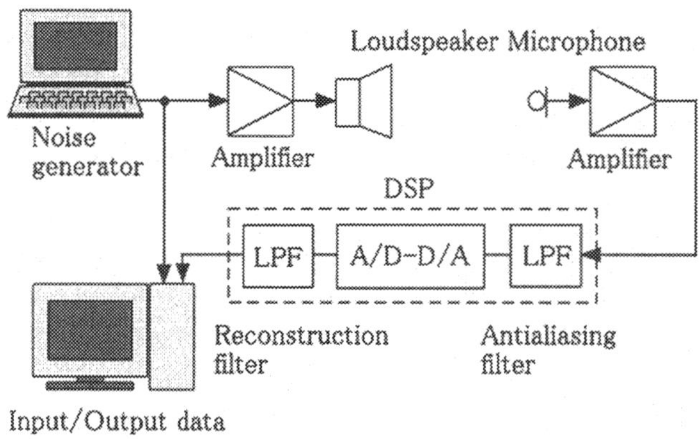


FIGURE 2.2. Setup of the system identification experiment

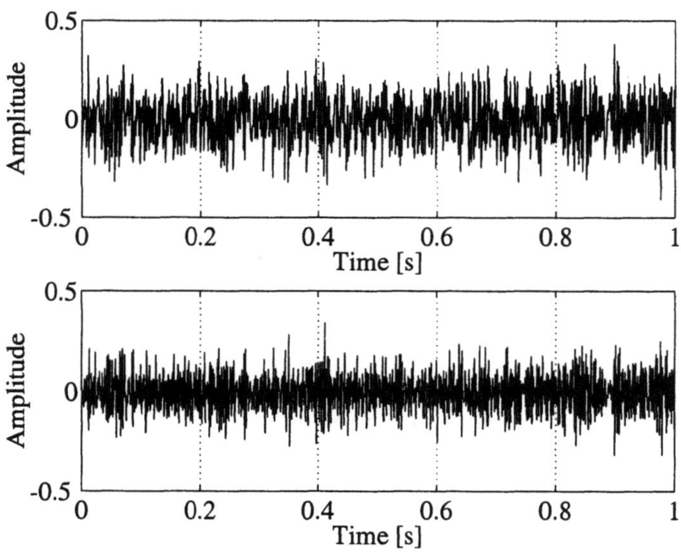


FIGURE 2.3. Input (bottom) and output (top) signals after decimation processing of the original data measured in the system identification experiment

the number of data for system identification to 20,000. The decimated input and output signals are shown in Figure 2.3.

A 64th-order FIR model was used for system identification, and the LMS algorithm was applied to the data. Bode plots of the identification result for $l = 0.2$ m are shown in Figure 2.4, in which it can be seen that the gain plot of the first-principle model (dashed line) is almost the same as the system identification result (solid line). However, the phase plot of the first-principle model differs greatly from the system identification result, because the first-principle model did not take the DSP dynamics of the implementation device into consideration. This is therefore modeled in the next step.

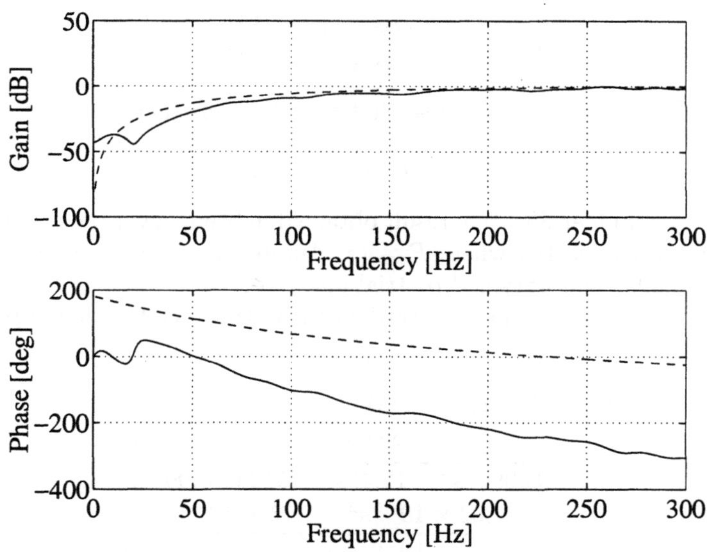


FIGURE 2.4. Comparison between the system identification result (solid line) and first-principle model (dashed line) for $l = 0.2$ m

*Step 3: Modeling of the implementation device*

The DSP dynamics can be approximated by an all-pass filter:

$$P_i(s) \approx \frac{-0.0008s + 1}{0.0008s + 1} \tag{2.6}$$

in frequency range below 500 Hz for the Active-5 DSP (Redec Co.) device that was used.

*Step 4: Parameter tuning*

The plant model can be written for Steps 1 to 3 as follows:

$$\bar{P}(s) = P_a(s)P_i(s) = \frac{1.02 \times 10^{-2}s^2}{9.03 \times 10^{-3}s^2 + 12.43s + 3512.6} \cdot \frac{-0.0008s + 1}{0.0008s + 1} \cdot \exp(-5.88 \times 10^{-4}s). \quad (2.7)$$

In general, it is preferable to describe the model by a rational transfer function when the feedback controller has been designed based on robust control theory. In order to approximate $\exp(-5.88 \times 10^{-4}s)$ by a rational function, Pade approximation of order 2,

$$P_d(s) = \frac{1 - \alpha s + (\alpha s)^2/12}{1 + \alpha s + (\alpha s)^2/12}, \quad (2.8)$$

is applied. Parameter $\alpha$ is tuned so that the phase characteristics of

$$P_{nom}(s) = P_a(s)P_i(s)P_d(s)$$

match the system identification result obtained in Step 2 over the frequency range from 40 to 500 Hz, where $P_{nom}(s)$ denotes a nominal model of the plant. $\alpha = 0.0035$ was obtained by trial and error.

As a result, the plant can be described by a 5th-order IIR model represented by

$$\begin{aligned} P_{nom}(s) &= P_p(s)P_d(s)P_f(s) \\ &= \frac{0.0203s^2}{0.00903s^2 + 12.43s + 3512.6} \frac{-0.0008s + 1}{0.0008s + 1} \\ &\quad \cdot \frac{1.02 \times 10^{-6}s^2 - 1.75 \times 10^{-3}s + 1}{1.02 \times 10^{-6}s^2 + 1.75 \times 10^{-3}s + 1}. \end{aligned} \quad (2.9)$$

A Bode plot of the model is shown by the solid line in Figure 2.5, in which the dashed line shows the system identification result based on the 64th-order FIR model. It is clear that the 5th-order IIR model matches well with the conventional 64th-order FIR model over the frequency range from 40 to 500 Hz. Since the proposed approach is based on first-principle modeling, a continuous-time model can be directly obtained, which is very suitable for feedback controller design.

*Step 5: Evaluation of the model uncertainty*

An evaluation of the model uncertainty is required to design an FB-ANC system by $\mathcal{H}_\infty$ control theory. The additive uncertainty can be evaluated by

$$\Delta_a(j\omega) = P_{id}(j\omega) - P_{nom}(j\omega) \quad (2.10)$$

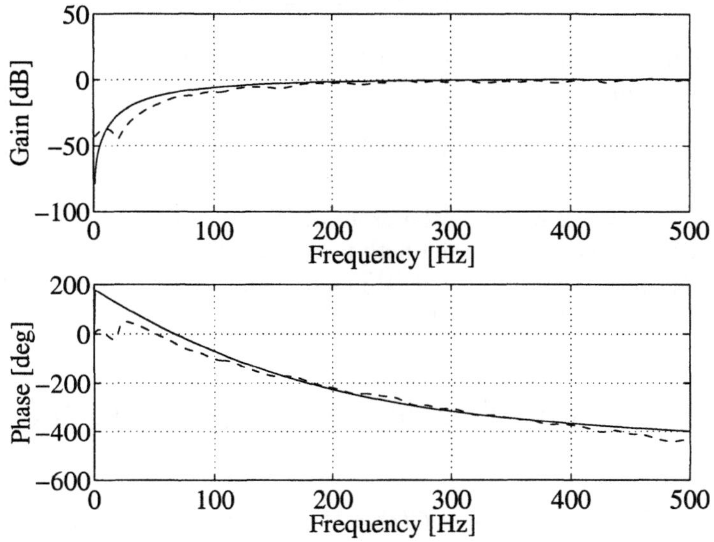


FIGURE 2.5. Proposed model result (solid line) and system identification result (dashed line) for $l = 0.2$ m

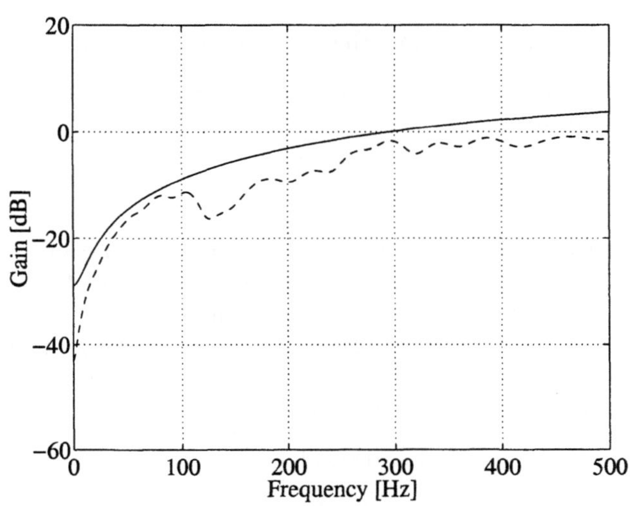


FIGURE 2.6. Additive model uncertainty $|\Delta_a(j\omega)|$ (dashed line) and corresponding frequency weighting function $|W_T(j\omega)|$ (solid line) for $l = 0.2$ m

where $P_{id}(j\omega)$ is a precise model which was obtained by system identification in Step 2 and $P_{nom}(j\omega)$ is the nominal model which was calculated in Step 4.

In order to cover the additive model uncertainty $|\Delta_a(j\omega)|$, a frequency weighting function

$$W_T(s) = \frac{2.9(s+62.8)}{s+5027} \tag{2.11}$$

was chosen for $l = 0.2$ m. Since the order of $W_T(s)$ influences the order of the designed controller, a 1st-order high-pass filter was employed. Figure 2.6 shows $|\Delta_a(j\omega)|$ (dashed line) and $|W_T(j\omega)|$ (solid line). $W_T(s)$ plays an important role in robust stabilization.

# 3 Feedback controller design incorporating $\mathcal{H}_\infty$ control

The design of feedback controller $C(s)$ in Figure 1.2 is based on the nominal model described by Equation (2.9) and frequency weighting function $W_T(s)$ for the additive model uncertainty.

The control specifications of the FB-ANC system are as follows:

S1: Stabilization of the closed-loop system

S2: Attenuation of the sensitivity function in the low-frequency range, especially around 150 Hz

S3: Robust stabilization against the additive model uncertainty.

In S2, sensitivity function $S(s)$, which is the transfer function from noise $d$ to error $e$, is defined by

$$S(s) = \frac{1}{1+P(s)C(s)}. \tag{3.1}$$

Also, we aim at attenuating noise level at 150 Hz, because low frequency road noise below 200 Hz is difficult to reduce by passive methods without increasing car weight in an actual car development.

$\mathcal{H}_\infty$ control was adopted for the feedback controller design method, because it can assign the appropriate attenuation performance according to the sensitivity function. The nominal performance can be described by

$$||W_S(s)S(s)||_\infty < 1 \tag{3.2}$$

where $W_S(s)$ is a frequency weighting function. To meet the control specification S2,

$$W_S(s) = \frac{Ks^2}{s^2 + 2\zeta\omega_n s + {\omega_n}^2} \tag{3.3}$$

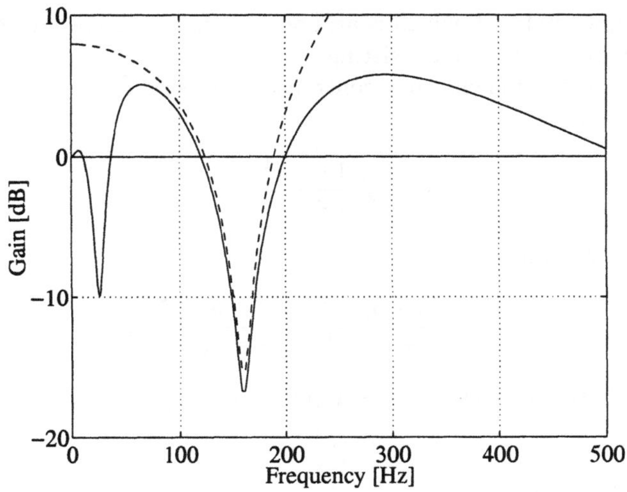


FIGURE 3.1. Sensitivity function $|S(j\omega)|$ (solid line) and frequency weighting function $|W_S(j\omega)|^{-1}$ (dashed line)

was applied where $\omega_n = 2\pi \times 150$. The other parameters $K$ and $\zeta$ were tuned by trial and error based on the shape of $|S(j\omega)|$ so as to attenuate the noise level at 150 Hz. In this experiment, $K = 0.6$ and $\zeta = 1/14$ were selected.

The design specification S3 is described by

$$||W_T(s)T(s)||_\infty < 1 \tag{3.4}$$

where $T(s)$ is the complementary sensitivity function defined by

$$T(s) = \frac{P(s)C(s)}{1 + P(s)C(s)} \tag{3.5}$$

which is the transfer function from noise $d$ to the plant output. Frequency weighting function $W_T(s)$ is given by Equation (2.11) in the previous section. The control problem considered here is known as a mixed sensitivity type.

The $\mathcal{H}_\infty$ controller was designed in continuous-time by the Glover and Doyle method [5]. The resulting sensitivity function is shown in Figure 3.1, indicating a noise attenuation by the feedback controller of about 15 dB at a frequency of 150 Hz. It can be seen from Figure 3.1 that there are amplification areas such as those from 40 to 120 Hz and from 200 to 500 Hz, where the sensitivity function is greater than 0 dB. This is an inherent drawback of the FB-ANC system, which is known as the water-bed effect, because the acoustic plant is a non-minimum phase system due to the time delay [14]. In this study, however, we aim at attenuating the noise in a particular frequency band, which is around 150 Hz in our specifications, at the cost of amplification areas existing.

The obtained continuous-time controller can be digitized by bilinear transformation as

$$s = 2f_s \frac{1 - z^{-1}}{1 + z^{-1}}.$$

The discrete-time controller

$$C(z) = \frac{b_0 + b_1 z^{-1} + \cdots + b_8 z^{-8}}{1 + a_1 z^{-1} + \cdots + a_8 z^{-8}}$$

can then be obtained in which the sampling rate is $f_s = 3$ kHz.

# 4 Experiments

The performance of the proposed feedback controller was tested by the experimental apparatus shown in Figure 4.1. Pseudo road noise was used to simulate the noise inside an automobile.

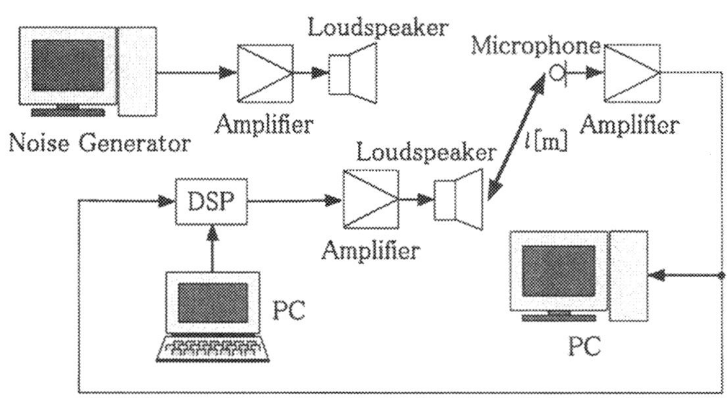


FIGURE 4.1. Configuration of the experimental apparatus

The plot of the A-weighted relative sound pressure level for $l = 0.2$ m is shown in Figure 4.2. It is clear that the FB-ANC system attenuates the noise level at 150 Hz by about 15 dB, this being the predicted attenuation level based on the sensitivity function.

It is known that the zone of quiet, within which the pressure is at least 10 dB below that due to the primary source, is a sphere with a diameter of about one-tenth of a wavelength [11]. Therefore, the zone of quiet in this experiment is expected to be a sphere with a diameter of 0.226 m, since the frequency of the target noise is 150 Hz.

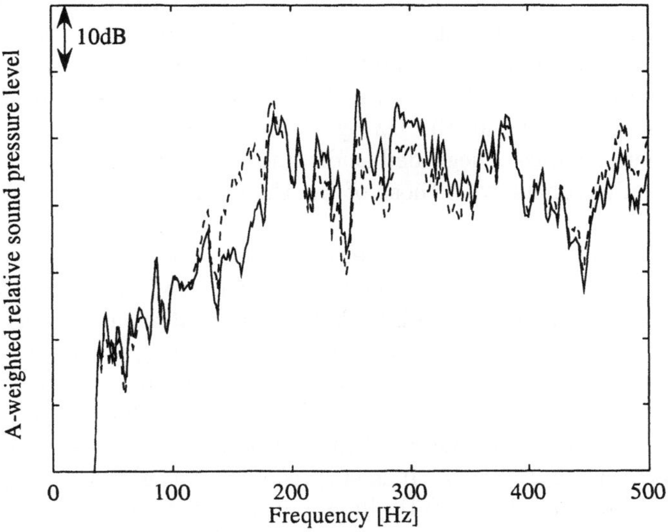


(a) Frequency range from 0 to 500 Hz

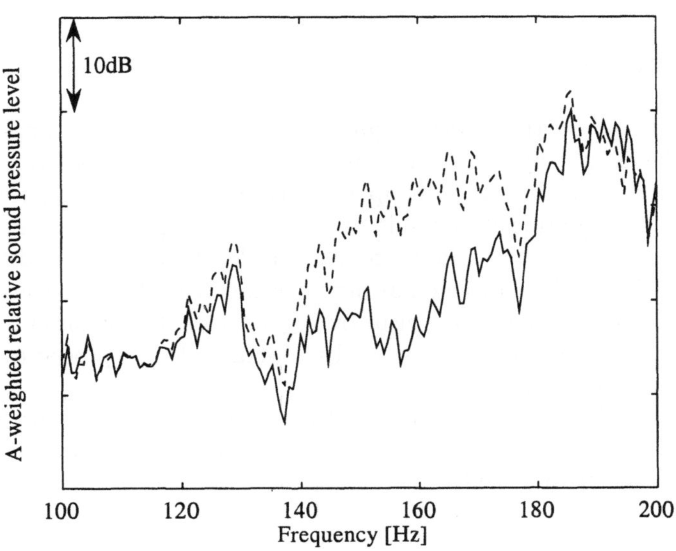


(b) Frequency range from 100 to 200 Hz

FIGURE 4.2. Sound pressure level of the FB-ANC system (solid line), where the dashed line indicates the original sound pressure level for $l = 0.2$ m

By a similar modeling and controller design procedure, an FB-ANC system for $l = 0.5$ m was designed, whose design specifications are the same as those for $l = 0.2$ m. A Bode plot of the obtained model is shown in Figure 4.3(a). It is clear that the proposed modeling procedure is capable of being applied to the acoustic plant with $l = 0.5$ m. The experimental result is shown in Figure 4.3(b). It can be seen that the designed system attenuates the noise level at 150 Hz by about 11 dB and that the noise attenuation performance is degraded as distance $l$ becomes longer. The experimental results concur with those of the theoretical study on FB-ANC systems [14].

An acoustic plant with $l = 1$ m was also modeled, the result being shown in Figure 4.4(a), where the dashed line represents the system identification result. It is clear from the dashed line that the natural acoustic modes are dominant in this acoustic plant, so the proposed modeling procedure could not be applied. Instead of this method, the curve fitting method in the frequency domain was applied such that the gain and phase characteristics below 200 Hz match the system identification result. The experimental result is shown in Figure 4.4(b). It is surprising that the designed system attenuates the noise level at 150 Hz by about 10 dB. However, the attenuation zone, i.e., the difference between the dashed line and solid one, has shrunk in comparison with the cases for $l = 0.2$ and 0.5 m.

# 5 Conclusions

The modeling and controller design procedure for an FB-ANC system for free sound fields is proposed. The acoustic plant was modeled as a low-order IIR model by the first-principle modeling method in conjunction with the system identification result under the assumption that the distance was sufficiently short relative to the wavelength of the sound to be controlled. A feedback controller was next designed to attenuate the noise level in the vicinity of the error sensor by $\mathcal{H}_\infty$ control theory based on the IIR model. Finally, the effectiveness of the proposed design procedure was demonstrated by experimental tests.

A free sound field has been assumed in this study. A systematic modeling and controller design procedure for an acoustic field with reverberation will be the subject of further study.

# Acknowledgments

The author is grateful to Dr. H. Sano of Honda R & D Center for cooperation in the experiments and his helpful advice.

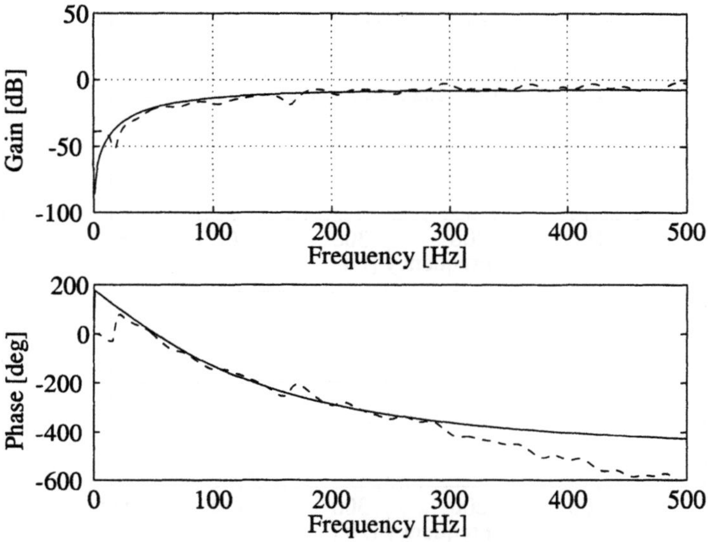


(a) Proposed modeling result (solid line) and system identification result (dashed line)

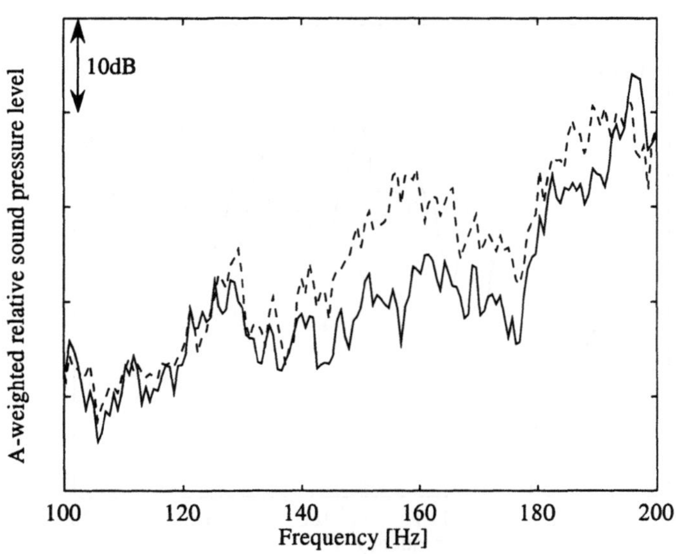


(b) Sound pressure level of the FB-ANC system (solid line), where the dashed line indicates the original sound pressure level

FIGURE 4.3. Experimental results for $l = 0.5$ m

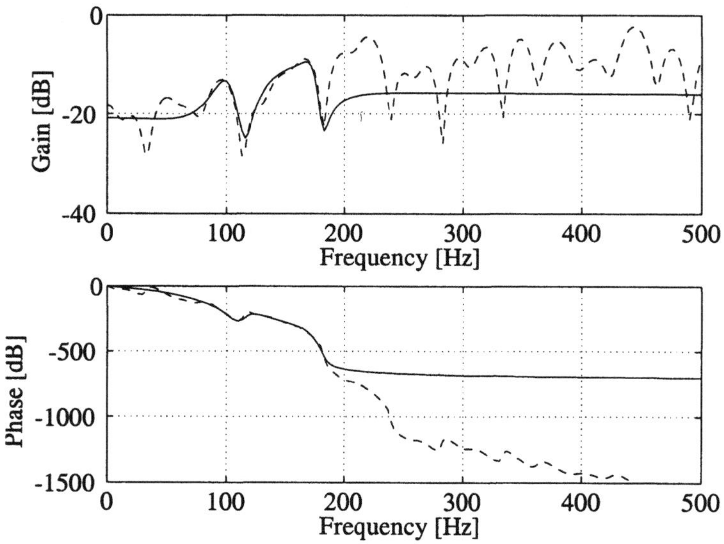


(a) Modeling result by curve fitting (solid line) and the system identification result (dashed line)

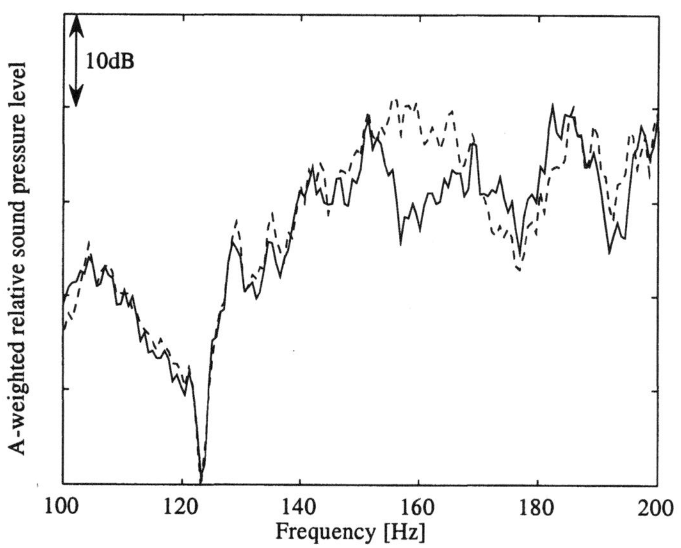


(b) Sound pressure level of the FB-ANC system (solid line), where the dashed line indicates the original sound pressure level

FIGURE 4.4. Experimental results for $l = 1$ m

## References


[1] S. Adachi and H. Sano, "Application of two-degree-of-freedom type active noise control using IMC to road noise inside automobiles," in *Proc. 35th IEEE Conf. Decision and Contr.*, 1996, pp. 2794–2795.

[2] R. J. Bernhard, "Road noise inside automobiles," in *Proc. Active95*, 1995, pp. 21–32.

[3] J. C. Doyle, B. A. Francis, and A. R. Tannenbaum, *Feedback Control Theory.* Macmillan, 1992.

[4] K. Eghtesadi, W. K. W. Hong, and H. G. Leventhall, "The tight-coupled monopole active attenuator in a duct," *Noise Contr. Eng. J.*, vol. 20, no. 1, pp. 16–20, 1983.

[5] K. Glover and J. C. Doyle, "State-space formulae for all stabilizing controller that satisfy an $\mathcal{H}_\infty$-norm bound and relations to risk sensitivity," *Systems & Control Letters*, vol. 11, pp. 167–172, 1988.

[6] B. Hassibi, A. H. Sayed, and T. Kailath, "Optimality of the LMS algorithm," *IEEE Trans. on Signal Processing*, vol. SP-44, no. 2, pp. 267–280, 1997.

[7] J. James, J. Akers, R. Venugopal, R. Lee, A. Sparks, P. Washabaugh, and D. Berstein, "Modeling, identification, and feedback control of noise in an acoustic duct," *IEEE Trans. Control Systems Technology*, vol. 4, no. 3, pp. 283–291, 1996.

[8] S. M. Kuo and D. R. Morgan, *Active Noise Control Systems — Algorithms and DSP Implementations.* John Wiley & Sons, 1996.

[9] S. M. Kuo and J. Tappia, "The implementation of modified leaky SHARF algorithm for the active noise cancellation," in *Proc. IEEE ASSP Workshop Applications of Signal Processing to Audio and Acoustics*, 1989.

[10] L. Ljung, *System Identification — Theory for the User*, 2nd ed. Prentice Hall PTR, 1999.

[11] P. A. Nelson and S. J. Elliott, *Active Control of Sound.* Academic Press, 1993.

[12] H. Sano, S. Adachi, and H. Kasuya, "Application of a least squares lattice algorithm to active noise control for an automobile," *Trans. ASME, J. of Dynamic Systems, Measurement, and Control*, vol. 119, no. 2, pp. 318–320, 1997.

[13] H. Sano and S. Adachi, "Two-degree-of-freedom active control of road noise inside automobiles," in *Proc. Active97*, 1997, pp. 543–546.

[14] H. Sano, S. Miyamoto, and S. Adachi, "Two-degree-of-freedom active noise control with internal model control in feedback control" (in Japanese), *Trans. of the Japan Society of Mechanical Engineers*, vol. 63, no. 608, pp. 1208–1213, 1997.

[15] T. J. Sutton, S. J. Elliott, and A. M. McDonald, "Active control of road noise inside vehicles," *Noise Contr. Eng. J.*, vol. 42, no. 4, pp. 137–147, 1994.

[16] B. Widrow and S. D. Sterns, *Adaptive Signal Processing.* Prentice-Hall, 1985.

Shuichi Adachi
Department of Electrical and Electronic Engineering
Utsunomiya University
7-1-2 Yoto, Utsunomiya, Tochigi 321–8585, Japan
adachis@cc.utsunomiya-u.ac.jp

# Visuomotor Architecture for High-Speed Robot Control

Koichi Hashimoto, Akio Namiki, and Masatoshi Ishikawa

*Dedicated to Professor Hidenori Kimura, on the occasion of his 60th birthday, for guiding the first author into the area of vision and control and for his continuous encouragement*

ABSTRACT A hierarchical control architecture is proposed on the basis of an interaction model between efferent and afferent information in brain motor control. The model has five levels: motoneurons, premotor interneurons, pattern generator, parameter selection and action planning. The effectors including biophysical properties receive the commands from motoneurons. In the proposed architecture, the premotor interneurons and motoneurons are implemented as a servo module; the pattern generator corresponds to the motion planner; and the parameter selection is realized by adaptation module. The afferent information is the feedback signal and the efferent information corresponds to motion command and parameter adaptation. Grasping and handling of a dynamically moving object are implemented on a DSP network with a high-speed vision, a dextrous hand and a 7 DOF manipulator. The results show responsive and flexible actions that exhibit the effectiveness of the proposed hierarchical modular structure.

**Keywords:** Vision, Control, Robot, Human brain architecture, High-speed vision, Grasping

## 1 Introduction

Human beings have high ability to achieve dynamical tasks based on visual information feedback. Ball games such as baseball, football and tennis are good examples that require vision-based dynamical tasks. Hitting a cockroach with a folded newspaper is a little more difficult because the dynamics of the object motion is unknown. To realize such ability it is essential to measure the dynamically changing environment and the moving object in real-time. The object motion model learned by training may be an important help but real-time feedback is necessary. (Model construction should be done through visual measurement.) Thus use of real-time sensory feedback of vision (as well as other sensors) is essential [10, 3].

Limiting the task to grasping, several applications of real-time sensory (mainly visual) feedback have been studied. Allen et al. realized grasping of a toy train based on the optic flow field [1]. Hong and Slotine realized grasping of a flying ball using stereo visual feedback [4]. Jägersand et al. realized grasping using visual servo with model estimation [6]. Namiki et al. developed a high-speed grasping system using visual feedback at the cycle time of 1 ms [8]. For the grasping tasks realized so far, the visual feedback is used only for position control of the gripper.

However, the visuomotor loop in the human brain controls not only hand position. At least, the shape of the hand and the directions of both eyes are simultaneously controlled. Moreover, the sensory outputs from other sensors (afferent information) and the information of arm motion command (efferent copy) interact in the visuomotor loop. A model of this motor control architecture is shown in Figure 1.1 [2]. It illustrates the many structural/functional levels on which sensory and motor pathways interact.

In this paper, a hierarchical parallel processing architecture is proposed based on the hierarchical efferent/afferent interaction model. The hierarchical structure of the proposed architecture is appropriate for multiple heterogeneous tasks. Grasping is a good example of a coordination of heterogeneous tasks such as tracking control of active vision, position control of the arm, preshaping of the hand. Thus we developed a real-time system that realizes very quick grasping tasks. Experimental results of high-speed tracking, grasping, manipulation and collision avoidance are presented to demonstrate the effectiveness of the proposed hierarchical parallel processing architecture.

## 2 Efferent/afferent interaction model [2]

The different functions incorporated in motor control structure define a hierarchy of control levels. Generation of activity in specific motor units lies at the bottom and task function is at the top. The signal from sense organs (afferent information) and the copy of the commands (efferent copy) for the pattern generator (and premotor interneurons) act on each level.

The lowest structural level denotes the reflex in which activity in sensory fibers (proprioceptors and exteroceptors) is passed directly onto motoneurons. The function of these connections ranges from selecting a response to modifying details of motor activity. These direct connections provide fast responses but offer limited complexity. Direct connections from sense organs to motoneurons which trigger actions are common in withdrawal reflexes where a fast response is advantageous for avoiding injury.

The premotor interneurons are introduced as an intermediate structural level between the oscillator or pattern generator and the motor neurons. This level is not present in all systems but it is useful for illustrating some kinds of temporal and spatial integration. Premotor interneurons can par-

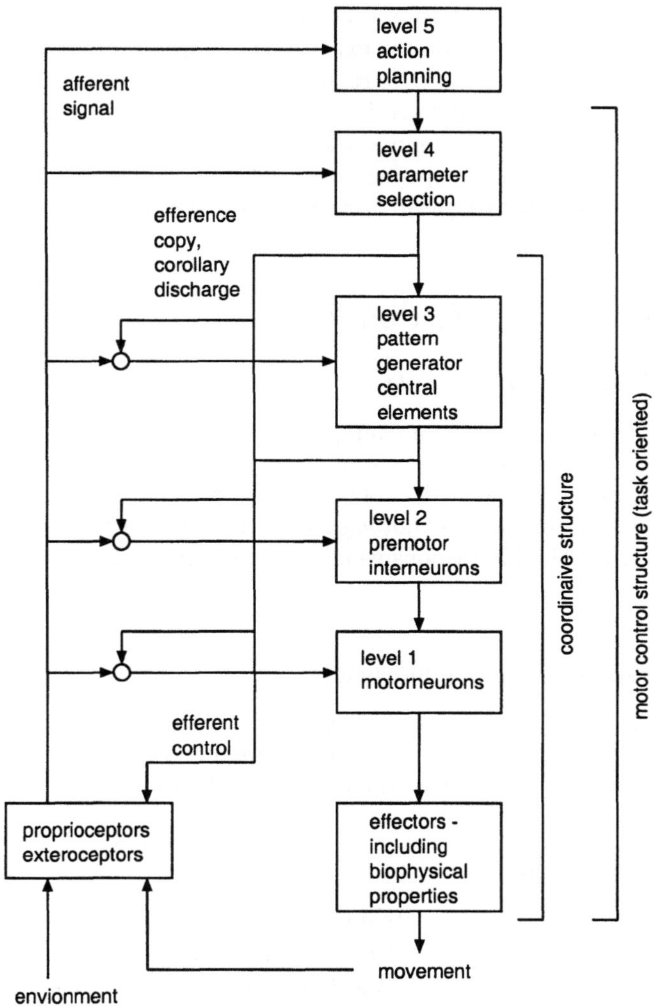


FIGURE 1.1. Efferent/afferent interaction model

ticipate in temporal integration by prolonging a response triggered by brief sensory inputs and by transforming sensory input in various ways. The introduction of interneurons is a simple way to increase the number of response combinations.

The pattern generator may incorporate complicated central networks as well as sensory influences. In many behaviors, sensory inputs can affect a rhythm and the sensory information on a pattern generator is used to determine the successful completion of a movement.

At higher levels of motor organization, sensory information is used to select appropriate motor control structures and to determine their parameters. This function depends on the extraction of appropriate sensory information about the environment. It requires sensory processing of variable complexity and may even involve internal maps or representation of the animal and its relation to its environment.

Note that the hierarchical representation of the control functions does not necessarily correspond to separate structural levels in the nervous system.

# 3 Usage of visual feedback

## 3.1 Feedback for servo control

Part (a) in Figure 3.1 shows an example of servo control in which a hand is controlled by observing the positional error through vision. Usually the vision sensor is used as an outer loop of the joint servo loop [3, 5] but in this example the joint loop is thoroughly controlled by a vision sensor. To achieve stable and highly responsive control, high-speed visual feedback is necessary.

## 3.2 Desired trajectory generation

Sensory feedback is used for generating desired hand trajectory. Part (b) in Figure 3.1 is an example in which the optimal arm trajectory against a moving obstacle is computed by using visual feedback.

## 3.3 Task switching

Sensory feedback is used for switching tasks according to the state of the environment. Part (c) in Figure 3.1 gives an example of task switching in which three subtasks (grasping, object handling, and collision avoidance) are switched according to the state of the environment. To select an optimal subtask in a dynamically changing environment, high-speed feedback is needed.

# 4 Hierarchical sensory feedback model

To simplify the problem, it is assumed that the state of the system is uniquely described by a state vector $\boldsymbol{z} \in \boldsymbol{R}^{m_z}$. A part of the state can be controlled, e.g., joint angle of the arm and direction of the eye. We assume the existence of a function $\boldsymbol{f}$ that selects the controllable state from the whole state $\boldsymbol{\theta} = \boldsymbol{f}(\boldsymbol{z})$. The controllable state $\boldsymbol{\theta} \in \boldsymbol{R}^{m_\theta}$ is called the motion parameter. We also introduce a state of sensor signal $\boldsymbol{s} \in \boldsymbol{R}^{m_s}$. It is also

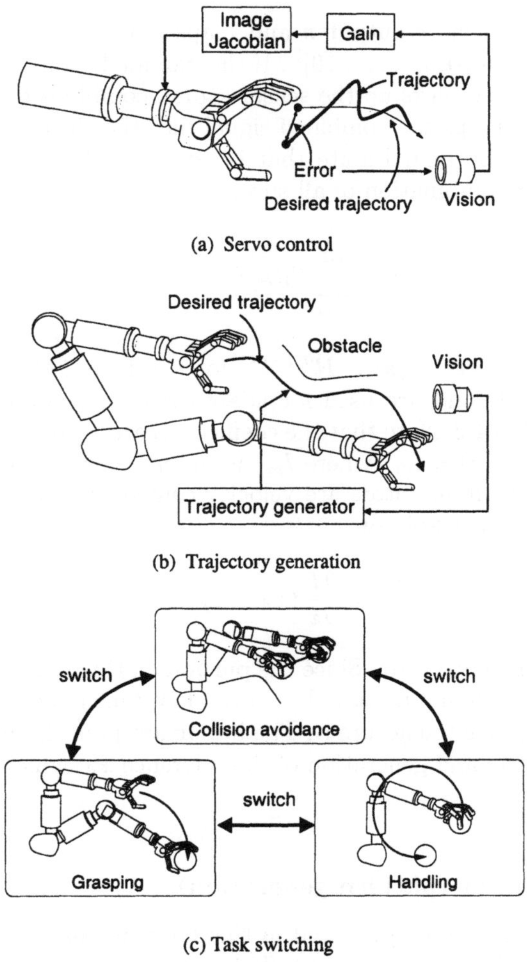


FIGURE 3.1. Use of real-time sensory feedback

assumed that the sensor signal is rich enough to estimate the whole system state.

## 4.1 Control layer

Corresponding to the usage of a vision sensor shown in Figure 3.1 (a), the servo control layer is implemented as levels 1 and 2 of the efferent/afferent interaction model.

Let the number of subtasks that need servo control be $M_z$ and let the state concerning each tasks be $\boldsymbol{z}_i(\boldsymbol{s}) \in \boldsymbol{R}^{m_z}$ for $(i = 1, \dots, M_z)$. For example, let the first subtask $i = 1$ correspond to the pan motion of the eye and the pan angle $q_p$ be the third element of the system state $\boldsymbol{z}$. Then $\boldsymbol{z}_1(s)$

selects $q_p$ from the sensor output $\boldsymbol{s}$ and maps it to the third element of the state, i.e., $\boldsymbol{z}_1 = [0, 0, q_p, 0, \ldots, 0]^T$. If the pan angle control needs other sensor output, $\boldsymbol{z}_1$ also picks up the sensor signal, combines the signal with the pan angle and maps the combined signal into the third element.

Then we define the actual state that is estimated from current sensor output as a linear combination of all subtasks

$$\boldsymbol{z} = \sum_{i=1}^{M_z} S_i(\boldsymbol{s})\boldsymbol{z}_i(\boldsymbol{s}). \tag{4.1}$$

The coefficient matrices $S_i(\boldsymbol{s}) \in \boldsymbol{R}^{m_z \times m_z}$ for $i = 1, \ldots, M_z$ are functions of $\boldsymbol{s}$, sensor output, and are used for adaptation. The adaptation law is described in Section 4.3. Note that the coefficients are semi-positive definite and satisfy $\sum_{i=1}^{M_z} S_i(\boldsymbol{s}) = I_{m_z}$ where $I_{m_z}$ is the $m_z \times m_z$ identity matrix.

We assume that all actuators are velocity-control type and we adopt a simple proportional control law

$$\dot{\boldsymbol{\theta}} = \frac{\partial \boldsymbol{f}}{\partial \boldsymbol{z}} (\boldsymbol{z}_d - \boldsymbol{z}), \tag{4.2}$$

where $\boldsymbol{z}_d$ is the reference state. Since the purpose of this paper is to present the architecture, we do not discuss the control law. One may use any kind of control law to improve the accuracy. The important point is the estimation of system state (4.1) and generation of the reference $\boldsymbol{z}_d$, which is discussed below.

### 4.2 Planning layer (pattern generation)

The reference trajectory is generated in this layer on the basis of the sensor usage described in Figure 3.1 (b). This layer corresponds to level 3 in the efferent/afferent interaction model. To deal with various subtasks, the desired trajectories (for eye direction, arm motion, finger shape and so on) are given as combinations of trajectories for all subtasks.

Let the number of subtasks that affects the control parameters of the system be $M_{dz}$ (which is not necessarily equal to $M_z$) and let the reference state for each subtask be $\boldsymbol{z}_{di}(\boldsymbol{s}) \in \boldsymbol{R}^{m_z}$ for $(i = 1, \ldots, M_{dz})$. Then the actual reference state (motion pattern) is generated as a linear combination of each reference state

$$\boldsymbol{z}_d = \sum_{i=1}^{M_{dz}} U_i(\boldsymbol{s})\boldsymbol{z}_{di}(\boldsymbol{s}). \tag{4.3}$$

The coefficients $U_i(\boldsymbol{s}) \in \boldsymbol{R}^{m_z \times m_z}$ for $i = 1, \ldots, M_{dz}$ are used for adaptation and they are semi-positive definite matrices that satisfy $\sum_{i=1}^{M_{dz}} U_i(\boldsymbol{s}) = I_{m_z}$.

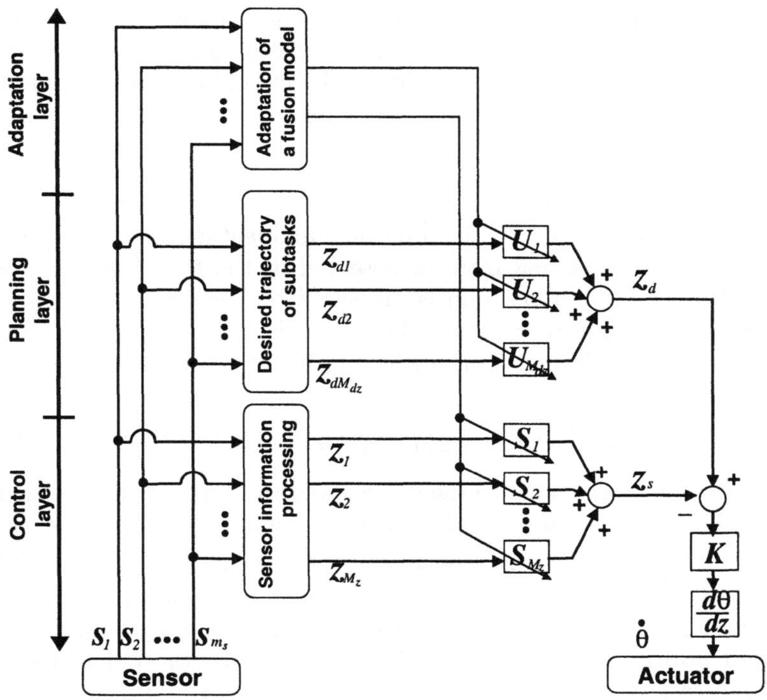


FIGURE 4.1. Hierarchical sensory feedback model

### 4.3 Adaptation layer (task selection)

The third layer is the adaptation. This corresponds to the sensor usage of Figure 3.1 (c). This layer is mainly for selection of the main task among various subtasks. The selection is realized by changing the coefficient matrices $S_i(\boldsymbol{s})$ and $U_i(\boldsymbol{s})$.

On the basis of these control strategies, a hierarchical parallel processing model is constructed as depicted in Figure 4.1. Since all feedback loops (including vision loop) have very high sampling rate (1 kHz), adaptation and task selection corresponding to the dynamic changes of the environment can be achieved. Moreover, this computation model has an advantage that the structural change of the control system, e.g., addition or deletion of sensory feedback, is relatively easy because the dependence between tasks and sensors are described explicitly in the definition of the state.

## 5 High-speed grasping algorithm

We assume that the system is composed by a 6-axis manipulator with a multi-finger hand and a monocular vision. Each joint of the finger has a force sensor. The contact force between the fingers and the object can be measured by using them. The task is a series of general manipulation processes, i.e., grasp the object, handle the object, and avoid other objects.

### 5.1 Control algorithm

Suppose that the joint angle vector is $\boldsymbol{\theta}^a \in \boldsymbol{R}^6$, the position and the orientation of the hand is $\boldsymbol{x}^a \in \boldsymbol{R}^6$, the position and orientation of the object is $\boldsymbol{x}^o \in \boldsymbol{R}^6$, and the force and moment observed by the force/torque sensor is $\boldsymbol{F}^a \in \boldsymbol{R}^6$. The object position is measured by vision and the force is measured by a force/torque sensor. Then, on the basis of a typical force control scheme, the desired joint angle velocity $\boldsymbol{v}_d^a \in \boldsymbol{R}^6$ is computed as follows:

$$\boldsymbol{v}_d^a = {J^a}^{-1} K^a (\boldsymbol{x}_d - \boldsymbol{x}_s) - K^{av}\dot{\boldsymbol{\theta}}^a + {J^a}^T K^{af}\boldsymbol{F}^a, \tag{5.1}$$

where $\boldsymbol{x}_d$ and $\boldsymbol{x}_s$ correspond to the desired and the actual hand positions, respectively. These vectors are defined in the following subsection. The matrix $J^a \equiv \frac{\partial \boldsymbol{x}^a}{\partial \boldsymbol{\theta}^a}$ is the arm Jacobian and the matrices $K^a$, $K^{av}$ and $K^{af}$ are positive definite diagonal gain. In this equation the first term denotes the position control based on a vision sensor ($\boldsymbol{x}_d$ and $\boldsymbol{x}_s$ are generated from the object position and the hand position observed by the vision sensor), the second term denotes the velocity feedback, and the third term is the force feedback using the force/torque sensors.

### 5.2 Task encoding

The variables that appeared in the first term of (5.1) are defined as follows:

$$\begin{aligned}\boldsymbol{x}_d \equiv\ & (I_6 - G^a)(I_6 - G^m)(I_6 - G^s)\,\boldsymbol{x}^o \\ & + (I_6 - G^a)(I_6 - G^m)\,G^s\,\boldsymbol{x}_d^a \\ & + (I_6 - G^a)\,G^m \boldsymbol{x}_d^o + G^a\,\boldsymbol{x}_d^c, \end{aligned}\tag{5.2}$$

$$\boldsymbol{x}_s \equiv (I_6 - C^m)\,\boldsymbol{x}^a + C^m\,\boldsymbol{x}^o, \tag{5.3}$$

where $\boldsymbol{x}_d$ is the reference trajectory generated by integrating desired trajectories of the subtasks and $\boldsymbol{x}_s$ is the actual controlled position. The block diagram of this control scheme is given in Figure 5.1. The reference is a function of the object position $\boldsymbol{x}^o$, the desired handling trajectory $\boldsymbol{x}_d^o$, the desired avoidance trajectory $\boldsymbol{x}_d^c$ and the desired reaching trajectory $\boldsymbol{x}_d^a$. The matrices $G^a$ and $G^m$ select the subtasks and the matrix $G^s$ divide the workspace into tracking and reaching subspaces. The vector $\boldsymbol{x}_s$ denotes the task dependent state of the system generated by combining the visual information $\boldsymbol{x}^o$ with the internal sensor information $\boldsymbol{x}^a$ (the fingertip position computed from the joint angle measurement $\boldsymbol{\theta}^a$). The matrix $C^m$ switches the controlled point. (If the hand is holding the object the controlled point becomes the object. Otherwise, the controlled point is the fingertip.)

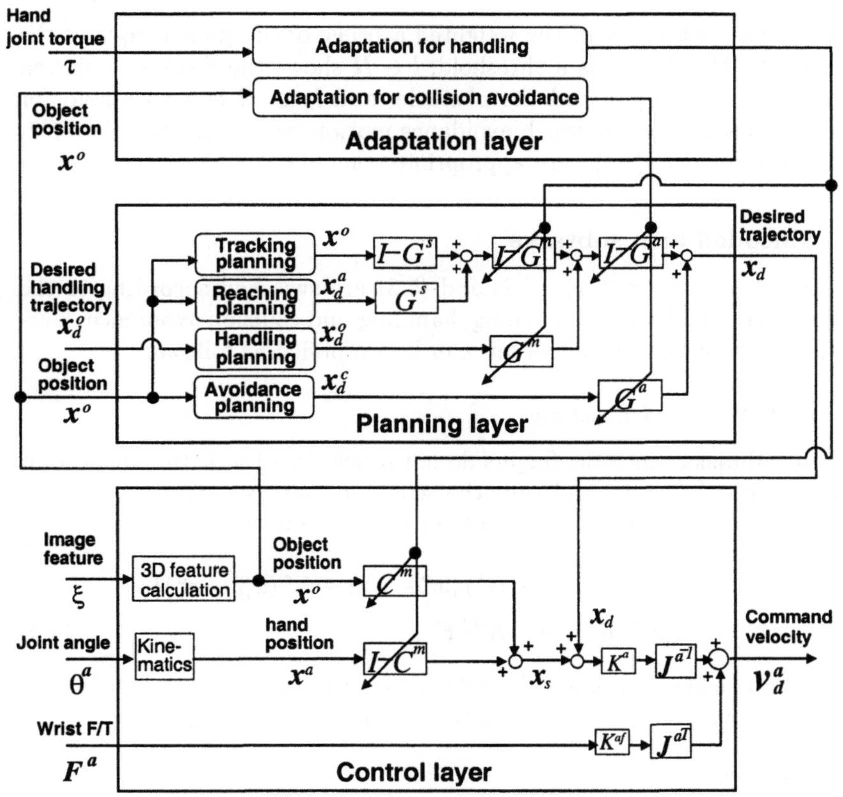


FIGURE 5.1. Block diagram

## 5.3 Selection matrices

The coefficient matrices that appeared in (5.2) and (5.3) are defined as follows:

$$G^s = \mathrm{diag}\,(1,\,0,\,0,\,0,\,1,\,1)\,,$$

$$G^m = \mathrm{diag}\,(\,g_i^m\,)\,,\quad i=1,2,\cdots,6, \tag{5.4}$$

$$g_i^m = \begin{cases} 1 & (1<\bar{g}_i^m), \\ \bar{g}_i^m & (0\le \bar{g}_i^m \le 1), \\ 0 & (\bar{g}_i^m<0), \end{cases} \tag{5.5}$$

$$\bar{g}_i^m = \int_{t'=0}^{t} \gamma_i^m \mathrm{sgn}(\tau^h(t')-\tau_o^h)\,dt', \tag{5.6}$$

$$\tau^h = \sqrt{\sum_{i=1}^{4} \boldsymbol{\tau}_i^{h\,T} H_i \boldsymbol{\tau}_i^h},$$

$$G^a = \mathrm{diag}\,(\,g_i^a\,)\,,\quad i=1,2,\cdots,6, \tag{5.7}$$

$$g_i^a = \frac{1}{1+\exp(\gamma_i^a(l-l_o))},$$

$$C^m = C^\alpha\, G^m, \tag{5.8}$$

where $\tau^h$ shows the size of the weighted average of the joint torque of the hand $\boldsymbol{\tau}_i^h \in \boldsymbol{R}^3$, $\tau_o^h \in \boldsymbol{R}$ is a threshold, $l \in \boldsymbol{R}$ shows the distance between the manipulated object and the obstacle, and $l_o \in \boldsymbol{R}$ is a threshold for adjusting the distance in which avoidance motion starts. The matrices $C^\alpha$, $H_i$ and the scalers $\gamma_i^m$, $\gamma_i^a$ are appropriate constants.

## 5.4 Encoded arm subtasks

The arm control laws (5.1), (5.2) and (5.3) are switched according to the subtasks, i.e., tracking and reaching; handling and collision avoidance. During these subtasks, the control law can be simplified as follows:

### *5.4.1 Tracking and Reaching*

In these subtasks, since the fingers do not touch the object, the finger joint torque is zero ($\tau^h = 0$) and thus the selection matrices satisfy $G^m = O_6$ and $G^a = O_6$. Therefore the control law (5.1) becomes

$$\begin{aligned} \boldsymbol{v}_d^a = {J^a}^{-1} K^a \left\{ (I_6 - G^s)\,(\boldsymbol{x}^o - \boldsymbol{x}^a) + G^s\,(\boldsymbol{x}_d^a - \boldsymbol{x}^a) \right\} \\ -K^{av}\dot{\boldsymbol{\theta}}^a + {J^a}^T K^{af} \boldsymbol{F}^a, \end{aligned} \qquad (5.9)$$

where the tracking motion $(I_6 - G^s)\,(\boldsymbol{x}^o - \boldsymbol{x}^a)$ is orthogonal to reaching motion $G^s\,(\boldsymbol{x}_d^a - \boldsymbol{x}^a)$, there is no interference between them.

### *5.4.2 Handling*

After grasping the object, the subtask is switched to handling mode. When the grasp is stiff, the switching function $g_i^m$ becomes unity. Thus the selection matrices are given by $G^m = I_6$ and $G^a = O_6$. Then the control law (5.1) is described by

$$\boldsymbol{v}_d^a = {J^a}^{-1} K^a\,(\boldsymbol{x}_d^o - \boldsymbol{x}_s^o) - K^{av}\dot{\boldsymbol{\theta}}^a + {J^a}^T K^{af} \boldsymbol{F}^a, \qquad (5.10)$$

where $\boldsymbol{x}_s^o \equiv (I_6 - C^\alpha)\boldsymbol{x}^a + C^\alpha \boldsymbol{x}^o$ is the controlled position generated by combining the visual information with the internal sensor information.

### *5.4.3 Collision avoidance*

After grasping, if other objects appear in the viewing area then the subtask is switched to collision avoidance mode. In this subtask the selection matrices become $G^m = I_6$ and $G^a = I_6$. Thus the control law (5.1) is given by

$$\boldsymbol{v}_d^a = {J^a}^{-1} K^a\,(\boldsymbol{x}_d^c - \boldsymbol{x}_s^o) - K^{av}\dot{\boldsymbol{\theta}}^a + {J^a}^T K^{af} \boldsymbol{F}^a. \qquad (5.11)$$

Note that the avoidance force is generated by the term $(\boldsymbol{x}_d^c - \boldsymbol{x}_s^o)$ which is obtained by combining the visual information with the internal sensor information. The desired trajectory for collision avoidance is defined by

$$\boldsymbol{x}_d^c \equiv \begin{cases} \boldsymbol{x}_s^o + \begin{bmatrix} \boldsymbol{n}^l (l_o - l) \\ \boldsymbol{0} \end{bmatrix} & \text{if } l_o - l > 0, \\ \boldsymbol{x}_s^o & \text{otherwise,} \end{cases} \tag{5.12}$$

where $\boldsymbol{n}^l \in \boldsymbol{R}^3$ is the unit vector in the direction from the obstacle to the manipulated object.

# 6 Experiment

## 6.1 Experimental system

The experimental system is the 1 ms sensory-motor fusion system [9]. This system consists of a hierarchical parallel processing system, an active vision, a 7-axis arm, and a 4-fingered hand. The hierarchical parallel processing system is a multiprocessor system with eleven general-purpose floating-point DSPs. The active vision is a 2-DOF actuator (tilt and pan) and a high-speed vision chip system SPE-256 (resolution: 16 $\times$ 16 pixel) [7] is mounted on it. The hand is a 4-fingered 14-DOF system that has force sensors in each joint. The cycle time of the whole system control is 1 ms.

## 6.2 Experimental result

We assume that the handling subtask is to keep the object at a specified position. Part (a) of Figure 6.1 shows the time response of object and hand trajectories. After grasping, obstacle trajectory is also shown. It illustrates that tracking, grasping, handling and collision avoidance subtasks are carried out. To show the subtask switching mechanism, the time responses of coefficients for adaptation $G^a$ and $G^m$ is given in part (b) and the time response of the grasping torque $\tau^h$ is given in part (c). Task switching from the reaching subtask to the object handling subtask is realized by the change of the coefficient matrix $G^m$ according to the grasping torque $\tau^h$. The task switching from the object handling subtask to the collision avoidance subtask is realized by the change of the coefficient matrix $G^a$ due to the approach of the obstacle. In Figure 6.2 the total manipulation process is shown as a continuous sequence of pictures every 0.3 [s].

# 7 Conclusion

In this paper, a hierarchical parallel processing model based on high-speed sensory feedback is proposed. On the basis of this model, grasping and

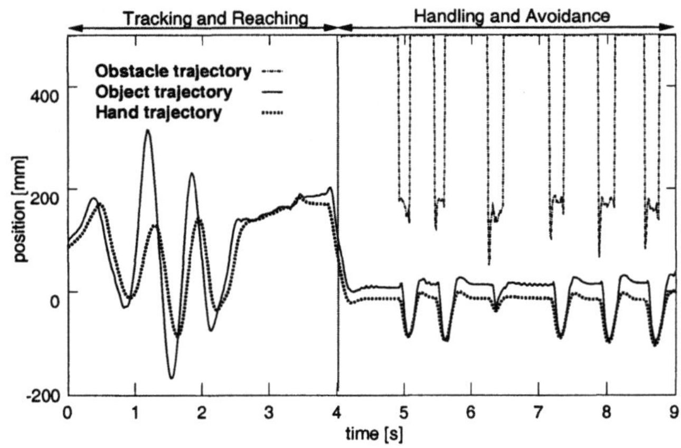


(a) Trajectory

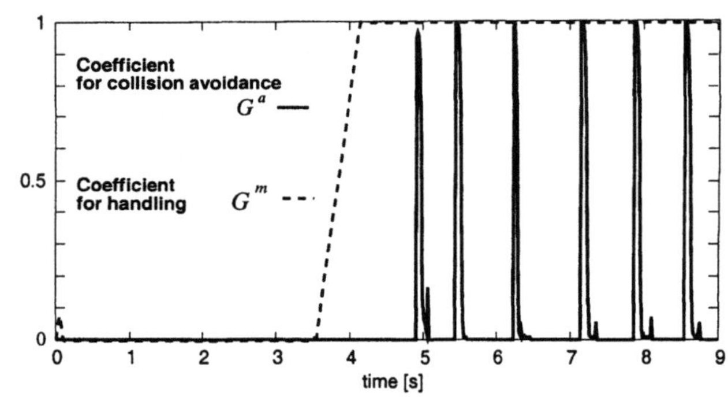


(b) Coefficient

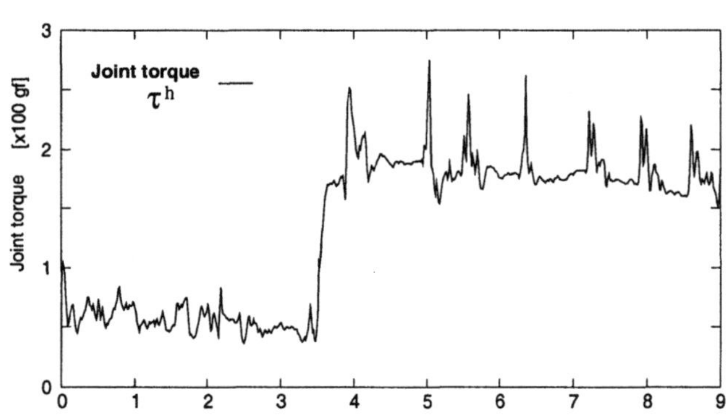


(c) Joint torque

FIGURE 6.1. Time response: task switching

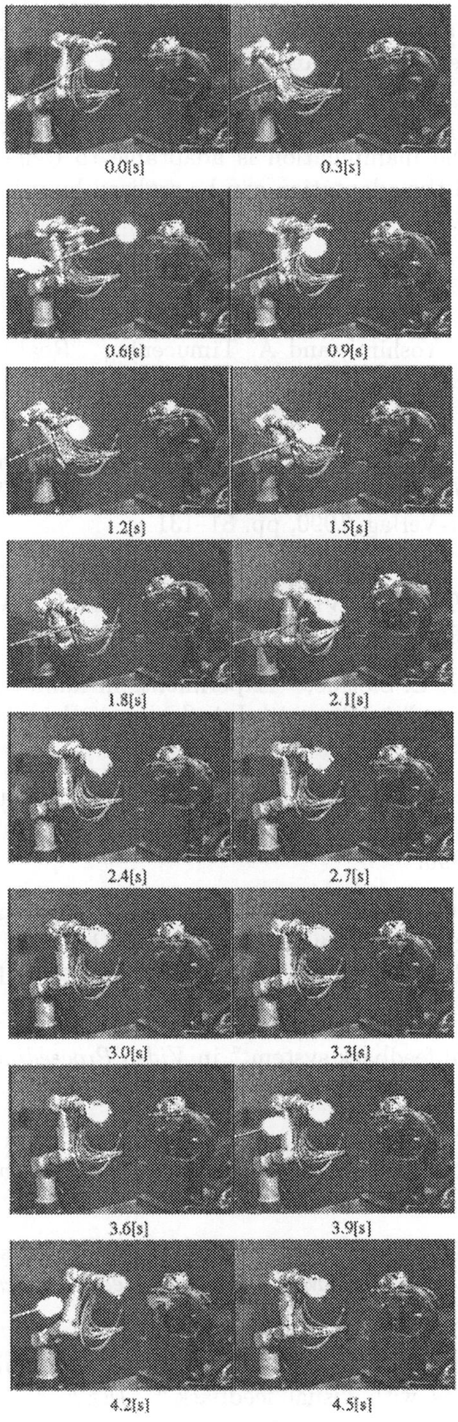


FIGURE 6.2. Experimental result: collision avoidance

manipulation tasks are realized. It is shown that the system is highly responsive and flexible to the dynamic changes of an environment. The proposed algorithm for multiple heterogeneous subtasks is well suited for a hierarchical efferent/afferent interaction model. The experimental results demonstrate that the manipulation is adaptable to dynamic changes of an environment and the adaptation can be realized by using the real-time sensory feedback.

## References

[1] P. K. Allen, B. Yoshimi, and A. Timucenko, "Real-time visual servoing," in *Proc. IEEE Int. Conf. Robotics and Automation*, 1991, pp. 2376–2384.

[2] J. Dean, "The neuroethology of perception and action," in *Relationships Between Perception and Action*, O. Neuman and W. Printz, Eds. Berlin: Springer-Verlag, 1990, pp. 81–131.

[3] K. Hashimoto, Ed., *Visual Servoing — Real-Time Control of Robot Manipulators Based on Visual Sensory Feedback.* World Scientific, Singapore, 1993.

[4] W. Hong and J. E. Slotine, "Experiments in hand-eye coordination using active vision," in *Proc. 4th Int. Symp. on Experimental Robotics (ISER'95)*, 1995.

[5] S. Hutchinson, G. D. Hager, and P. I. Corke, "A tutorial on visual servo control," *Trans. on Robotics and Automation*, vol. 12, no. 5, pp. 651–670, 1996.

[6] M. Jägersand, O. Fuentles, and R. Nelson, "Acquiring visual-motor models for precision manipulation with robot hands," in *Proc. of 4th European Conf. on Computer Vision*, 1996, pp. 603–612.

[7] Y. Nakabo, I. Ishii, and M. Ishikawa, "High speed target tracking using 1ms visual feedback system," in *Video Proceedings of IEEE Int. Conf. Robotics and Automation*, Minneapolis, 1996.

[8] A. Namiki, Y. Nakabo, I. Ishii, and M. Ishikawa, "High speed grasping using visual and force feedback," in *Proc. IEEE Int. Conf. on Robotics and Automation*, 1999.

[9] A. Namiki, Y. Nakabo, I. Ishii, and M. Ishikawa, "1ms sensory-motor fusion system," *IEEE/ASME Trans. on Mechatronics*, vol. 5, no. 3, pp. 244–252, 2000.

[10] L. Weiss, A. Sanderson, and C.P. Neuman. "Dynamic sensor-based control of robots with visual feedback," *IEEE J. of Robotics and Automation*, vol. RA-3, no. 5, 1987.

Koichi Hashimoto
Department of Information Physics and Computing
Graduate School of Information Science and Technology
The University of Tokyo
Hongo 7-3-1, Bunkyo, Tokyo 113-8656, Japan
koichi@k2.t.u-tokyo.ac.jp

Akio Namiki
Department of Information Physics and Computing
Graduate School of Information Science and Technology
The University of Tokyo
Hongo 7-3-1, Bunkyo, Tokyo 113-8656, Japan
namik@k2.t.u-tokyo.ac.jp

Masatoshi Ishikawa
Department of Information Physics and Computing
Graduate School of Information Science and Technology
The University of Tokyo
Hongo 7-3-1, Bunkyo, Tokyo 113-8656, Japan
ishikawa@k2.t.u-tokyo.ac.jp

MIX
Papier aus verantwortungsvollen Quellen
Paper from responsible sources
FSC® C105338

If you have any concerns about our products,
you can contact us on
**ProductSafety@springernature.com**

In case Publisher is established outside the EU,
the EU authorized representative is:
**Springer Nature Customer Service Center GmbH**
**Europaplatz 3, 69115 Heidelberg, Germany**

Printed by Libri Plureos GmbH
in Hamburg, Germany